Evolution, the Logic of Biology

Evolution, the Logic of Biology

John S. Torday

*Departments of Pediatrics, and Obstetrics and Gynecology
Harbor-UCLA Medical Center, Torrance, California, USA
Evolutionary Medicine Program
UCLA, Los Angeles, California, USA*

Virender K. Rehan

*Department of Pediatrics
Harbor-UCLA Medical Center
Torrance, California, USA*

Registered Office
John Wiley & Sons, Inc., 111 River Street, Hoboken, NJ 07030, USA

Editorial Office
111 River Street, Hoboken, NJ 07030, USA

For details of our global editorial offices, customer services, and more information about Wiley products visit us at www.wiley.com.

Wiley also publishes its books in a variety of electronic formats and by print-on-demand. Some content that appears in standard print versions of this book may not be available in other formats.

Library of Congress Cataloguing-in-Publication data applied for

ISBN: 9781118729267

Cover design: Wiley
Cover image: © Man_Half-tube/Gettyimages

Set in 10/12pt Warnock by SPi Global, Pondicherry, India
Printed and bound in Malaysia by Vivar Printing Sdn Bhd
10 9 8 7 6 5 4 3 2 1

Dr. Torday dedicates this book to his wife Barbara, his children Nicole Anne and Daniel Philip Torday, his daughter-in-law Dr. Erin Kathleen Torday, his granddaughters Abigail Eve and Delia Rose Torday, his parents Steven and Maria Torday, and his mentor Mary Ellen Avery.

Dr. Rehan dedicates this book to his parents Sain Das and Nirmala Rehan, his wife Yu Hsiu and children Amit and Anika Rehan, and his brother (the late) Dr. Sudhir Rehan.

Contents

Preface

> We shall not cease from exploration
> and the end of all our exploring will be
> to arrive where we started
> and know the place for the first time.
> *T.S. Eliot, Four Quartets*

This book is a sequel to our first publication on the cellular-molecular basis for vertebrate evolution, which was entitled *Evolutionary Biology, Cell-Cell Communication, and Complex Disease*. In it we showed the utility of a cellular-molecular approach to understanding the evolution of complex physiology from the unicellular state. In the current book, the Ur hypothesis is that all complex physiology has evolved from the cell membrane of unicellular organisms, offering a functional integration of all physiologic properties intersecting structurally and functionally in the unicellular "singularity." This combined holistic-reductionistic perspective provides a fundamental insight to the logic of biology never before available.

By tracing the emergence and contingence of novel evolutionary traits backwards in the history of the organism using ontogeny and phylogeny as guide posts, we have been able to deconvolute the lung as an archetype for understanding how and why physiologic traits have evolved from their unicellular origins – how cholesterol evolved from the sterol pathway of bacteria to facilitate oxygenation, metabolism, and locomotion in primitive eukaryotes, only to later recur molecularly, cellularly, structurally, and functionally as the swim bladder of fish, and subsequently as the lung of amphibians, reptiles, mammals, and birds. That "arc" has provided a way of connecting the evolutionary dots to other physiologic traits – skin, kidney, skeleton, brain – by tracing their evolution in tandem with the lung ontogenetically and phylogenetically, in combination with pathophysiologic data to provide the fullest picture for such complex, arcane interrelationships. Such interconnections become more apparent during times of stress like the water-to-land transition (see Chapter 2, e.g.). The specific causes of the gene mutations and duplications that occurred during that phase of vertebrate evolution, when seen in the context of having to adapt to terrestrial life, become self-evident when factored into the prevailing physiologic constraints. This is particularly true of the developmental and phylogenetic properties that are mediated by soluble growth factors and their cognate receptors.

In brief, Chapter 1, "Introduction," sets the stage for a paradigm shift in the way we think about the evolution of vertebrate physiology, based on mechanisms of cell–cell interactions. Chapter 2, entitled "On the fractal nature of evolution," shows the value added in the cellular-molecular approach to evolution, beginning with the origins of life. It is predicated on the balancing selection of calcium and lipids as the fundament of vertebrate evolution. We suggest that the cell is the ultimate physiologic fractal, expressing the "first principles of physiology." Biology mimics the Big Bang of physics, generating its own internal pseudo-Universe through the entraining of calcium by lipids, which is vertically integrated to generate complex physiology. Chapter 3, entitled "The historic perspective on paracrinology and evolution as lead-ins to a systems biology approach," traces the history of cell culture and its influence on our insights to the role of the cell in evolution. That realization was a game-changer for our understanding of embryology, homeostasis, and repair as a functional continuum. Chapter 4, entitled "Evolution of adipocyte differentiation related protein, or 'Oh, the Places You'll Go' – Theodore Geissel, aka Dr Seuss," describes the discovery of neutral lipid trafficking within the lung alveolus for stretch-regulated surfactant production. The realization of the paracrine regulation of surfactant phospholipid substrate provided seminal cellular-molecular insights to the evolution of the lung and other complex physiologic mechanisms. Chapter 5, entitled "Evolutionary ontology and epistemology," provides the rationale for reconfiguring the logic of biology and evolution. In this chapter, we delve into philosophical aspects of the cellular-molecular approach to evolution. The identification of homeostasis as the underlying, overriding mechanism of evolution fundamentally affects the way in which we think of the process. Chapter 6, entitled "Calcium-lipid epistasis: like ouroboros, the snake, catching its tail!," delineates the epistatic balancing selection for calcium and lipids as the operating principle for vertebrate biology and evolution. We have gone into detail with regard to calcium/lipid epistatic balancing selection as the founding principle of eukaryotic physiology and evolution. Chapter 7, entitled "The lung alveolar lipofibroblast: an evolutionary strategy against neonatal hyperoxic lung injury," describes the evolution and role of the lipofibroblast in lung development, homeostasis, repair, and evolution. The lipofibroblast is emblematic of lipid homeostatic balance in eukaryotic physiologic regulation. We exemplify the vertical integration from the molecular to the physiologic/pathophysiologic using the lung alveolus as a model. Chapter 8, entitled "Bio-logic," is a further exposition on the cellular approach to understanding the logical basis of physiology as a continuum from unicellular to multicellular organisms. The chapter looks at complex physiology as a "vertically synthetic," internally consistent, scale-free process, both developmentally and phylogenetically. Chapter 9, entitled "Cell signaling as the basis for all of biology," focuses on cell–cell signaling as the mechanistic basis for all of the principles of physiology. Chapter 10, entitled "Information + negentropy + homeostasis = evolution," explains the functional interrelationships between physical chemistry and biology, catalyzed by circumventing the second law of thermodynamics. Chapter 11, entitled "Vertical integration of cytoskeletal function from yeast to human," examines the roles of the mechanical "superstructure," the cytoskeleton and skeleton, as organizing principles for integrated physiology as an existing form that relates all the way back to the origins of life itself. In Chapter 12, entitled "Yet another bite of the 'evolutionary' apple," we go through the reverse logic of physiology emanating from the unicellular state. It may seem redundant, but because of the counterintuitive nature of the approach it is helpful to see this viewpoint in multiple ways. Chapter 13, entitled "On eliminating the subjectivity from biology: predictions," addresses the value added in seeing physiology and evolution from their origins. Chapter 14, entitled "The predictive value of the cellular approach to evolution," recapitulates the concept that by starting from the cellular origins of life the underlying principles of seemingly complex, indecipherable physiologic principles can be understood and expanded to all of physiology. Chapter 15, entitled "Homeostasis as the mechanism of evolution," provides a mechanistic integration for the how and why of evolution. Beginning with the protocell, homeostasis acts as the integrating principle on a scale-free basis. Chapter 16, entitled "On the evolution of development," takes the dogma of development and shows how it becomes part of the continuum of evolution using the principles provided in the previous chapters. Chapter 17, entitled "A central theory of biology," provides the first comprehensive perspective on the "first principles of biology." By utilizing the unique view provided by cell biology as the common denominator for ontogeny and phylogeny, biology can be seen as having a logic. Chapter 18, entitled "Implications of evolutionary physiology for astrobiology," demonstrates how the principles of physiology and evolution on Earth can provide a logical way of thinking about extraterrestrial life. Chapter 19, entitled "Pleiotropy reveals the mechanism for evolutionary novelty," provides a way of thinking about how the return to the unicellular state during the life cycle offers the opportunity for the reallocation of genes to generate novel physiologic traits. Chapter 20, entitled "Meta-Darwinism," provides examples of the power of the cellular-molecular approach to evolution.

The content of this book constitutes a novel, mechanistic, testable, refutable approach to the questions of 'how and why' evolution has occurred. This book is dedicated to that sea change.

John S. Torday
Virender K. Rehan

1

Introduction

There are these two young fish swimming along, and they happen to meet an older fish swimming the other way, who nods at them and says, "Morning, boys. How's the water?" And the two young fish swim on for a bit, and then eventually one of them looks over at the other and goes "What the hell is water?"
David Foster Wallace, Kenyon College Commencement Speech, 2005

The premise of this book is that the Big Bang of the Universe gave rise to inorganic and organic compounds alike. Both are formed by bonds, the former constituted by inertness, the latter doing quite the opposite by giving rise to life itself. Organic chemistry provided the physical space within which negentropy, chemiosmosis, and homeostasis all acted in concert to form the first primitive cells. Single-celled organisms dominated the Earth for the first 3 or 4 billion years, followed by the generation of multicellular organisms as exaptations. How and why this occurred provides the mechanism for the emergence of human biology, starting with the "first principles of physiology." Such a rendering is way overdue, since the human genome was published more than a decade and a half ago. Without such an effectively predictive working model for physiology, such information is of little value.

Mind the Gap

Let us go then, you and I,
When the evening is spread out against the sky
Like a patient etherized upon a table...
T.S. Eliot, The Love Song of J. Alfred Prufrock

The Michelson–Morley experiment (1887) refuted the notion of luminiferous aether, a theorized medium for the propagation of light, making way for novel thinking about the fundamental principles of physics at the close of the nineteenth century and the beginning of the twentieth. This second scientific revolution was crowned by Relativity Theory (1905), equating energy and mass, a counterintuitive relationship that changed not only the way we see the world around us, but also how we see ourselves. The understanding of the inner workings of the Bohr atom similarly gave insights to physics and chemistry

Evolution, the Logic of Biology, First Edition. John S. Torday and Virender K. Rehan.
© 2017 John Wiley & Sons, Inc. Published 2017 by John Wiley & Sons, Inc.

that were previously inaccessible and inconceivable. The twenty-first century has been declared the "age of biology," given our foreknowledge of the genetic makeup of humans and an ever-increasing number of model organisms. Yet the promise of the human genome – the cure for all of our medical ills – has not transpired 15 years hence. We contend that this is symptomatic of our not having attained the level of knowledge in biology that the physicists had reached at the dawn of the second scientific revolution... we are still mired in the sticky, sludgy, stodgy "aether" of descriptive biology. Deep understanding of the inner workings of the cell, particularly as they have facilitated the evolution of multicellular organisms, will herald such breakthrough science. The way in which the cell acts at the interface between the external physical and internal biological "worlds," authoring the script for Life, is a reality play without an ending, reiterative and reinventive. Thus life is formulated to continually learn from the ever-changing environment, making use of such knowledge in order to sustain and perpetuate it.

Duality, Serendipity, and Discovery

The field of biomedical research is characterized by paradoxes, serendipitous observations, and occasional discoveries. This is due to the lack of a central theory of biology, DNA notwithstanding. It is also the reason why we have been unable to solve the challenging "puzzle" of evolution. In lieu of guidelines and principles, we collect anecdotes and make up "Just So" stories based on associations and *a posteriori* reasoning. This book was written to elucidate how to understand biology based on its origins in unicellular life, evolving in the forward direction of biologic history, both ontogenetically (short-term history) and phylogenetically (long-term history). We use the figure-ground image (Figure 1.1), made popular by gestalt psychology, as a way to express the inherent problem in seeing biologic phenomena as dualities: inorganic-organic, genotype-phenotype, proximate-ultimate, structure-function, health-disease, synchronic-diachronic, ontogeny-phylogeny. It is the latter duality that was the breakthrough for us, realizing that ontogeny and phylogeny, looked at from a cellular perspective, are actually one and the same process, only seen from different perspectives. With that issue put behind us, we could address the "first principles of physiology," beginning with the plasmalemma of protists as the homolog for all of the subsequent traits expressed by multicellular organisms.

Figure 1.1 Figure-ground "faces."

Biology as "Stamp Collecting"

As working scientists, the authors of this book have been involved in studies of developmental physiology for many decades. One of us (J.T.) was first introduced to the concepts of cell biology in reading Paul Weiss, one of the founders of the discipline, when he took advanced placement biology in high school. It was Weiss who admonished us not to ask "how or why" questions, but merely to describe biologic phenomena. That attitude prevailed in biology until the advent of molecular biology in the 1960s, which demanded that we ask how biologic mechanisms functioned, having finally "reduced" the problem to its smallest functional unit, the cell, like the Bohr atom in physics. Yet this reductionist approach has not solved some of the remaining fundamentals of physics, hence string theory, "branes," and multiverses. By analogy, we are of the opinion that we must think in terms of the cell as the smallest functional unit of biology. Conversely, stripping away billions of years of biologic information to focus on DNA is a systematic error that is misleading and misguided, in our opinion. This book is intended to demonstrate how the cell-molecular approach to evolutionary biology provides novel insights to the how and why for the evolution of form and function.

The "why" question has emerged from the New Synthesis of evolutionary biology, particularly after it had embraced developmental biology as evo-devo. But even at that, the evolutionists were not delving into the mechanisms of evolution, seemingly content with random mutation and population selection as the mechanisms of evolution. For working scientists like ourselves, studying how organs develop across species, this didn't seem like a reasonable process since we could see the common denominator between ontogeny and phylogeny at the cellular level, suggesting (to us) that some underlying organizing principle was at large. Not to mention that the ongoing serendipitous, anecdotal nature of both biology and medicine, even in the post-Human Genome Project era, was frustrating given that science is ultimately supposed to be predictive.

Then in 2004 Nicole King published her ground-breaking paper demonstrating for the first time that the complete multicellular genomic "toolkit" of sponges was expressed in the unicellular free-swimming amoeboid form during the life cycle. That reversed everything in biology because up until that point in time physiology was described based on biologic traits in their extant form, not based on how they evolved from the unicellular state. That perspective precipitated our hypothesis that the complete phenotypic toolkit was present in the plasmalemma of unicellular organisms, and raised the question as to how the genome determined complex physiology ontogenetically and phylogenetically, from unicellular organisms to invertebrates and vertebrates.

That question was made all the more pertinent because we had published a seminal paper on the cell-molecular basis of alveolar physiology that had emerged from decades-long study of how the fetal lung develops at the cell-molecular level. Those studies were catalyzed by two landmark observations – the physiologic acceleration of lung development by glucocorticoids, and the observation that parathyroid hormone-related protein (PTHrP) was necessary for alveolar formation. The linking of those two phenomena through the serial paracrine interactions between the lung endoderm and mesoderm culminated in our fundamental understanding of the physiologic principle of ventilation-perfusion matching – essentially how the distension of the alveolar wall molecularly coordinated surfactant production and alveolar capillary perfusion to maintain both local and systemic homeostasis.

The experimental evidence for the coordinating effects of cell stretching on PTHrP, leptin, and their cognate cell-surface receptors on surfactant production and vascular perfusion led to the first scientific documentation of the physiologic continuum from development to homeostasis. More importantly, it begged the question as to how these specific cell types evolved the mammalian lung phenotype, given that the molecular ligand and receptor intermediates involved were highly conserved, deep homologies that could be traced at least as far back as the origins of vertebrate phylogeny in fish, if not all the way back to the unicellular state. If that "story" could be told, it would provide insight to both lung physiology and pathophysiology based on first principles – a counterintuitive idea predicted by this cell-molecular approach.

More importantly, the functional genomic linkage between lung evolution in complex climax organisms – mammals and birds – and homologous mechanisms in emerging unicellular eukaryotes, formed the basis for fundamental insights into the evolution of all visceral organs. The advent of cholesterol, which Konrad Bloch referred to as a "molecular fossil," was critical for the evolution of eukaryotes from prokaryotes. The insertion of cholesterol into the cell membrane of eukaryotes enabled vertebrate evolution by facilitating endocytosis (cell eating), increased gas exchange (due to the thinning of the eukaryotic cell membrane), and increased locomotion (due to increased cytoplasmic streaming). And vertebrate evolution is founded on those three biologic traits – metabolism, respiration, and locomotion. Therefore, all of the visceral organs – lung, kidney, skin, skeleton, brain, and so forth – likely evolved from the plasmalemmae of unicellular organisms, providing a unified, common homolog for all of these organs. The existence of vertebrate physiologic mechanisms based on functional cell-molecular homologies, rather than on the tautologic "Just So" stories for physiologic structure and function that currently prevail, would no longer hamper forward progress in understanding the "how" and "why" of biology and medicine.

Up until now, the void between descriptive and mechanistic physiology has been filled by either top-down descriptive physiology, or bottom-up abstract philosophy and mathematics. With the insights gained from the "middle-out" ligand-receptor approach we have employed to understand lung evolution, we are now enabled for a paradigm shift from post-dictive to predictive physiology and medicine.

Historically, physicists became actively involved in biology after the Second World War as an alternative source of employment, having successfully developed and deployed the atomic bomb. The Greek philosophers understood the unity of life intellectually as far back in written history as the fifth century BCE, but had no scientific evidence for it. Beginning with quantum mechanics, physicists felt empowered to comment on the meaning of life, given that they had discovered the operating principle behind the atom and had unleashed its power. Bohr was the first modern physicist to address the question of "what is life" by applying the conceptual principle of the duality of light to biology in his Como lecture in 1927. He went on to explain that this seeming duality was a technical glitch due to the different ways in which the wave and packet forms of light were measured, a phenomenon he referred to as complementarity. This was a metaphor that poets such as Robert Frost (in his poem *The Secret Sits*) have reconciled more facilely, in our opinion:

> We dance round in a ring and suppose,
> But the Secret sits in the middle and knows.

Or for that matter, the glib comment in Robert Frost's published notebook: "Life is that which can mix oil and water." Erwin Schrödinger later wrote a monograph, entitled *What is Life?*, in which he tried to apply physical principles to the question of the vital force. Others followed, such as the Nobelist Ilya Prigogine, and the polymath Michael Polanyi, who expressed their considered opinions that biology was "irreducible." More recently, in his Nobel Prize acceptance speech, Sydney Brenner stated that the problem of biology is soluble, citing his CELL project to map all of its intracellular pathways. Of course, the greatest of all physicists, Albert Einstein, kept the problem of "life" at arm's length, yet it was his intuitive insight that led him to $E = mc^2$, transcending the stigma of descriptive physics by equating mass and energy. He had already seen the "forest for the trees" at 16 years of age, dreaming that he was traveling in tandem with a light beam (see *Einstein* by Walter Isaacson). Like scientific feng shui, he was able to conceive of the fundamentals of the physical world – Brownian movement, the photoelectric effect, and relativity theory – all in his wunderjahr of 1905. Of course, he famously said that "G_d does not play dice with the Universe," so he would not have agreed with the conventional stochastic approach to evolution.

But perhaps the solution to the evolution puzzle is not based on random mutation and population selection – biology and medicine are on the threshold of a conceptual breakthrough on a par with such breakthroughs as heliocentrism and relativity theory. By subordinating descriptive biology to cell-molecular signaling in development, homeostasis, and regeneration as the essence of evolution, the fundamental mechanism of life, we will be able to understand the "Inner Universe" of physiology, starting from its origins, resynthesizing biology from the bottom up. Embracing this approach to physiology would be a game changer.

Galileo is widely considered to be the father of modern science because his ideas were not derived solely from thought and reason. He was guided by experimentation and observation. This was a revolutionary change in science – observational experience being the key method for discovering Nature's rules. By understanding how celestial bodies moved through the night sky, Galileo wrested power from the astrologers, placing it in the hands of the astronomers and their scientific heirs. That changed the course of human history.

A Cellular-Molecular Model of Evolutionary Biology

The formulation of Quantum Theory launched modern physics, providing a mechanistic explanation for the origin and formation of the Universe, beginning with the Big Bang. This represented a major transition from Newtonian mechanics, which was deterministic, to relativity theory, which is probabilistic. Yet biology remains wedded to descriptive mechanisms, not to the underlying mechanicism of biology. Evolution is the mechanism underlying all of biology, whereas natural selection, survival of the fittest, and descent with modification are all metaphors. In this book, we provide the biologic analog of Quantum Mechanics by showing how unicellular organisms gave rise to derivative multicellular organisms through the self-organizational cellular intersection of negentropy, information acquisition, and homeostasis. All of these components have been addressed in the biologic literature at one time or another over the course of the past 50 years, yet they have not been formulated as a mechanistic process that would explain evolution, especially as it applies to non-fossilized tissues and organs. The advent of cell–cell interactive cellular-molecular embryology and phylogeny offers the opportunity to model vertebrate evolution. Herein, we provide a unique perspective on evolutionary biology.

The Evolutionary Continuum from Development to Physiologic Homeostasis, Repair and Reproduction

The central concept of this book is that there is a biologic continuum mediated by cell–cell communication that starts with the zygote, generating multicellular organisms, facilitating reproduction, culminating in the next generation of the organism, et cetera, et cetera. The experimental breakthrough in this way of thinking about biology occurred in 1978, when George Todaro discovered that cultured cells produce soluble growth factors. These substances allow cells to communicate with one another in space-time through specific signaling receptors, mediating the generation of structure and function. At birth, many of these signaling mechanisms become the agents that determine physiologic homeostasis and repair.

Evolution is said to be "emergent and contingent." Moreover, newly evolved biologic traits are known to have their antecedents in earlier phylogenetic forms of the same species, what is referred to as preadaptation or exaptation. If you follow that precept back far enough, you end up in the unicellular state, which King, Hittinger, and Carroll have shown to be the origin of multicellular organisms.

And if we trace the evolution of eukaryotes all the way back to the single-celled form, we see homologies with prokaryotes, such as negentropy, bioenergetics, and homeostasis, that are common to both. In fact, bacteria express sterol-like hopanoids that are under control by oxygen, as is the case for sterols in eukaryotes, yet they do not serve the same fundamental purpose of rendering the cell membrane interactive with the environment. Instead, bacteria possess a hard exterior cell wall that isolates the internal milieu. In contrast to this, eukaryotes possess a compliant cell membrane that facilitates metabolism, respiration, and locomotion. And the endomembranes that derive from the cell membrane provide the cell with the ability to form compartments in which factors derived from the external environment that would otherwise have been harmful – oxygen, nitrogen, ions, heavy metals, gravity – are exploited for physiologic functions. These structural-functional relationships between prokaryotes and eukaryotes offer an opportunity to consider the possibility that we are still part of that continuum, particularly in light of the recent recognition that the microbiome is an integral component of the hologenome.

Resonance at the Denouement of this Book

If we may wax a bit poetic, once we determine the mechanisms that resonate throughout biology, we will truly be able to understand our origins, and predict our future as a species. Like the hierarchical relationships generated by the Big Bang, biology shares a oneness with the Universe because of its fundamental mimicking of the external environment, not unlike a great literary or musical composition, though that falls short because it is a creation of the human mind, unlike evolution. Biology resonates from unicellular to multicellular life, derived from the Cosmos, as George Gurdjieff and the monk Teilhard de Chardin had proposed, culminating in the human mind and what Raymond Maslow referred to as "peak experiences." That being said, once we determine the interrelationship between the physical world and the human mind, we will truly be able to act in harmony with our environment, organic and inorganic alike.

Why is Physiology?

It should be obvious to even the most casual observer that there are orthogonal patterns of inheritance, like those of physiology and physiognomy, that is, things just seem to "fit together." After all, the physical Universe was generated by the Big Bang, creating an informational hierarchy by distributing the elements based on their mass. The biologic "Universe" was generated by mimicking that physical Universe. Even at the molecular level, we see that genes "fit together" into clusters, or cassettes. Perhaps the most striking examples are the homeobox genes, which align on the chromosome in the same sequence as they are expressed during embryology – how could this have happened by chance? I maintain that this cannot merely be the result of random variation and phenotypic selection, whereby over eons such associations occurred adaptively, so natural selection favored those organisms that expressed them as such.

Alternatively, this may have been the net result of ongoing interactions between internal and external selection, internal selection resulting from physiologic stress, generating radical oxygen species that are known, for example, to cause gene duplications closely associated with evolutionary change. The epitome of internal selection is Haeckel's Biogenetic Law (Figure 1.2) – ontogeny recapitulates phylogeny – dismissed by contemporary evolutionists for lack of "hard" fossil evidence but rejuvenated by molecularists providing evidence for ghost lineages due to epigenomic inheritance mediated by genetic "marks." We have suggested that perhaps the process of phylogenetic recapitulation during embryonic development is necessary in order to ensure that any given genetic change, particularly those due to epigenetic inheritance, is internally consistent with the subsequent homeostatic adaptations fostered by such an event – we see this repeatedly when experimentally manipulating genes, resulting in either some phenotypic developmental change, or embryonic lethality, which we refer to as either "proof of principle" in the case of the former, or casually dismiss as being incompatible with life in the case of the latter, without further delving into the underlying mechanisms involved.

Starting with the generation of unicellular organisms, life forms have internalized and mimicked the external physical environment, "fueled" by negentropy, perpetuated and molded by homeostasis, structurally realized through endomembrane compartmentation...or perish. For example, early in the evolution of primitive cells, cytoplasmic calcium had to be drastically reduced in concentration to avoid the denaturing of proteins, nucleotides, and lipids. This was achieved by cholesterol forming calcium channels in the cell membrane, fostering integrated physiology, from unicellular to multicellular eukaryotes.

We have observed and documented this phenomenon for the evolution of the lung, genetically annotating the processes of lung ontogeny and phylogeny at the cellular-molecular level. As a result, we have documented a seamless, alternating pattern of internal and external selection for specific functional traits, accommodating both

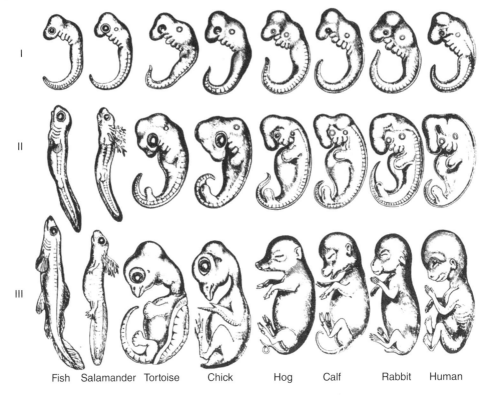

I

II

III

Fish Salamander Tortoise Chick Hog Calf Rabbit Human

Figure 1.2 Haeckel's Biogenetic Law. Note the similarity between the early stages of all the organisms pictured. Such similarities are lost as these species grow and differentiate. From Blancke, 2010. Reproduced with the permission of John Wiley and Sons.

external and internal selection pressure. Since this mechanism of lung morphogenesis is determined by cell–cell interactions, the physical compartmentation of the genes expressed by the endoderm and mesoderm determines the ontogeny and phylogeny of the lung, providing a self-referential, self-perpetuating interactive mechanism for constantly monitoring and gleaning information from the environment, resulting in optimalized physiologic adaptation on an ongoing basis.

Experimental evidence for such an interactive cellular-molecular process (Figure 1.3) comes from studies to determine the effect of gravity on fundamental phenotypic cellular expression. In organs as disparate as the skeleton (fossil evidence shows that vertebrates attempted adaptation to land on at least five separate occasions), lung, and uterus, PTHrP signaling is mechanosensitive; those reiterative attempts to breach land must have concomitantly caused physiologic stress on the visceral organs, resulting in microvascular shear stress within the capillaries, generating radical oxygen species known to cause gene duplications; and perhaps even the PTHrP receptor and β-adrenergic receptor gene duplications that occurred during the water–land transition. We hypothesized that the resultant amplification of PTHrP signaling promoted the various organ-level adaptations for land habitation – lung, skin, bone, kidney – as exaptations of ancient biologic adaptations to gravity, which is the most ubiquitous, unidirectional, unrelenting environmental effector of vertebrate biology. For example, microgravity

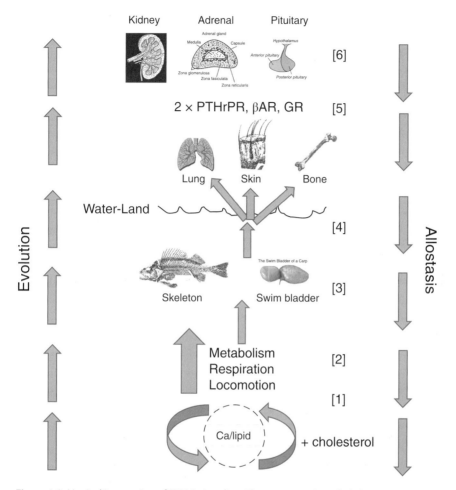

Figure 1.3 Vertical integration of PTHrP signaling. The ontogenetic and phylogenetic integration of calcium-lipid homeostasis, **[1]** from protocell incorporation of cholesterol into the plasmalemma for **[2]** metabolism, respiration, and locomotion, the foundation of vertebrate evolution. **[3]** Coevolution of the skeleton and swim bladder of physostomous ray-finned fish **[4]** under physiologic stress of the water–land transition gave rise to the lung, skin, and bone phenotypes of land vertebrates, **[5]** largely through the duplication of the parathyroid hormone-related protein (PTHrP), beta-adrenergic (βA), and glucocorticoid (G) receptors. **[6]** Expression of PTHrP in the pituitary, adrenal cortex, and kidney glomus gave rise to endothermy and the glomerulus.

causes both cellular depolarization and failure to form buds in yeast. Cell polarity is necessary for integrated physiology, whereas budding is the means by which yeast reproduce. Hence, there are deep homologies between the simplest and most complex eukaryotic traits that are mechanistically linked through the effects of gravity.

In the lung, mechanotransduction via fluid distension of the airways *in utero* determines the epithelial-mesenchymal interactions underlying gene regulatory mechanisms for alveolar barrier formation, including surfactant biosynthesis, intracellular matrix production, and tubular myelin formation. When lung alveolar epithelial cells are cultured under $0 \times g$ conditions the cells round up, losing their polarity, and their PTHrP signaling capacity declines, reaching a newly established, stable baseline of expression several orders of magnitude less than in unit gravity several hours later, resulting in the loss of cell–cell signaling for lipid metabolism. As noted above, uterine PTHrP expression is also stretch-regulated during pregnancy, regulating the rate of calcium flux between the mother and fetus. Hence, there appear to be highly conserved mechanisms common to cellular polarity, metabolic homeostasis, and reproduction alike that have evolved in adaptation to gravity. Even a simple eukaryotic yeast will lose both its polarity and ability to duplicate, or bud, in $0 \times g$. In retrospect, this should not be surprising, given that actin and tubulin are common to mitosis, meiosis, and homeostasis alike.

These physical adaptations may have originated from the cytoskeleton, since it is homologous to both unicellular organisms and to bone, the latter regulating calcification through PTHrP in compliance with Wolff's law. That is to say, the skeletal adaptation to the water–land transition may have created the selection pressure for PTHrP signaling in other visceral organs, fostering the phenotypic adaptations seen in the lung, skin, and bone alike.

In support of this hypothesis (see Figure 1.3), PTHrP is also expressed in the anterior pituitary, where it regulates Adrenocorticotropic Hormone (ACTH), and in the adrenal cortex, where it controls corticoid synthesis. The net effect of such a PTHrP-dependent hypothalamic-pituitary-adrenal (HPA) axis would be increased adrenaline production by the adrenal medulla, due to the corticoid stimulation of the rate-limiting step in catecholamine synthesis in response to physiologic stress. The most potent naturally occurring physiologic stressor is hypoxia, which would have occurred transiently due to fluctuations in atmospheric oxygen during the Phanerozoic eon (Figure 1.4) as the lung evolved from the swim bladder of bony fish to the lungs of amphibians, reptiles, mammals, and birds; that is, the gradual increase in lung surface area occurred in response to metabolic demand, and would periodically have been constrained, resulting in periods of hypoxia, stimulating the HPA and adrenaline production, and releasing the bottleneck on the lung by increasing surfactant production ad hoc. Therefore, there appears to have been a mechanism of physiologic evolution arising from external environmental stress, from unicellular to multicellular organisms, contingent on cellular compartmentation of genetic expression.

Using this cellular approach, even the evolution of the central nervous system can be traced from unicellular to multicellular organisms. Both paramecia and neurons communicate environmental information through calcium flux, in the case of the former as hyperpolarization, in the case of the latter as polarization and depolarization. Therefore, the evolution of the human brain can be seen as an arc of consciousness, beginning with the simplest organisms, culminating in creativity – "first there were bacteria, now there is New York" (Simon Conway Morris in *Complexity and the Arrow of Time*).

As independent corroboration of this foundational conceptualization for the origins of physiology, Jack Szostak has shown that lipid-based protocells can generate nucleotides, whereas nucleotides cannot produce lipids. This indicates that micelles preceded DNA-RNA, which may have evolved in supporting the evolutionary effort by providing a "long-term" memory for the adaptational strategy. That is to say, the nucleotide "memory mechanism" would have put those organisms that evolved DNA at an advantage due to their "reference library" for adaptive strategies rather than random selection. The obvious efficiency of the former would have put such organisms at an advantage, ultimately becoming the mode of life to the exclusion of non-DNA-based life forms.

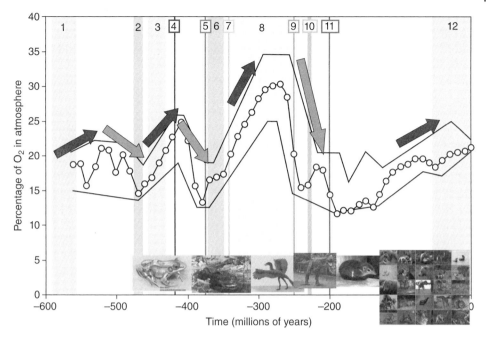

Figure 1.4 Phanerozoic oxygen, hypoxia and physiologic stress. Over the course of the last 500 million years, called the Phanerozoic eon, atmospheric oxygen levels have varied between 15% and 35%, fluctuating fairly drastically. The increases have been shown to foster metabolic drive, but the physiologic effects of the decreases have been overlooked. Hypoxia is the most potent physiologic agonist known, so the episodic decreases in atmospheric oxygen would have fostered remodeling of the internal organs. (*See insert for color representation of the figure.*)

These ontologic insights to the origin, creation, and perpetuation of physiology suggest a long-awaited approach to elucidating mechanisms for what Aristotle referred to as *entelechy*, a factor that directs the individual regularities of organisms; what Rupert Riedl referred to as *burden*, the responsibility carried by a feature or decision; what Wallace Arthur called *bias*; and what L.L. Whyte referred to as *coordinative conditions*: "The coordinative conditions hold the clue to the relation of physical laws to organic processes and to the unity of the organism." Many have invoked the existence of an internal principle as the basic problem of evolutionary theory without providing a mechanistic understanding. Yet, we know that evolution is not a conscious process. The power of Darwinian evolution is in usurping the vitalists, yet they are encroaching yet again as creationists and intelligent designers!

The supporters of Synthetic Theory, on the other hand, hold that no causal mechanism has yet been proven. They are also concerned that the search for it would allow for such unproven phenomena as finality and entelechy. Entelechy, a concept taken from Aristotelian metaphysics, is assumed to be a factor that directs the individual regularities of organisms, specifically, their orderliness, harmony, plan, or goal. Entelechy would arise from the pre-established harmony of living organisms, which is outside the realm of science. Instead, these biologists tend to discount the problem, which is unfortunate. In some cases, they even claim that there is no place in the modern synthesis for an internal principle. However, it is important to note that even the authorities on this viewpoint, such as Theodosius Dobzhansky, Curt Kosswig, and Ernst Mayr, acknowledge that the epigenetic system confers a fundamental, but not fully understood, ordering effect. They also ask whether this pattern of mutual gene effects will ever be understood, because of its complexity.

We maintain that the cellular-molecular, ontogenetic/phylogenetic approach provides the solution to this age-old quandary. Our working hypothesis is that the unicellular stage of the life cycle is the level of selection, going from zygote to zygote. This perspective has several advantages. It is based on the functional origin of life from the unicellular state, which naturally would integrate genotype and phenotype. And the iterative return to the unicellular state by metazoans during reproduction would provide the "filter" for newly acquired epigenetic marks from the environment, using ontogeny to recapitulate phylogeny as a means of determining the fidelity of any given mutation – reprising Haeckel's Biogenetic Law (see Figure 1.2) after laying fallow for more than 200 years.

Additionally, all three states of unicellular organisms – homeostatic, mitotic, and meiotic – are determined by signaling from the cytoskeleton, which is ubiquitous in eukaryotes. Therefore, the unicellular state is self-organizing and autonomous.

In this Chapter, we have set the stage for a paradigm shift in the way we think about the evolution of vertebrate physiology based on mechanisms of cell–cell interactions. Chapter 2, entitled "On the Fractal Nature of Evolution," will attempt to show the value added in the cellular-molecular approach to evolution, beginning at the origins of life. It is predicated on the balancing selection of calcium and lipids as the fundament of vertebrate evolution.

Selected Reference

King N, Hittinger CT, Carroll SB. (2003) Evolution of key cell signaling and adhesion protein families predates animal origins. *Science* **301**:361–363.

2

On the Fractal Nature of Evolution

Life is that which can mix oil and water
Robert Frost, The Notebooks

The mathematician Benoit Mandelbrot has described physiology as fractal. That is to say, it is self-similar, due to the underlying, integrative mechanisms of cellular ontogeny, phylogeny, and homeostasis. This process is self-referential, harking back in vertebrate phylogeny to unicellular eukaryotes. What are the integrating mechanisms that would account for the evolution of multicellular organisms from unicellular life? If they were known, they would provide fundamental insights to the mechanisms of evolution.

Elsewhere, we have argued that the "first principles of physiology" can be determined since they are the mechanistic basis for ontogeny and phylogeny, which are one and the same process when seen from the cellular perspective, occurring on different timescales, that is, they are diachronic (as discussed later). Given that, we maintain that the innate organizing principle of physiology – homeostasis – is fractal. We hypothesize that the epistatic balancing selection between calcium and lipid homeostasis was essential to the initial conditions for eukaryotic evolution, starting the process of vertebrate evolution, continuously perpetuating and embellishing it from unicellular organisms to metazoans in all phyla.

We would like to use a classic misapprehension of the mechanism of evolution to illustrate the difference between Darwinian mutation and selection, on the one hand, and cellular-molecular evolution from unicellular organisms on the other. In their paper, entitled "The spandrels of San Marco and the Panglossian paradigm," Stephen J. Gould and Richard Lewontin used the metaphor of the spandrels that were used in Byzantine Venetian architecture to fill in the gaps in the mosaic design of the central dome of St Mark's Cathedral, to disabuse their audience of the notion that everything in biology is purposeful, and that therefore there must be an underlying selection mechanism for them as well. However, if metazoans are derived from unicellular organisms, perhaps *all* metazoans are actually spandrels, derived from the organizing principles of unicellular life. For example, it is commonplace in the evolution literature for authors to point to evolved traits as "preadaptations," the inference being that the trait pre-existed in an ancestral form. Carried to its logical extension, this would have led to the prediction that metazoans originated from unicellular organisms. In turn, if metazoans are derived from unicellular organisms, then by determining the homologies that interconnect unicellular and multicellular organisms perhaps we can determine the underlying mechanisms involved.

Evolution, the Logic of Biology, First Edition. John S. Torday and Virender K. Rehan.
© 2017 John Wiley & Sons, Inc. Published 2017 by John Wiley & Sons, Inc.

Fractal Physiology: How and Why?

Like relativity theory, the biology of multicellular organisms is also due to the interrelationships between space and time. In both cases, the Big Bang radiated out from its point of origin to give rise to both the physical and the biologic world. In the case of the physical world, it generated and then scattered the elements throughout the Universe, as evidenced by the so-called redshift – when visible light or electromagnetic radiation from an object moves away from the observer, increasing in wavelength and shifting to the red end of the light spectrum – as well as creating the background radiation of the Universe detected by radio telescopes. The scattering of the elements based on their mass renders a structural hierarchy of "information." By contrast, the biologic "Big Bang" was caused by the mimicking of that informational hierarchy of the physical Universe, biology generating an internal "pseudo-Universe" through the dynamic interactions generated by cellular information gathering, negentropy, and homeostasis. Biologic "relativity" is due to the interactions between cells of different germline origins – ectoderm, endoderm, and mesoderm. The germlines within the organism develop through interactions between one another, via soluble growth factors that signal to their cell-surface receptors residing on neighboring cells expressed in a different germlines. Those signaling motifs, or gene regulatory networks (GRNs), are largely expressed as a function of their specific germlines, providing physical reference points for the physiologic internal environment to orient itself to the external physical environment. The genes take their "cues" from the interaction between the germline cells to generate form and function relative to the prevailing environmental conditions.

As a result, the organism can evolve in response to the ever-changing environmental conditions, the genes of the germline cells "remembering" previous iterations under which they (by definition) successfully mounted an adaptive response. Then, by recapitulating the germline-specific GRNs under newly encountered conditions, they may form novel, phenotypically adaptive structures and functions by recombining and permutating the old GRNs. This process is conventionally referred to as "emergent and contingent." It explains how and why the same GRN can be exploited to generate different phenotypes as a function of the history of the organism, both as ontogeny (short-term history) and phylogeny (long-term history), the germlines orienting and adapting the internal environment to the external environment by expressing specific genetic traits.

For example, seen from the traditional perspective, this phenomenon is described as pleiotropy. However, once we realize that this is actually the consequence of an ongoing process, and not merely a chance occurrence, the causal relationships become evident, offering the opportunity to understand how and why form and function have evolved. Such interactive, cellular-molecular mechanistic pathways project both forward and backward in time and space, offering the opportunity to understand our unicellular origins, and the functional homologs that form the basis for the "first principles of physiology."

In the Beginning

Life likely began with the formation of liposomes through the agitation of lipids in water (Figure 2.1), bearing in mind that the Moon broke away from the Earth 100 million years after the Earth's formation, offering billions of years for wave action to fashion life in

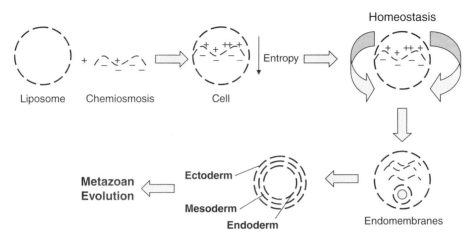

Figure 2.1 On the evolution of form and function. Starting with the formation of micelles and chemiosmosis, life began as the reduction in entropy, maintained by homeostasis. The generation of endomembranes (nuclear envelope, endoplasmic reticulum, Golgi) created intracellular compartments; compartmentation gave rise to the germlines (ectoderm, mesoderm, endoderm), which monitor environmental changes and facilitate cellular evolution accordingly. This is the basis for "fractal physiology." (*See insert for color representation of the figure.*)

this way. That interaction was chemiosmosis generating bioenergy, aiding and abetting negentropy, maintained by homeostasis. Endomembranes such as the nuclear envelope, endoplasmic reticulum, peroxisome and Golgi apparatus formed intracellular compartments; compartmentation gave rise to the germlines (ectoderm, mesoderm, endoderm) that monitor environmental changes and facilitate cellular evolution accordingly. These are the elements that fostered "fractal physiology."

Chemiosmosis

Peter Mitchell proposed his chemiosmotic hypothesis in 1961, that adenosine triphosphate (ATP) synthesized in aerobic cells is generated by the electrochemical gradient formed across the inner membranes of mitochondria using the energy of nicotinamide adenine dinucleotide (NAD) and flavin adenine dinucleotide (FAD) hydroquinone produced by the breakdown of glucose.

Glucose is catabolized to acetyl CoA, which is oxidized in the mitochondrial matrix, where it is chemically coupled to the reduction of such carrier molecules as NAD and FAD. These carriers shunt electrons to the electron transport chain (ETC) within the inner mitochondrial membrane, transferring them on to other proteins in the ETC. The energy of the electrons pumps protons from the mitochondrial matrix across the inner mitochondrial membrane, storing the energy in a transmembrane electrochemical gradient. The protons are shunted back across the inner mitochondrial membrane via the enzyme ATP synthase. The passage of protons back to the mitochondrial matrix by means of ATP synthase allows ADP to bond with inorganic phosphate to form ATP. The electrons and protons subsequently react with oxygen to form water.

On the Evolution of Calcium Homeostasis

With the advent of the first cell, the biologic world segregated into intra- and extracellular spaces, immediately controlling the ion content of the cytoplasm. The first cell was defined by a membrane composed of ion-conducting pores and ion pumps in combination with chemiosmosis – the movement of ions across a selectively permeable membrane – as the source of energy to maintain entropy far from equilibrium. Failure of any of these components would have flooded the cytosol with calcium, testing the viability of the molecular machinery, potentially obviating the possibility of life.

The ion carrier characteristics were determined by the chemical composition of the primordial oceans. Only a few ions were present at the time – sodium, chloride, magnesium, calcium, potassium, and other trace ions. Calcium ions were the most problematic because they universally denature proteins, lipids, and nucleic acids. At unphysiologic concentrations, calcium aggregates proteins and nucleic acids, disrupts lipid membranes, and precipitates phosphates. High cellular calcium levels are incompatible with life; from bacteria to eukaryotes, excess cellular calcium levels are cytotoxic. Thus, prototypical life forms needed calcium regulation to maintain intracellular calcium at physiologic concentrations, which were orders of magnitude lower than the extracellular levels. Even the most primitive bacteria have cell membrane calcium pumps.

High calcium gradients have typified life ever since its inception. Maintenance of calcium homeostasis demands substantial energy consumption, so it is no wonder that evolution has led to adaptation of calcium regulatory systems whose initial function was to protect against the otherwise-overwhelming calcium pressure, leading to ever more complex physiology.

On the Nature of Calcium Channels

Prokaryote calcium channels likely acted as signaling mechanisms. The ubiquitous transmembrane calcium gradients referred to above produced substantial currents upon channel opening, though facilitated repolarization still was not necessary since the calcium signal could be stopped by just closing the calcium channel, allowing for dispersal of the calcium. This is correlative of the observation that voltage-gated sodium channels are virtually absent from bacteria and less advanced animals. Yet certain bacteria concomitantly express a sodium-permeable analog of the calcium channel. Since this analog is structurally similar to L-type calcium channels, but is selective for sodium, it may represent a precursor of voltage-gated sodium channels. Calcium-dependent potassium channels, which are sensitive to both voltage and intracellular calcium, likely evolved in prokaryotes. Such calcium-dependent potassium channels were fortuitous, as the rise in intracellular calcium mediated through voltage-gated channels also activated a repolarization mechanism that rapidly inhibited calcium influx. The presence of the calcium-dependent potassium channels in prototypical organisms suggests their early evolution. Having few cell membrane channels and calcium pumps, prokaryotes established prototypical calcium signaling systems. Intracellular calcium flux became useful for regulating bacterial locomotioin, chemo- and phototaxis, survival maneuvers, and sporulation.

The Inception of Calcium-Storing Organelles

The fundamental difference between prokaryotic and eukaryotic cells is that the former do not have a nucleus or other intracellular organelles that are membrane bound, subdividing the cytoplasm into specialized compartments with distinct mechanisms of calcium regulation. The appearance of the nuclear envelope marked the inception of this endomembrane system. Prototypical eukaryotes evolved endosymbiotic relations with protobacterial rickettsiae, which eventually evolved into mitochondria. These rickettsiae already possessed calcium channels and sodium-calcium ion exchangers, enabling them to act independently as calcium-regulating and -storing/releasing organelles. And since uniporter-mediated calcium entry into the mitochondria regulated the activity of oxidative enzymes, and hence ATP synthesis, an increase in cytosolic calcium was now capable of linking cell activity with energy production. The origin of the secondary calcium storage source, the endoplasmic reticulum, is unclear. Tubular cisternal endoplasmic reticulum-like structures may have developed from yet another endosymbiotic event with microtubule-forming protoorganisms, or else have been due to specific invaginations of the plasmalemma, or to primordial vacuoles. Whatever their origin, it is indisputable that the endoplasmic reticulum and its endogenous enzymatic systems appeared very early in eukaryotic development. Even prototypical protozoa, which are devoid of mitochondria, express the endoplasmic reticulum and its indigenous heat-shock proteins. In any case, the advent of the endoplasmic reticulum generated an extended, omnipresent organelle for protein synthesis, posttranslational protein modification, and endogenous calcium transport. Significantly, the endoplasmic reticulum has evolved its own endogenous mechanism of calcium homeostasis. In contrast to the intracellular space, where maintenance of low concentrations is existential, the calcium concentration in the endoplasmic reticulum is comparable to that in the extracellular space. Such a high endoplasmic reticulum calcium concentration enabled a second level of calcium signaling complexity, since ionic calcium could enter the intracellular space from either the external space or from an intracellular source. This represents yet another example of the cellular strategy of using otherwise toxic environmental factors for endogenous biologic adaptation. Calcium efflux from endoplasmic reticulum stores is now known to be controlled by several sets of intracellular channels, namely the calcium-gated ryanodine receptors (RyRs), inositol-trisphosphate-calcium-gated receptors (InsP$_3$Rs) and nicotinic acid adenine dinucleotide-controlled receptors. The phylogeny of the intracellular channels is poorly characterized, yet very early on eukaryotes already possessed G-protein-coupled receptors and inositol trisphosphate-associated second messenger systems that stimulate intracellular calcium mobilization by means of inositol trisphosphate receptors. However, inositol trisphosphate signaling appears to be more important in more complex eukaryotes. The ryanodine receptors appeared later in evolution. With the appearance of calcium stores, a new calcium influx pathway controlled by calcium within the endoplasmic reticulum became physiologically relevant.

And Calcium-Binding Proteins

Prototypical eukaryotes manifested cellular polarity early in evolution, requiring precise localization and generation of calcium signals. Some primitive eukaryotes had survived by forming colonies from aggregates of single-celled organisms that are the ancestors of multicellular organisms. Such developments increased the need for signaling systems, particularly in their encoding and targeting. These demands were met by the appearance of a versatile class of "calcium effector/sensor" or EF motif-containing proteins. The EF motif of EF-hand proteins refers to a helix–loop–helix structure that specifically binds calcium ions. Linking two EF motifs greatly enhanced their calcium affinity, forming the main functional unit for EF-hand proteins. Functionally, when calcium binds to EF motifs it modifies their stereospecific protein conformation, facilitating their function in response to calcium binding. The acquisition of these properties likely established EF-hand calcium binding proteins as functional intracellular calcium sensors, ensuring evolutionary success. The EF-hand calcium sensors are virtually absent from bacteria, are abundant in yeast, and are omnipresent in multicellular organisms. Representatives of such calcium-binding proteins include calmodulin, troponin C, calpains, and all of the calcium buffering proteins. The appearance of the EF-hand proteins and their ability to act as sensitive calcium sensors conferred signaling specificity par excellence to a previously dull signaling system.

Calcium Signaling Facilitates the Transition from Polarized Cells to Multicellular Organisms

Cell polarity required more efficient calcium signaling pathways since EF-hand protein function had not yet fully evolved. Such binding proteins rapidly mop up any calcium ions passing through the plasmalemma, slowing their cytoplasmic diffusion. Strong calcium buffering tends to sequester cytoplasmic calcium signals, promoting spatial precision and coding for calcium signals, as characterized as calcium microdomains. Concomitantly, cytoplasmic calcium binding retards broader cytoplasmic calcium diffusion in polarized cells. The challenge to broader calcium signaling was reconciled by the unusually effective calcium management properties of the endoplasmic reticulum. The endoplasmic reticulum maintains high intracellular calcium concentrations, and contains calcium-binding proteins with affinities in the 10^{-3} M range. Consequently, calcium readily diffuses through the endoplasmic reticulum compared to the cytoplasm. This relationship offers an opportunity for long-distance calcium transport through intra-endoplasmic reticulum calcium channels. The endoplasmic reticulum intracellular calcium release channels also convey excitation properties to the endomembrane, so that any regenerative recruitment of ryanodine receptors/inositol trisphosphate receptors generates propagating calcium waves. The appearance of multicellular aggregates and proto-multicellular organisms also brought about the mechanism of apoptosis, or programmed cell death: Some cells had to die in a regulated manner to allow other cells to live and thrive. Here too, calcium played a fortuitous role, since rapid disruption of calcium homeostasis is the most efficient mechanism for cellular apoptosis. Thus calcium-dependent death motifs began regulating tissue homeostasis and development. Cell polarization was an initial step in fostering cell specialization and functional multicellular organisms. Obviously, multicellular organisms required intercellular signaling and highly evolved mechanisms for encoding and deciphering. There are two distinctly different mechanisms of cell–cell communication – directly, through intercellular contacts (such as gap junctions), and indirectly, through the release of active chemicals such as hormones and neurotransmitters. The chemical mechanism was the most primitive since even protocells needed to sense changes in the molecular composition of their environment. As a result, diffuse chemical signaling was widespread in ancient multicellular organisms. The utilization of such signals required a system of regulated release of packaged hormones, this being achieved by regulated exocytosis. As vesicular release acquired calcium sensitivity, the regulation of exocytosis became yet another important function of calcium. Diffusion of hormones, however, represents a problem for complex multicellular and multi-tissue organisms, since they have a rather non-specific global action, which may be beneficial for general regulation, but is useless for specific local information transfer. Therefore, there was a pressing need to develop focal mechanisms for the release of discrete amounts of chemicals that could exert their action in a very local space. Once again, this was achieved by calcium-regulated exocytosis, operational in the context of spatially and temporally limited release sites. Brief episodes of calcium entry through plasmalemmal channels generate short-lived, high-magnitude calcium microdomains, controlling local, rapid release of neurotransmitters. A complementary mechanism is the localized expression of high-level calcium signaling complexes, including the receptors, to generate highly responsive cellular microdomains. Life, initially confronted with the existential necessity for precise control over cellular calcium concentrations and movements, has developed a relatively limited number of calcium handling systems, including membrane calcium channels (both plasmalemmal and intracellular), cytoplasmic calcium buffers, and several sets of membrane calcium transporters, represented by calcium pumps and calcium exchangers. These systems appeared early in evolution, and became conserved phylogenetically. Most importantly, however, various combinations of the components of calcium homeostatic systems offer almost endless permutations and combinations for creating "calcium signaling toolkits," which in turn determine the versatility and individuality of calcium signaling events in various cell types under varied conditions. Indeed, calcium signals resulting from cell activation appear in multiple forms, from very local microdomains, which, for example, regulate neurotransmitter release, post-synaptic plasticity, or cell process guidance, to global calcium signals, which control excitation–contraction or excitation–secretion coupling, and regulate gene expression and tissue development, to propagating calcium waves, accomplishing integration in complex multicellular structures, such as in the pan-glial syncytia in the brain. At the same time, the calcium signaling machinery is responsible for the control of cell survival, and numerous calcium-dependent pathways trigger various modes of cell death, which are indispensable for both normal tissue development and homeostasis. At their very core, however, calcium-homeostatic systems remain deeply rooted in the life process, and their damage is universally detrimental, being intimately related to the etiology of pathologic processes.

Why is Physiology? a Vertical Integration of Ca, Lipid, and Homeostasis

There are patterns of inheritance, like those of physiology and physiognomy, that is, things just seem to "fit together." Even at the molecular level, we see that genes "fit together" into clusters, or cassettes. The most striking example of this are the home-obox genes, which align in the chromosome in the same sequence as they are expressed during embryogenesis – how could that have happened? I maintain that this cannot be explained merely by the neutral theory, namely that over eons such associations occurred and were adaptive, so natural selection favored organisms that expressed them as such. Alternatively, this may have been a consequence of ongoing interactions between internal and external selection, starting with the formation of primitive unicellular organisms, mimicking and internalizing the external physical environment, "fueled" by negentropy, perpetuated and molded by homeostasis...or perish.

For example, early in the evolution of primitive cells, cytoplasmic calcium had to be evacuated to avoid the denaturing of proteins, nucleotides, and lipids. This was achieved by cholesterol forming calcium channels in the cell membrane. The arc of this interrelationship between calcium and cholesterol is seen all the way from the origins of life to highly integrated physiologic processes. For example, endoplasmic reticulum stress (ER stress) causes the breakdown of lipid metabolism, which is manifested in many organs.

We have observed and documented this phenomenon in the evolution of the lung, genetically annotating the processes of lung ontogeny and phylogeny at the cellular-molecular level. As a result, we have documented the seamless, alternating pattern of internal and external selection for specific functional traits relevant to external and internal selection pressure. Since this process is determined by cell–cell interactions, the physical compartmentation of the genes expressed by the endoderm and mesoderm determines the ontogeny and phylogeny of the lung, providing a self-referential, self-perpetuating interactive mechanism for constantly gleaning information from the external environment in order to optimize physiologic adaptation.

Experimental evidence for such an interactive cellular-molecular process comes from studies to determine the effect of gravity on fundamental phenotypic cellular expression. In organs as disparate as the skeleton (fossil evidence showing that aquatic vertebrates attempted adaptation to a terrestrial environment on at least five occasions), lung, and uterus, parathyroid hormone-related protein (PTHrP) signaling is mechanosensitive; those iterative attempts to breach land concomitantly caused physiologic stress on the internal organs, with consequent microvascular shear stress, generating radical oxygen species known to cause gene duplications, and perhaps even the PTHrP receptor gene duplication, which is known to have occurred in the water–land transition. We hypothesized that this amplification of PTHrP signaling promoted the various visceral organ adaptations to land – lung, skin, bone, kidney – as exaptations of ancient biologic adaptations to gravity, which is the most ubiquitous, unidirectional, and relentless environmental effector of vertebrate biology.

In the lung, mechanotransduction via fluid distension of the airway *in utero* determines the epithelial–mesenchymal interactions underlying gene regulatory mechanisms for alveolar barrier formation, including surfactant biosynthesis, intracellular matrix production, and tubular myelin formation. When lung alveolar epithelial cells are cultured under microgravitational conditions the cells round up and lose their polarity. In addition, their PTHrP signaling mechanism declines, reaching a newly established and stable baseline of expression, which results in the loss of cell–cell signaling for lipid metabolism. As noted above, uterine PTHrP expression is also stretch-regulated during pregnancy, regulating calcium flux between the mother and fetus. Interestingly, when yeasts are cultured under microgravitational conditions, they also lose their polarity, and they cannot form buds, which is their means of reproduction. Hence, there appear to be highly conserved mechanisms for determining cellular polarity, metabolic homeostasis, and reproduction that have evolved in adaptation to gravity. These physical adaptations may have originated from the cytoskeleton since the latter is common to both unicellular organisms and to bone; in the latter calcification is regulated via PTHrP in compliance with Wolff's law. That is to say, the skeletal adaptation to the water–land transition may have created the selection pressure for PTHrP signaling in other internal organs, fostering the phenotypic adaptations seen in the lung, skin, and bone alike.

In support of this hypothesis, PTHrP is also expressed in the anterior pituitary, where it stimulates ACTH, and in the adrenal cortex, where it stimulates corticosteroid synthesis. The net effect of such a PTHrP-dependent hypothalamic-pituitary-adrenal (HPA) axis would be increased adrenaline production under physiologic stress. The most potent naturally occurring physiologic stressor is hypoxia, which happened periodically as the fish swim bladder evolved into the lungs of amphibians, reptiles, mammals, and birds; that is, the progressive increase in lung surface area occurred in compliance with the demand for more efficient metabolism, and would periodically have resulted in hypoxia, stimulating the HPA axis and adrenaline production, and temporarily releasing the evolutionary physiologic constraint on the lung by increasing surfactant production ad hoc. This scenario would have been accommodated by natural selection for those members of the species able to upregulate their PTHrP signaling mechanism, and/or through local mutation of PTHrP signaling elements due to micro-vascular shear stress giving rise locally to radical oxygen species. In further support of this hypothesis, the other gene duplication that occurred during the water–land/fish–amphibian transition was that of the gene for the β-adrenergic receptor (βAR), which mediates the effect of adrenaline at specific physiologic sites, particularly the alveolus of the lung, and the formation of the chambers of the heart. This was certainly fortuitous, but may not have been a chance event since the βAR activates the PTHrP receptor by phosphorylating it, providing a functional genomic mechanism for the evolution of this exaptational trait. Hence, there appears to have been a continuous mechanism of physio-logic evolution arising from external environmental stress, from unicellular to multicellular organisms, contingent on cellular compartmentation of genetic expression.

In further support for the origins of physiology, Jack Szostak has shown that lipid-based protocells can generate nucleotides, whereas nucleotides cannot produce lipids. This indicates that micelles preceded DNA-RNA, which may have evolved in supporting the evolutionary effort by providing a "long-term" memory for the adaptational strategy. That is to say, the nucleotide "memory mechanism" would have put those organisms that evolved DNA at an advantage due to their "reference library" for adaptive strategies, rather than to random selection. The obvious efficiency of the former would have put such organisms at an advantage, ultimately becoming the mode of life to the exclusion of non-DNA-based life forms.

These ontologic insights to the origin, creation, and perpetuation of physiology suggest a long-awaited approach to elucidating physiologic mechanisms for Aristotle's *entelechy*, Rupert Riedl's *burden*, Wallace Arthur's *bias*, and L.L. Whyte's *coordinative conditions* (see Chapter 1). Several theorists have ascribed an internal principle to evolution but without a mechanism for an acknowledged active process. The power of Darwinian evolution is in usurping the vitalists, but they are encroaching yet again as creationists and intelligent designers!

Adherents of the "synthetic theory" maintain that no causal mechanism has been provided, at the same time worrying that the search would support such phenomena as finality and entelechy. Entelechy is thought to direct the individual regularities of organisms, arising from the pre-established harmony of living organisms, which is outside the realm of science. It is important to note that even the recognized authorities on this viewpoint – Theodosius Dobzhansky, Curt Kosswig, and Ernst Mayr – acknowledge that the epigenetic system confers an ordering effect, though they are skeptical that this pattern of mutual gene effects will ever be understood because of its complexity. We maintain that the cellular-molecular, ontogenetic-phylogenetic approach provides the solution to this age-old quandary. The key is in recognizing the central role of the cell as the "author" of evolution, making it the primary form of life, to the exclusion of all other forms, thus simplifying what otherwise is seen as highly complex. The following will attempt to show the value added in this approach.

In the current Chapter we have suggested that the cell is the ultimate physiologic fractal, expressing the "first principles of physiology." Biology mimics the Big Bang of physics, generating its own internal pseudo-Universe through the entraining of calcium by lipids, which is vertically integrated to generate complex physiology. Chapter 3, entitled "The Historic Perspective on Paracrinology and Evolution as Lead-ins to a Systems Biology Approach," traces the history of cell culture and its influence on our insights to the role of the cell in evolution.

3

The Historic Perspective on Paracrinology and Evolution as Lead-ins to a Systems Biology Approach

In order to understand and fully appreciate the empiric origins of the cellular-molecular approach to evolution taken in this book, it is instructive to consider the history of cell biology. The German biologist Matthias Schleiden was the first to formally propose that cells lived schizophrenic lives – an autonomous life only relevant to itself, and a life relevant only to the organism. This view characterizes how thinking about cells is complicated by the context in which they are perceived. It is reminiscent of Niels Bohr's explanation for the seeming duality of light as both a particle and a wave. In his Como Lecture of 1932 he explained that this was an experimental artifact, which he termed complementarity, resulting from the different methods used to determine the properties of light. The same can be said for the genotype and phenotype, which appear to be divergent properties of the cell, but in reality are different aspects of the same entity measured by different criteria, creating a seeming dialectic. The purpose of this book is in large part to eliminate this misconception, among others emanating from the failure to realize the significance of the single cell.

Historic Perspective

How we conceive of cells has always been as important as how we observe and study them. Scientific discussions of cells are replete with mechanistic language and imagery. Cells are commonly described in industrial terms, such as "chemical factories," composed of "protein machines," regulated by "genetic programs." But there is another way of perceiving cells that characterizes them as minute automata making "decisions" about what to do and what type of cell to become, living in rich social environments involving other cells.

Whereas Schleiden and Schwann's cell theory of the early nineteenth century depicted cells as discrete "building blocks" from which a more complex plant or animal is constructed, by the mid-nineteenth century cells were commonly regarded as autonomous "elementary organisms," and the plant or animal as a "society of cells" or "cell-states." In 1855, the highly influential Rudolf Virchow came to the conclusion that all cells came from pre-existing cells, completing the classical cell theory. The principle of the division of labor and its role in the organization of modern political states provided biologists with an analogy for thinking about the functional relationship between the organism as a whole and its constituent parts. But insight was still lacking into how cells were able to arrange themselves in a hierarchical system of tissues and organs in

Evolution, the Logic of Biology, First Edition. John S. Torday and Virender K. Rehan.
© 2017 John Wiley & Sons, Inc. Published 2017 by John Wiley & Sons, Inc.

which physiologic responsibilities were divided up. Cells were said to subordinate their individual interests to the greater good of the organism, or cell society, absent an actual mechanism for how and why they would do so, and what unified them as an integral whole at the organismal level.

Cell theory has undergone significant revision since the 1960s, with the advent of readily available cell culture plasticware and commercially available cell culture media. A major obstacle to the widespread study of isolated cells in culture was the availability of plastic that would support cell growth – it was the Falcon Company of Oxnard, California, that incorporated polylysine in their tissue culture plasticware, creating a negatively charged surface that solved this technical obstacle. The isolation and cultivation of cells in culture gave rise to concepts of cell-cell communication, since it was discovered empirically that epithelial cells required their neighboring connective tissue cells to maintain their differentiated state. The subsequent discovery that cells actively secreted growth factors that bound to cognate receptors on target cells marked a revolution in biomedicine, finally providing a mechanism for the long-anticipated Spemann organizer. The further elucidation of how such growth factors activated cell-signaling pathways, describing how "signals" received on the surface of or through the cell membrane are internally processed to result in changes in cell behavior and morphology, provided a mechanism for embryogenesis for the first time. This enlightened understanding of cells as communication hubs helped to explain how a cluster of genetically and morphologically similar blastocyst cells, with no division of physiologic labor, became differentiated and organized into complex systems of tissues and organs, ultimately resulting in the formation of a highly integrated organism. It also provided further support for the view that cells and their derivative organisms resulted from social forces and arrangements.

This social perspective provides a way of thinking about organisms and cells as distinct from, yet complementary to, the predominant mechanistic and reductionist perspectives that scientists have in thinking about the causal details of how cells manage to function. Cells are viewed as individuals, but with the proviso that they must be understood within a social context, whether that context involves a dispersed community, a loosely organized colony, a tissue or organ system, or a tightly integrated multicellular organism as a whole.

Parts and Wholes

Questions about biologic individuality and the relationships between parts and wholes were central to much of biologic theory during the nineteenth century. During that era, scientists used divergent styles of reasoning to understand the relationship between the biologic organism and its component parts. What was referred to as "analysis:synthesis" was associated with the idea that plants and animals are compound organisms, aggregates of more elementary organisms. Advocates for this point of view tended to compare compound organisms to human societies and their constituents. Herbert Spencer described the animal body as a "commonwealth of monads." The publication of Darwin's *On the Origin of Species* in 1859 promoted further speculation about the origin of complex organisms from more ancient and simpler ones. Ernst Haeckel was one of the first to merge cell theory and the theory of evolution to search for clues to the origin of multicellular organisms from unicellular organisms. His hobby was the detailed drawing of radiolaria, which are protozoans that produce intricate mineralized cytoskeletons; they exhibit a great deal of phenotypic heterogeneity, and piqued Haeckel's curiosity. The cell theory has popularized the idea that multicellular animals and plants are "cell-states" in which individual cells are analogous to the citizens of a higher social order. Cells were characterized as specialized individuals, arranged into various professions and classes within the greater cell society. Haeckel speculated on how complex cell-states like our own bodies might have evolved from more primitive sorts of cell societies. Increased division of labor would result in more specialized and differentiated cells, eventually leading to the point at which none of them could survive independently, having become dependent upon the specialized functions of their neighbors. This "colonial theory" of multicellular origins was closely aligned with Haeckel's Biogenetic Law of recapitulation ("ontogeny recapitulates phylogeny"), providing a more mechanistic approach to twentieth century biology. But this social construct for multicellular organisms fell into disfavor for lack of scientific evidence. The subsequent reductionist focus on genetics, and ultimately on molecular biology, relegated talk of the body as a "society of cells" to a lesser status, in favor of more mainstream DNA-based mechanisms and informatics.

The Advent of Cell Communication Theory

Political and economic metaphors (division of labor, cell state) dominated social metaphors for the body in the nineteenth century, whereas communication metaphors subsequently emerged. By the mid-nineteenth century, the nervous system was commonly analogized to telegraph wires carrying electrical signals throughout the body, but cells in the body communicating with one another was unsuspected. Even as late as the first decade of the twentieth century, it could only be hinted that the cells of a developing embryo might be in "communication" with one another by means of the slender "protoplasmic bridges" seen connecting them. Experiments performed during that era showing the disruptive effects of killing or removing early blastula cells, and the inductive influence of tissues like Spemann's organizer on developmental processes, all pointed to some form of coordination between the developing structures within an embryo. The "how" of it was nebulous, although some chemical basis was suspected. Physiologists made strides in understanding how organ functions are coordinated by means of chemical communications. At the turn of the twentieth century, Ernest Starling and William Bayliss proposed that chemical molecules called "hormones" acted as "messengers" to orchestrate various bodily functions, and fostered intense efforts to identify the body's chemical messengers. In the late 1950s, Karlson and Butenandt coined the term "pheromone" to describe chemical messengers acting externally between individuals of various species of social insects. X-ray crystallography, radioisotope and fluorescent labeling, and nuclear magnetic resonance techniques provided insights to hormonal activity at the cellular level. The theory of cell-cell communication and intracellular signaling emerged in the 1950s, 60s, and 70s through the creative conceptual work of people like Sutherland ("second messengers"), Gorski (estrogen receptor), and Rodbell ("signal transduction"), all of whom were trying to understand the bioactivity of hormones at the cellular level, in association with Grobstein, who had first suggested the existence of soluble growth factors in the process of embryonic development. It would be another two decades before George Todaro's breakthrough discovery that cultured fibroblasts actively produced soluble growth factors. The theory of cell-cell communication caused a sea change in classical cell theory – cells were no longer just building blocks, but sentient social organisms in constant communication with one another by means of chemical and physical signaling.

Cell Sociology

While biochemists in the early twentieth century were preoccupied with understanding the internal chemical dynamics of cells as parts of larger tissue, organ, and organismal systems, others in the field were using the newly developed techniques of cell and tissue culture to study the behavior of cells both as individuals and in groups. As early as 1931, the French surgeon Alexis Carrel called for the creation of a "new cytology" that would transcend the established methods of studying the remains of dead cells with stains and dyes in favor of live cells in culture, whose behavior could be studied using the new technology of microcinematography. This would allow for what Carrel called the study of "cell sociology." "Cell colonies, or organs," he wrote, "are events which progressively unfold themselves. They must be studied like history. A tissue consists of a society of complex organisms..."; its physiological properties belong to "the supracellular order, and are the expression of sociological laws." This concept was reiterated by the French pathologist Albert Policard. It was not until the 1970s that a comprehensive attempt was made to develop a theory of "cell sociology," unencumbered by data. Beginning in 1976, the French biologist Rosine Chandebois formally proposed a theory of development she termed "cell sociology."

Evolutionary developmental biologist Brian Hall has applied Chandebois' notion of "cell sociology" to modularity in biologic systems to explain that, while all cells in a developing embryo communicate, it makes a difference whether the signals in question are exchanged between cells *within* a similar group (a homotypic interaction in a localized condensation of cells) or *between* dissimilar groups (a heterotypic inductive signal between separate layers of cells, for example). Signals exchanged among similar cells within a homotypic context foster the emergent features known as group or community effects, such as the upregulation of certain tissue-specific genes. Since these effects are not attainable by isolated cells or numbers of cells beneath a minimal density, they behave like social phenomena. Recently, the idea of cell sociology has also been applied by Shirasaki to the immune system to highlight the complex and dynamic interactions of the various cell types in the body's host defense system. The communication between cells within a particular group (a condensation, a tissue, an organ, etc.) helps to explain the emergence and integration of these new levels of biologic organization. The emphasis on group or community speaks to the organizational structure of the social interactions occurring between cells of a developing embryo, and reflects the significance of population structure for understanding transitions in evolutionary individuality more globally.

In addition to cell "sociology," some have more recently alluded to "cell sociobiology" or "socio-microbiology." These sociobiological approaches to the study of cell behavior are distinguished by their preoccupation with the evolution of cooperative behavior among predominantly single-celled organisms such as bacteria and slime molds, and the employment of inclusive fitness and evolutionary game theory, which are missing from Chandebois' strictly developmental focus. This is already a well-established effort by Richard Michod operating under the heading of "evolutionary transitions in individuality."

Given all these various "cell sociologies," one might question the overall significance of such cell-cell communication mechanisms. The pre-eminent cell biologist Paul Weiss admonished us not to ask "how or why" questions, yet contemporary cellular-molecular biology behooves us to do so, given that we now have all the "parts." The core of this book is to focus on the cell as the fundament, driving force, and perhaps the *raison d'être* for all of biology and its evolution.

Discovery of Neutral Lipid Trafficking: *A Priori* Experimental Evidence for Cellular Evolution

The methods that have been devised for routine laboratory use of cell culture have an interesting history. Investigators like Alexis Carrell were attempting to grow tissue fragments in glassware at the beginning of the twentieth century, with the help of such luminaries as Charles A. Lindberg, who took a personal interest because he had a sister with heart disease, so he wanted to figure out how to grow a heart in culture. Interesting experiments were conducted, particularly by developmental biologists like Rudnick, Wessels, and Spooner keen to learn about the properties of the embryo independently of its mother. Beginning in the late 1950s, Clifford Grobstein performed a provocative series of experiments in which he propagated tissues in culture to study their properties – particularly those of the lung. He discovered that the embryonic lung could continue developing in culture, but that if he separated the mesodermal and endodermal tissue layers, the isolated tissues could no longer develop, instead balling up into amorphous clumps of cells. However, when he recombined these same tissue layers, development recommenced; of particular interest was his seminal observation that even if he interposed a semipermeable membrane between the two tissue layers, development would continue "normally," the inference being that there must be low molecular weight molecules that were able to pass through the pores in the membrane, mediating the process of organogenesis. That observation remained fallow for a decade, until in 1978 de Larco and Todaro discovered that Swiss 3T3 fibroblasts actively secreted soluble growth factors in cell culture, a discovery that revitalized the field of molecular embryology, which had crashed and burned earlier in that same decade with the demise of the chalone theory, rendering the field of embryologic development fallow for a number of years.

Following up on the milestone observation that glucocorticoids produced by the maternal and fetal adrenal cortex could faithfully accelerate lung surfactant production in fetal sheep, our laboratory (Torday) had shown that cortisol could accelerate fetal lung surfactant synthesis in mixed lung cell culture as well, documenting the direct effect of the hormone on lung development in preparation for birth. Unexpectedly, the hormone did not directly affect the isolated alveolar epithelial type II cells that produce lung surfactant; instead, it stimulated the neighboring interstitial fibroblasts, which produced a low molecular weight peptide, which, in turn, stimulated the alveolar type II cells to synthesize surfactant. That paracrine effect was underscored by the fact that when we studied such cultures isolated at advancing stages of lung development, known to produce progressively more surfactant, the baseline rates of surfactant synthesis increased as a function of the gestational age of the donor, but failed to progress in culture spontaneously; that indicated that the process of lung development was determined by extrinsic factors such as the exposure to circulating levels of cortisol (but was in conflict with Grobstein's observations regarding the sustained development of the lung in explant culture mediated by intrinsic factors). This was a fundamental observation for the central role of paracrine cell–cell interactions in lung embryogenesis. Barry Smith called the fibroblast-secreted substance fibroblast-pneumonocyte factor (FPF). Three decades later, the Torday laboratory determined that FPF was probably leptin, which is secreted by lung lipofibroblasts (which, like homologous adipocytes, produce leptin in response to cortisol), has a molecular weight equivalent to FPF, and stimulates lung surfactant phospholipid synthesis by alveolar type II cells mediated by the cell-surface leptin receptor.

The other factor known to affect lung development is lung fluid distension. The "father" of fetal physiology, Alfred Jost, and the pathologist Alfred Policard had demonstrated that the lung actively secreted lung liquid in 1948. They demonstrated that the fluid within the fetal lung arose from the lung itself and did not, as had been thought, represent aspirated amniotic fluid. At that time investigators speculated that the fluid was a transudate derived from the vasculature. However, experiments in the 1960s demonstrated that the fetal lung liquid had an unusually high chloride ion concentration, and that the movement of liquid into the developing fetal lung lumen occurred as a result of the active transport of ions by the alveolar epithelium. The active role played by lung liquid in the process of lung development was discovered serendipitously in the 1970s when investigators studying the production of fetal lung surfactant *in utero* were draining lung liquid from the lungs of fetal lambs in the womb using an endotracheal tube. They initially exteriorized the tube through the mother's flank into a receptacle to collect the lung liquid, only to find that the rate of lung development had been retarded by this procedure; by leaving the tracheal fluid collection bag inside the maternal peritoneum, this intriguing effect was alleviated. Progress in our understanding of the physiologic effect of fluid distension on the rate of lung development was slow, finally being systematically studied by Moessinger *et al.* two decades later. The lack of progress in this area of study was largely discouraged by the absence of a stretch-regulated mechanism for the control of surfactant production. That changed with a series of studies from the Torday laboratory starting in 1994, when we first reported that the expression of parathyroid hormone-related protein (PTHrP) by a stretch-regulated gene in alveolar type II (ATII) cells of the lung stimulated lung fibroblast differentiation by a receptor-mediated paracrine mechanism.

A key study done in support of the role of fluid distension harkened back to the earlier comment about the failure of cultured lung fibroblasts to progress beyond the stage at which they were harvested from the donor lung. In contrast to this, organ culture of the developing lung did paradoxically exhibit spontaneous maturation in culture, offering the opportunity to determine how and why this occurred. We hypothesized that the formation of fluid within the alveoli was driving the maturation of the lung. We tested this hypothesis by blocking the fluid formation chemically, finding that this inhibited the spontaneous maturation of the lung epithelial and mesenchymal cells molecularly, confirming the role of fluid distension in the process of lung development.

The specific intermediate mechanisms involved in the stretch-activated increase in lung surfactant production were not fully elucidated until 1995 (Figure 3.1), when the

Torday laboratory discovered the principle of *neutral lipid trafficking*, which mediates the transfer of lipid from the alveolar microcirculation to the interstitial lipofibroblast (LIF), and from the LIF to the alveolar type II cell for facilitated surfactant production. Prior to those findings, it was merely assumed that lipid substrate entered the lung tissue passively by mass action. We found that cultured fetal lung fibroblasts actively accumulate serum neutral lipid from the surrounding medium, but interestingly, that they could not release it; in contrast to this, cultured alveolar epithelial type II cells were unable to take up the neutral lipid. However, when the lipid-filled fetal lung fibroblasts were co-cultured with the lung epithelial cells, the neutral lipid from the fibroblasts was rapidly and specifically incorporated into surfactant phospholipids by the epithelial type II cells. The transfer of the neutral lipid was found to be cortisol-stimulated, substantiating the hypothesis that this was a regulated mechanism. The paradoxical active transfer of lipid from LIFs to alveolar epithelial type II cells, but the inability of the LIFs to release the lipids in cell culture, was resolved empirically when it was discovered that exposing the LIFs to culture medium containing the secretions of the epithelial type II cells (referred to as cell-conditioned medium) caused release of the neutral lipid, indicating that some soluble factor(s) secreted by the epithelial cells caused the release of

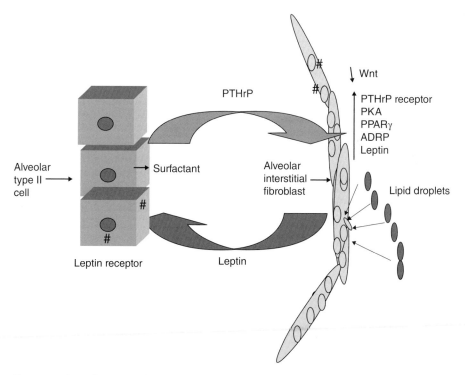

Figure 3.1 Stretch-activated increase in lung surfactant production. Parathyroid hormone-related protein (PTHrP), secreted by the alveolar type II (ATII) cell, binds to its receptor on the adjoining alveolar interstitial fibroblast, activating the protein kinase A (PKA) pathway, which actively downregulates the default Wnt pathway and upregulates the adipogenic pathway through the key nuclear transcription factor, PPARγ, and its downstream regulatory genes ADRP (adipocyte differentiation-related protein) and leptin. Lipofibroblasts in turn secrete leptin, which acts on its receptors on ATII cells, stimulating surfactant synthesis. (*See insert for color representation of the figure.*)

the neutral lipid from the LIFs. Subsequent chemical analysis of the secretions of the alveolar epithelial type II cells revealed that they secreted a lipid-soluble substance that caused the release of the triglyceride from the LIFs. On further examination, it was found that these cells produced prostaglandin E_2, a highly biologically active mediator of lipid secretion by adipocytes, and that there were specific prostaglandin E_2 receptors present on the LIFs. Experimentation revealed that the prostaglandin E_2 produced by the alveolar epithelial type II cells caused the active secretion of the neutral lipid from the LIFs. This cell signaling from the epithelium to the fibroblast for lipid substrate mediated by prostaglandin E_2 took on greater developmental physiologic significance when it was found that the production of prostaglandin E_2 (PGE$_2$) by alveolar epithelial type II cells increased several-fold as the time of birth approached, peaking just prior to delivery, and that both cortisol and the stretching of the type II epithelial cells increased PGE$_2$ synthesis and secretion. Taken together, these observations suggested cellular cooperativity for actively recruiting neutral lipids from the lung alveolar microcirculation, and their storage by the LIFs. The neutral lipids could subsequently be actively "trafficked" to the alveolar epithelial type II cells for surfactant phospholipid synthesis in preparation for air breathing at the time of birth. These findings, combined with the earlier observation that cortisol stimulated the overall transfer of lipid from the LIF to the alveolar epithelial type II cell, provided evidence for such a regulated, cell–cell interactive process.

We subsequently discovered (Figure 3.2) that the developing LIFs obtain the neutral lipid from the circulation by producing adipocyte differentiation-related protein (ADRP), a ubiquitous molecule that is necessary for the uptake and storage of lipid by all cells. ADRP encapsulates the lipid droplets for storage and secretion, and is required for the subsequent uptake and incorporation of the lipid into surfactant phospholipids by the alveolar epithelial type II cells. The neutral lipid trafficking mechanism for regulation of alveolar surfactant provided phenotypic functional genomic markers for our subsequent study of the developmental mechanisms that determine stretch-induced, on-demand surfactant production. We discovered that PTHrP (see Figure 3.1) is a stretch-regulated protein produced by lung alveolar epithelial type II cells, which stimulates surfactant synthesis through a cell–cell interactive mechanism: PTHrP binds to its G-protein coupled cell surface receptor, increasing both (1) lipid uptake by stimulating ADRP and

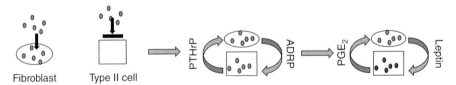

Fibroblast   Type II cell

Figure 3.2 Experimental evidence for neutral lipid trafficking. Monolayer cultures of lung fibroblasts actively take up neutral lipids, but do not release them unless they are co-cultured with type II cells (right two images). Type II cells cannot take up neutral lipid (second image from left) unless they are co-cultured with lung fibroblasts. Fibroblast uptake of neutral lipid is determined by parathyroid hormone-related protein (PTHrP) from the type II cell, which stimulates adipocyte differentiation-related protein (ADRP) expression by the lung fibroblast; stretching co-cultured lung lipofibroblasts and type II cells increases surfactant synthesis by coordinately stimulating prostaglandin E_2 (PGE$_2$) production by type II cells, causing release of neutral lipid by the lipofibroblasts, and leptin secretion by the lipofibroblasts, which stimulates surfactant phospholipid synthesis by the type II cells. (*See insert for color representation of the figure.*)

(2) leptin, the secretory paracrine product of the fibroblast that stimulates alveolar epithelial type II cell surfactant synthesis. Expression of both leptin and the PTHrP receptor by the LIF, and of PTHrP, PGE_2, and the leptin receptor by the alveolar epithelial type II cell are all stretch-regulated mechanisms. This results in the neighboring fibroblasts and type II cells coordinately mediating the increase in lung surfactant production through integrated, stretch-regulated increases in PTHrP and leptin specifically and coordinately signaling through their mutual cell surface receptors. This mechanism is further facilitated by the stretch-stimulated synthesis and release of PGE_2 by the alveolar epithelial type II cells; secreted PGE_2 binds to its receptor on the LIF, causing release of the neutral lipid from the LIFs, ensuring the availability of lipid substrate for stretch-mediated surfactant phospholipid synthesis. Moreover, PTHrP is a potent blood vessel dilator that stimulates alveolar capillary perfusion. Taken together, the coordinate stretch-mediated effects of PTHrP, leptin, and PGE_2 and their physiologically complementary target cell receptors account for the mechanism of ventilation/perfusion, or V/Q, matching, which is the descriptive physiologic principle for alveolar homeostasis.

Neutral Lipid Trafficking: Insights to the Evolution of the Lung

It is hard to imagine how such an exquisitely integrated mechanism for the coordinate physiologic regulation of surfactant phospholipid could have occurred merely by chance, (1) given the variety of cell types involved (endoderm, mesoderm), and the genetically determined ligand–receptor-mediated signaling mechanisms, (2) their close proximity to one another, (3) their molecular regulation by both endocrine hormones and locally produced paracrine factors, (4) all regulated by the effects of stretch in support of alveolar homeostasis for gas exchange, and (5) facilitating the increase in surface area-to-blood-volume ratio for maximally evolved efficient gas exchange. If you were to calculate the probability that these events occurred by chance, you would multiply the length of time it took to form the specialized LIFs of the mammalian lung by the length of time it took to evolve the coelom-like epithelium-lined alveoli. Even by the crudest of estimates, mammals took more than 4 billion years to evolve, and coelomic cavities lined with epithelia similarly took more than 4 billion years to evolve. Multiplying 4×10^9 by $4 \times 10^9 = 16 \times 10^{18}$ years, which is older not only than the age of the Earth, but also than the estimated age of the Universe itself!

Alternatively, it is more than reasonable to hypothesize that this mechanism was the result of positive phylogenetic selection pressure for progressively efficient surfactant production, mediated by cell–cell interactions. In order for the alveoli to have evolved a progressively smaller diameter, increasing the gas exchange surface area:blood volume ratio, surfactant production also had to become progressively more efficiently regulated, based on the law of Laplace, which states that the surface tension of a sphere (i.e., alveolus) is inversely related to its diameter. It also formed the basis for the structural evolution of the alveolar wall through the modifications of both the mesenchymal fibroblast and epithelial cell populations, as discussed above. This evolutionary concept is supported by the consistent observation that surfactant produced by the epithelial lining cells of the gas exchangers of fish, amphibians, reptiles, mammals, and birds increases progressively in response to hormonal and stretch regulation over both phylogenetic and developmental time in close association with the progressive decrease in alveolar diameter.

Mechanistically, the transition from the regulation of surfactant by constitutive genes to the cell–cell interactive mechanisms just described required the sequential evolution of *cis* regulatory control of surfactant synthesis – the PTHrP, leptin, and PGE$_2$ signaling pathways all act via second messengers such as cyclic AMP and inositol phosphate (IP$_3$), which interact with nuclear transcription factors to regulate surfactant synthesis and secretion. Transition from constitutive to *cis* regulatory mechanisms in adapting to the environment is a recurrent theme in evolutionary biology, although these mechanisms haven't been looked at longitudinally and diachronically (i.e. at multiple timescales) in the way that we have, systematically elucidating the cellular-molecular mechanisms of lung evolution.

Those observations led to a seminal experiment demonstrating the central role of lung fluid distension in the molecular development of the lung. It had long been known that when fetal lung explants were cultured they would develop relatively normally, both structurally and functionally as they do *in utero*. In contrast, isolated lung cells in monolayer culture remain developmentally "fallow," as described above. This left open the experimentally testable question: What is the innate mechanism for determining lung development? We hypothesized that the fluid distension of the alveolus-like cysts that characterize explanted lung tissue was mediating fetal lung development. We demonstrated that by inhibiting active fluid secretion we were able to stop the spontaneous development of lung explants in culture; treating the lung tissue with PTHrP overrode the inhibition due to the lack of lung fluid distension. Thus, for the first time, there was a mechanistic understanding of the effect of fluid distension on the process of lung development mediated by PTHrP. These observations were key to our insight to cell-cell signaling and evolution because PTHrP is a gravisensor, expressed at least as far back in vertebrate evolution as fish. It is expressed in the swim bladder, which is the fish homolog of the lung, yet in the swim bladder it mediates gas exchange in adaptation to buoyancy, a derivative of gravity, not for metabolic oxygenation per se but for feeding efficiency, providing insight to the fundamental nature of the evolutionary process – the same genes can be reallocated for a different phenotype, depending upon the "history" of the organism, that is, where, when, and how the gene regulatory network was used in previous generations determines the network's utility in subsequent generations. It's what the French biochemist François Jacob described as "tinkering," but now providing a mechanism in the context of ontogeny and phylogeny as a way to understand how and why the swim bladder evolved into the lung.

Lessons from the Hepatocyte

During that same era, George Michalopoulos discovered the importance of the collagen matrix in maintaining the metabolic activity of hepatocytes in monolayer culture. Other epithelial cell types such as the lung alveolar type II cell, and those of the mammary gland, pancreas, and prostate have also been shown to require their investing matrix proteins for normal structure and function *in vitro*, inferring their functional significance *in vivo*, both in health and disease. Such findings were serendipitous, like most discoveries in biology and medicine, due to the lack of a central theory. In retrospect, the bioactivity of the connective tissue cells and their active participation in the morphogenesis and differentiation of form and function should not have come as a surprise – Nature doesn't do anything unnecessarily. It was the systematic determination of the role of the mesoderm in alveolar growth and differentiation that led to the first demonstration of an integrated physiologic process mediated by cell–cell interactions, complete with stretch-regulated, hormonally responsive growth factors that mediate the physiologic development, homeostasis, and regeneration of the lung alveolus. This was accomplished by reverse-engineering the phylogenetic and ontogenetic mechanisms that determine the production of lung surfactant.

Hepatocytes have characteristics similar to ATII cells regarding lipid homeostasis. They require neighboring hepatic stellate cells (HSCs) to maintain their differentiated structure and function. The HSCs take up and store neutral lipids that are shuttled to the hepatocyte for metabolism. HSCs produce various growth factors necessary for hepatocyte growth and differentiation, and hepatocytes produce prostaglandins that cause lipid release from HSCs for uptake by hepatocytes. And like the pathophysiologic effect of overdistension on the alveolus, overdistension of the common bile duct causes liver fibrosis due to decreased PTHrP production by tubule epithelium.

Another such example is the mammary gland, which produces milk in a way homologous to surfactant production by the alveolus. In fact, both lung and mammary have alveoli. The mammary epithelial cells that synthesize milk are dependent on the adepithelial fibroblasts for both their growth and differentiation, particularly under hormonal regulation. Like the ATII cell, the mammary epithelium secretes ADRP, which surrounds the milk globules secreted into the alveolar space. The ADRP-lipid complexes are taken up by the gut epithelium for transit to the circulation.

In the mouse embryo, mammary buds are formed between embryonic days 10 (E10) and E11. The size of the bud slowly increases until a circumscribed epithelial "ball" is formed within the epidermis by E13. Continued proliferation until the end of gestation leads to the formation of a small ductal tree that consists of 10 to 15 branches arising from the single duct that emanates from the nipple. Mammary rudiments explanted *in vitro* follow the same timing of the outgrowth of the primary sprout, demonstrating that it is determined by an intrinsic program and not by systemic factors.

At the time of bud formation, the mesenchyme underlying the mammary bud appears indistinct from the rest of the dermis. The mesenchymal cells bordering the mammary buds orient themselves slowly around the epithelium. By E14, the epithelial gland buds are surrounded by a slightly denser mesenchyme consisting of several concentric layers of fibroblasts oriented around the epithelium. They are not set apart from the more distant dermis as is the case, for example, in the salivary gland.

The first gene expressed specifically in the epidermis at the site of mammary gland formation is *Lef1*, a gene encoding a transcription factor of the HMG-box family. *Lef1* is activated during the formation of the mammary bud at E11.0, even before it becomes morphologically distinct. Interestingly, *Lef1* is expressed in all epidermal thickenings that initiate development of skin appendages, including whisker, hair, and (ectodermally derived) teeth, but its expression is not restricted to the epidermal compartment.

By E12, the mammary epithelium has begun expressing PTHrP. Expression continues throughout embryonic development, but no information is available on the earliest appearance of these transcripts. At E13/E14, the epithelial bud expresses the transcription factor genes *Msx1*, *Msx2*, and *Lmx1b*.

Although tissue recombination experiments have produced unequivocal evidence for multiple reciprocal epithelial-mesenchymal interactions in the embryonic mammary gland, they do not provide evidence unequivocally showing which of the two tissues initiates mammary development. Tissue recombination experiments in the rabbit embryo, where the milk line appears on E13, suggest a leading role for the mesenchyme. At E12, that is, before visible mammary differentiation, combinations of mesenchyme from the (prospective) mammary region with heterotopic epidermis of the head or neck yielded mammary structures, whereas reciprocal combinations (of prospective mammary epidermis with non-mammary mesenchyme) did not. By E13, however, even just before the appearance of the milk line, mammary epidermis did form buds in association with mesenchyme from other regions. While this early study only used morphologic criteria for mammary development, Cunha and coworkers later showed that E13 (mouse) mesenchyme was able to induce rat midventral or dorsal epidermis to form functional mammary epithelium, as determined by immunohistochemical detection of casein and α-lactalbumin. Though these studies clearly show that the primary mammary mesenchyme can induce mammary development in the epidermis, they do not address the question of the initiation of mammary development. Conceivably, even the inducing mammary mesenchyme of the E12 rabbit embryo may already have received decisive cues from the overlying ectoderm, as recently shown for the initiation of limb development by Y. Yamamoto-Shiraishi and co-workers.

While the processes initiating mammary development remain unclear, there is fairly good evidence that formation of the primary mammary mesenchyme is governed by signals coming from the epithelial bud. When isolated epithelial gland buds are placed on mesenchyme that had not previously been in contact with mammary epithelium, they induce the same mesenchymal responses, including elevated [^{3}H]uridine and [^{3}H] glucosamine incorporation, elevated production of tenascin-C, and the synthesis of androgen and estrogen receptors. The induction of steroid receptors was shown to be a specific property of the mammary epithelium, not shared by other epithelia (epidermis, salivary gland, pancreas). Moreover, mesenchymal cells experimentally associated with mammary epithelium not only express androgen receptors, but also are capable of executing the androgen response, that is, they cause destruction of the epithelial anlage when exposed to testosterone. Although experimental evidence is not available for other marker genes of the primary mammary mesenchyme, it appears likely that their expression is also induced by the epithelial gland bud.

The endoderm of the early gut tube is uniform in its morphology along the anterior-posterior length of the primitive gut tube. There are no morphologic differences between the portions of tube formed by elongation of either the anterior or caudal intestinal portal, and no distinctions in regions that eventually form the antero-posterior portions of the gut: foregut, midgut, hindgut, and their adult phenotypes of esophagus/stomach, intestine, and colon. The primitive gut tube is lined by a single layer of a cuboidal/columnar endoderm/epithelium and encircled by a thin layer of splanchnic mesoderm. As the mesoderm grows and differentiates into smooth muscle, the gut tube alters its gross morphology, resulting in clear demarcations among the foregut, midgut, and hindgut. These distinctions can be made by gross morphology, histology, function, and the presence of the demarcating structures that separate these regions. The luminal epithelial morphology lags significantly behind the gross gut pattern in its regionally specific differentiation. In the chick, gross morphologic regional distinctions among the three primary AP subdivisions of the gut are evident by E3–5, but epithelial differences are not well developed until E10, and not clearly distinct until near hatching (E18–21). In some vertebrates, the gut epithelium continues to be plastic, often undergoing functional differentiation after birth, before forming the adult phenotype. The gut has the remarkable ability of continued epithelial growth and differentiation throughout the life of the organism along its radial axis. It is this axis in which the regionalization of the gut is often distinguished, as morphologic differences are easily discernible.

It has been known for decades that the gut cannot develop normally without an interaction between the endoderm and mesoderm. The direction of these endoderm–mesoderm interactions has been the focus of much investigation. Cultures of primitive foregut endoderm cannot differentiate without being co-cultured with mesodermal tissue. There is a developmental window after which the primitive gut endoderm, although still morphologically undifferentiated, is committed and develops into its regionally specified epithelium when cultured with a variety of tissues, including the vitelline membrane. However, at an earlier developmental time, the ultimate differentiation of primitive endoderm will depend on the AP region of its adjacent mesoderm. For example, early gizzard endoderm can differentiate into proventricular epithelium if co-cultured with proventricular mesoderm. Many studies have confirmed that the mesoderm directs the ultimate epithelial pattern in the gut, but the endoderm also has inductive capacities. Definitive endoderm co-cultured with somite mesoderm stimulates smooth muscle (splanchnic or visceral) rather than skeletal muscle development as assayed by histology, and by induction of visceral mesodermal proteins, for example, tenascin and smooth muscle actin. The mesodermal influence on endodermal patterning primarily involves specification of morphology that may not include all of the epithelial cytodifferentiation. Most of the endodermal gut regions studied appear plastic to influence from mesoderm in both morphologic and cytologic differentiation, except for the midgut region. Some midgut-specific epithelial cytodifferentiation appears to have cell autonomous/cell-specific features. Specific midgut epithelial expression of digestive enzymes is maintained even when influenced by non-gut tissues. This difference between the ability of the midgut and foregut endoderm to undergo complete heterologous differentiation may be an endogenous characteristic of the endoderm. Midgut endoderm must have some epithelial cell-autonomous features. Some of the molecular controls of early endodermal-mesodermal events have been described. Sonic hedgehog (Shh), a vertebrate homolog of *Drosophila* hedgehog (hh), encodes a signaling molecule implicated in mediating pattern formation in several regions of the embryo, including the limb bud, somite, and neural tube. Shh is expressed in the endoderm of the gut, and its derivatives, and is a candidate for an early endodermally derived inductive signal in gut morphogenesis because its earliest endodermal expression is restricted to the endoderm of the AIP and CIP prior to invagination. Shh is not the signal that initiates the invagination of the AIP or CIP, because murine null mutants for Shh develop a gut, although severe foregut abnormalities are present. These mutants have malformed esophagi with enlarged lumens and disorganized or absent subjacent mesoderm. This finding suggests that the endodermally derived signal from Shh is involved with mesodermal development, recruitment, or other aspects of mesodermal foregut patterning. Indeed, Shh must act as a signal from endoderm to mesoderm, because its receptor is present only in the gut mesoderm, and overexpression of Shh in the early primitive gut leads to a mesodermal (not endodermal) phenotype. In each organ in which the endoderm-derived tissue expresses Shh, there is closely associated mesenchymal mesoderm that expresses a homolog of *Drosophila* decapentaplegic (dpp). There are two vertebrate homologs of dpp expressed in the gut, bone morphogenetic proteins (Bmp) 2 and 4. In the primitive hindgut, at the earliest time Shh expression can be detected in the CIP region (even before invagination is apparent), Bmp4 is expressed in the subjacent mesenchymal mesoderm. In misexpression studies, Shh ectopically induces Bmp4 in the splanchnic mesoderm of the developing gut. A model has been proposed in which an endodermal role of Shh is to induce Bmp4 expression in the splanchnic mesoderm, which then controls aspects of smooth muscle development in the gut. Early during gizzard morphogenesis (the region of the chick stomach with the thickest smooth muscle layer), Bmp4 cannot be detected, nor can Shh induce its expression. With ectopic expression of Bmp4 early in primitive gizzard development, a thinning of the smooth muscle layer results. These findings suggest that Bmp4 may negatively regulate gut smooth muscle. Bmp4 may affect the mesoderm by negatively regulating growth and hypertrophy, or facilitating differentiation to smooth muscle. Such studies were performed during very early stages of anterior-posterior (AP) pattern formation in the primitive gut, before the initial gut tube is formed. Later in gut patterning, these same factors may play a different patterning role. Bmp2 is expressed at later stages, in a short time window, restricted to a region of the stomach mesoderm in the chick. At this time, Bmp2 appears to be important in mesoderm–endoderm interactions needed for proper glandular formation in the stomach epithelium. This is an example of the later function of the gut mesoderm in directing regionally specific epithelial patterning along the AP axis.

In this Chapter we have shown how the advent of cell culture empirically revealed the fact that complex physiology is mediated by cell-cell communication. That realization was a game changer for our understanding of embryology, homeostasis, and repair as a functional continuum. Chapter 4, entitled "Evolution of ADRP, or 'Oh the places you'll go,'" describes the discovery of neutral lipid trafficking within the lung alveolus for stretch-regulated surfactant production. The realization of the paracrine regulation of surfactant phospholipid substrate provided the seminal insights to the evolution of the lung and that of other complex physiologic mechanisms.

Selected Reading

Grobstein C. Morphogenetic interaction between embryonic mouse tissues separated by a membrane filter. *Nature* 1953;172(4384):869–70.

4

Evolution of Adipocyte Differentiation Related Protein, or "Oh, the Places You'll Go" – Theodore Geissel, Aka Dr Seuss

Neutral Lipid Trafficking Mediates Lung Alveolar Evolution

The discovery that neutral lipid trafficking mediates the well-recognized on-demand property of lung surfactant production (see Chapter 3) has provided important insights to the specific mechanisms that regulate lung alveolar development, homeostasis, and repair. But more importantly, this heretofore unknown underlying mechanism allowed the identification of novel functional genes that link physiology together as deep evolutionary homologies, for the first time tying principles of respiration and nutrition together all the way back to the advent of unicellular eukaryotes (see Chapter 2). The primary cell–cell signaling pathway interconnected parathyroid hormone-related protein (PTHrP) expression by the alveolar epithelial type II cell with the adepithelial mesenchymal cell, inducing the lipofibroblast (LIF) phenotype, an adipocyte-like homolog that facilitated land vertebrate lung evolution. Experimental investigation of that phenomenon revealed the involvement of leptin, the soluble paracrine secretory product of the LIF, signaling from the mesenchymal fibroblast to the alveolar type II cell for the regulation of surfactant phospholipid production. Leptin was the "missing link" in our search for the paracrine mediator of the PTHrP effect on lung surfactant production, since PTHrP itself had no direct empiric effect on alveolar type II cell surfactant synthesis (see Chapter 3), not to mention that the alveolar type II cell does not express the PTHrP receptor gene. We had generated experimental evidence that PTHrP stimulated surfactant synthesis by the alveolar type II cell in situ by explanted lungs in culture, but the effect turned out not to directly affect the alveolar epithelium; alternatively, we hypothesized that its mechanism of action was indirectly through its paracrine effect on the mesenchymal fibroblast. The determination that the LIF expressed and secreted leptin like its adipocyte homolog closed the paracrine loop, since leptin had already been shown by Victoria Funanage *et al.* to stimulate surfactant production by the alveolar type II cell. The sensitivity and specificity of this mechanism were ensured by the identification of the leptin receptor on the alveolar epithelium. And the dynamic interaction between the epithelium and mesenchyme for increased surfactant production was resolved by the demonstration that the coordinated effects of PTHrP–PTHrP receptor, and leptin–leptin receptor signaling are stretch-regulated mechanotransductive mechanisms.

Evolution, the Logic of Biology, First Edition. John S. Torday and Virender K. Rehan.
© 2017 John Wiley & Sons, Inc. Published 2017 by John Wiley & Sons, Inc.

In the Process of Lung Evolution, Homologies Run Very Deep

It is notable that the physiologic relevance of the LIF to alveolar homeostasis has led to deep homologies with the peroxisome, which is thought to have evolved in response to the otherwise pathologic effects of endoplasmic reticulum stress in unicellular organisms. According to the de Duve hypothesis, oxidant stress-induced disruption of intracellular calcium homeostasis caused selection pressure for peroxisome evolution, counterbalancing calcium dyshomeostasis using lipids.

This same scenario applies to the alveolar extracellular space, in which calcium and lipid homeostasis are of critical importance in controlling the alternating assembly and breakdown of tubular myelin (Figure 4.1), the structure that reduces surface tension in the alveolar hypophase. Tubular myelin is a mesh-like structure composed of lipids and proteins, among them surfactant proteins and antimicrobial peptides, packaged together as lamellar bodies within the alveolar type II cell, and actively secreted into the alveolar space. The identical mechanism of lamellar body secretion is observed in the stratum corneum of the skin, where the lipid-antimicrobial peptide gemish protects the skin against both infection and transudation of fluids; lung surfactant serves these same purposes in the alveolus, demonstrating the functional homology between the lung and skin as barriers against the environment, likely due to their common origin in the plasmalemmae of unicellular organisms.

Prostaglandin E$_2$ Mediates Neutral Lipid Secretion from the Lipofibroblast

One piece of the cell–cell interaction mechanism for facilitated surfactant production that remained unresolved was how lipids are actively recruited from the LIF. We had observed that lung fibroblasts actively took up neutral lipid from the surrounding culture medium, but were unable to release it into the surrounding medium in the absence of either the alveolar type II cell itself, or the secretions produced by them (referred to as "conditioned medium"). This suggested that the alveolar type II cell

(a) (b)

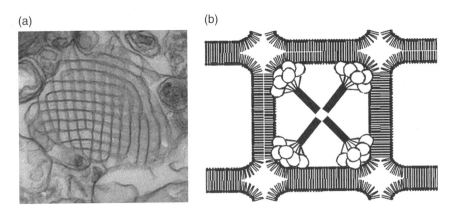

Figure 4.1 Tubular myelin. **a)** Electron photomicrograph of tubular myelin. **b)** A schematic showing the lipids "framing" the central protein structure of tubular myelin.

produced a factor(s) that caused the release of neutral lipids from the lung LIF. That riddle was solved by the discovery that alveolar type II cells secrete both prostaglandin E_2 (PGE$_2$) and PGF$_2$, but only the former mimicked the stimulated release of neutral lipid from the lung mesenchymal fibroblast. This is not a novel property of adipocytes since prostaglandins cause the release of fatty acids from these cells systemically. The physiologic relevance of this secretory mechanism was further validated by the fact that alveolar distension stimulates both alveolar type II cell PGE$_2$ secretion and PGE$_2$-specific EP2 receptor activity on the surface of the neighboring lung LIF, providing a physiologic, cell-specific mechanism for stretch-induced alveolar surfactant production. Thus, a complete mechanism for the facilitated uptake, release, and incorporation of neutral lipid for surfactant synthesis evolved in the mammalian lung. The one piece of this intriguing physiologic puzzle that was missing was the specific molecular agent responsible for the uptake, storage, and release of the neutral lipid by the fibroblast, and the specific mechanism facilitating the uptake of the neutral lipid for surfactant synthesis by the alveolar type II cell.

Adipocyte Differentiation Related Protein and Surfactant Homeostasis: The Plot Thickens

Adipocyte differentiation related protein (ADRP) is a member of a relatively newly discovered family of so-called trafficking proteins that form the limiting membrane around lipids stored in cytoplasmic droplets. It is expressed in adipocytes, where it facilitates the uptake and storage of neutral lipid. Since lung LIFs are modified adipocytes, we hypothesized that ADRP might also mediate neutral lipid trafficking between the LIF and alveolar type II cell. We first determined that ADRP is expressed in the lung LIF, where it surrounds the stored lipid vacuoles. Mechanistically, PTHrP signaling from the alveolar type II cell upregulates LIF ADRP, providing an active, stretch-mediated process for the recruitment of neutral lipid from the alveolar microcirculation, since ADRP is necessary for neutral lipid uptake by fibroblasts. We subsequently showed that PTHrP stimulates leptin expression by LIFs, and that leptin stimulates the uptake of ADRP-neutral lipid by the alveolar type II cell, both *in vitro* and *in vivo*. The identification of ADRP as the physiologic entity responsible for neutral lipid trafficking provides a portal to deep vertebrate homologies relevant to ancient mechanisms of lipid metabolism as early in phylogeny as fungi. Therefore, an in-depth perusal of the biology of ADRP and its cousins would provide "portals" for novel functional homologies in vertebrate physiology.

ADRP Mediates Alveolar Neutral Lipid Trafficking: Going Deeper and Wider

The perilipin-ADRP-TIP47, or PAT, family of lipid droplet proteins is comprised of five members in mammals:

- perilipin
- ADRP

- tail-interacting protein of 47 kilodaltons (TIP47)
- S3-12
- OXPAT.

These family members are also present in organisms as evolutionarily diverse as insects, slime molds, and fungi. All PAT proteins share sequence homologies and the functional ability to bind intracellular lipid droplets, either constitutively or in response to metabolic stimuli, such as increased lipid flux into or out of lipid droplets. Found at the surfaces of lipid droplets, PAT proteins mediate access by other proteins (e.g. lipases) to the lipid esters within the lipid droplet core, and can interact with cellular machinery that mediates lipid droplet biogenesis. Variations in the gene for the best characterized of the mammalian PAT proteins, perilipin, have been associated with abnormal metabolic phenotypes such as type 2 diabetes and obesity. The following is a description of how the PAT proteins regulate cellular lipid metabolism in both mammals and model organisms.

Lipid Droplets and Metabolism of Cellular Lipids

Lipids are sequestered in droplets within the cell cytoplasm for future use as signaling molecules, membrane constituents, or as fuel. Experiments have revealed that lipid droplets are actually highly regulated organelles containing a neutral lipid core enrobed in a phospholipid monolayer, coated with specific PAT proteins. Proteomic analyses have determined that these regulatory proteins control such varied cell functions as lipid metabolism, intracellular trafficking, signaling mechanisms, RNA metabolism, and the configuration of the cytoskeleton. The protein coat surrounding lipid droplets is highly variable, and can differ between droplets within a cell, between metabolic conditions, and between cell-types. Such variations correlate with morphologic changes and lipid droplet intracellular localizations that occur during metabolic alterations or development.

Perilipin and the PAT Family of Lipid Droplet Proteins

Constantine Londos discovered perilipin to be a lipid droplet protein that is phosphorylated in response to signals that stimulate breakdown of triacylglycerol (TAG) stored in adipocytes. This prompted interest in lipid droplet biology since it suggested the existence of intracellular regulatory mechanisms for the control of lipid storage. Subsequently, studies in the Londos laboratory determined that perilipin was the regulator of adipocyte lipid storage. Studies in both cultured cells and *in vivo* have shown that perilipin can either constrain or facilitate access to lipid droplets by enzymes that hydrolyze lipid esters (lipases) depending on the metabolic conditions. The members of the PAT family are distinct from one another with respect to their size, tissue expression, lipid droplet affinity, half-life when unbound from lipid droplets, and in their transcriptional regulation. Such heterogeneity would suggest that these proteins have different cellular functions, yet they all probably regulate the interface between lipid droplets and their cellular environment. The PAT proteins can be parsed into those expressed in a tissue-restricted manner (perilipin, S3-12, and OXPAT) versus a ubiquitous manner (ADRP and TIP47), and into those that are constitutively bound to lipid droplets (perilipin and ADRP), and those that demonstrate exchangeable lipid droplet binding (TIP47, S3-12, and OXPAT). As for their phylogenetic history, the PAT proteins are ancient, being found in many fauna and flora ranging from mammals and amphibians to insects, fungi, and slime molds.

The fundamental importance of the PAT proteins in regulating intracellular lipid stores is underscored by the highly conserved nature of the PAT proteins among mammals and insects, extending to other proteins such as receptors, enzymes, and cofactors, offering tremendous potential for the use of model organisms to identify evolved mechanisms for the control of intracellular lipid storage and utilization. Commonly occurring human pathologic conditions ranging from atherosclerosis to cardiomyopathy, obesity, type 2 diabetes, and non-alcoholic fatty liver disease are associated with abnormal cellular lipid metabolism – how the PAT proteins regulate cellular lipid stores might be involved in such diseases, and could lead to novel approaches to treating human diseases.

Perilipin

Perilipin expression is indicative of adipocyte differentiation and is therefore used as a reporter for identifying regulators of adipogenesis. It is largely under the control of peroxisome proliferator activated receptor gamma (PPARγ), though estrogen receptor related receptor alpha (ERRα) has also been implicated in perilipin regulation. In lipid droplets, perilipin has a half-life of 40 hours, whereas in the unbound state perilipin is rapidly degraded; for example, ubiquitinated perilipin accumulates in the presence of proteasome inhibitors in Chinese hamster ovary cells, and lysosomes have been implicated in perilipin breakdown since it is blocked by lysosomal protease inhibitors. The fact that all of the perilipin isoforms are stable only when they are bound to lipids indicates that neutral lipid droplets determine the cytoplasmic levels of perilipin protein.

Protein Kinase A Phosphorylation of Perilipin

The primary protein kinase A (PKA) substrate for lipid storage droplets is perilipin A, which can be phosphorylated at as many as six PKA sites in mice. Phosphorylation of perilipin A causes a change in function from storage to mobilization of stored neutral lipid, and it is dephosphorylated by protein phosphatase 1.

The Role of Perilipin in Triacylglycerol Metabolism

Functional studies of perilipin have primarily focused on the role of mouse perilipin A. Using cell culture models to study perilipin functionality has its strengths and weaknesses. On the one hand, 3T3-L1 fibroblasts synthesize perilipin A and many other adipocyte-specific proteins; on the other hand, 3T3 fibroblasts and Chinese hamster ovary cells lack many of these same proteins, but are relatively easy to culture and transfect. Moreover, since perilipin activity is determined by protein–protein interactions, certain proteins necessary for its function may or may not be present in all cell-types.

And results may vary among host cells derived from different species. Thus, caution should be used in interpreting studies in which perilipins of one species are expressed in a host cell of another species, and when using model cell systems that do not reproduce the intracellular environment in which perilipins are endogenously expressed. With these precautions in mind, such cell-based studies of perilipin have yielded insights to its roles in cellular lipid metabolism.

Stored lipids are utilized for the synthesis of membranes and for mobilization of metabolic substrates through lipolysis of fatty acyl esters. Fatty acids are esterified with glycerol to form mono-, di-, and triacylglycerols for energy storage. Catecholamines signal through β-adrenergic receptors and the G-protein coupled signaling cascade to stimulate intracellular cyclic AMP (cAMP) in adipocytes. cAMP activates cAMP-dependent protein kinase A (PKA), which phosphorylates the perilipins and hormone-sensitive lipase; the latter translocates from the cytoplasm to the surface of the lipid droplet.

Experimental evidence from transfected cells has implicated perilipin as an organizational motif for the control of lipolysis in adipocytes. Perilipin A prevents lipases from accessing neutral lipids in the droplet core in the hyperphosphorylated state. Consequently, hypophosphorylated perilipin reduces triacylglycerol hydrolysis; however, by recruiting hormone sensitive lipase to the droplet surface, phosphorylated perilipin A facilitates lipase action. Consistent with these findings, mice deficient in perilipin due to gene deletion exhibit increased basal and decreased stimulated lipolysis. Perilipin A possesses six PKA consensus sites, reflecting the complexity of perilipin A regulation. Phosphorylation of perilipin A is necessary for maximal PKA-stimulated lipolysis – for example, based on experiments using fibroblasts cultured from perilipin-deficient mice. Excess stimulation of PKA leads to a change in adipocyte lipid droplet conformation in that relatively few, large perinuclear lipid droplets disperse into thousands of microscopic droplets, each coated with perilipins throughout the cytoplasm. Such fragmentation and dispersion of large lipid droplets provides a mechanism for lipases gaining greater surface area, achieving higher levels of lipolysis.

Interaction of perilipin with other proteins is important for regulation of lipolysis in adipocytes, and is dependent on the phosphorylation state of perilipin. At baseline, perilipin A associates with CGI-58, an activator of the acyl hydrolase activity identified as a lipid droplet protein through proteomic studies. CGI-58 is found in lipid droplets from cells that express perilipin A, but not in cells that only express perilipin B.

Perilipin acts as a scaffold for the lipolytic machinery once it is phosphorylated by PKA; phosphorylated hormone sensitive lipase, for example, requires phosphorylated perilipin for optimal activity and localization to the lipid droplet. Hormone sensitive lipase then interacts with fatty acid binding protein, revealing the sequence of events involved in the lipolytic stimulation of adipocytes or adipocyte-like cell lines. Catecholamines bind to β-adrenergic receptors, which signal through a heterotrimeric G-protein to activate adenyl cyclase, elevating cAMP levels and activating PKA. The latter phosphorylates both perilipin A and hormone sensitive lipase, and CGI-58 simultaneously dissociates from the lipid droplet. Furthermore, prolonged lipolysis will also cause ADRP coating of the surfaces of larger lipid droplets.

Perilipin and Physiology

Mice lacking perilipin have normal body weights, but their adipose mass is significantly reduced. They have a normal appetite and resist diet-induced obesity, but exhibit increased peripheral insulin resistance at a young age. Adult animals have normal circulating glucose levels, but their insulin levels are twice those of wild-type littermates, indicating that perilipin-deficient mice are insulin resistant. Metabolic studies of perilipin-deficient mice indicate that they utilize more oxygen than control mice, probably due to increased thermogenesis. Treating control mice with the β-3 receptor agonist CL316243 causes an increase in oxygen utilization and a decrease in the respiratory exchange ratio, indicating increased adiposity instead of carbohydrate catabolism. By comparison, perilipin-deficient mice exhibit an increase in oxygen consumption and a decreased fall in the respiratory exchange ratio, consistent with elevated basal and decreased stimulated lipolysis. As further validation of these observations, lipolysis in adipocytes isolated from perilipin-deficient mice exhibits both elevated basal lipolysis and diminished stimulated lipolysis.

The abnormal fat distribution, development, or deterioration characteristic of lipodystrophy may occur in humans due to genetics or it may be acquired. Mouse models of lipodystrophy commonly have severe fatty liver disease, whereas perilipin-deficient mice are much healthier. There are two feasible explanations for this observation. In perilipin-deficient mice, leptin in circulation does not correlate with fat pad mass, and the perilipin-deficient and control animals have similar leptin levels, indicating that adipocytes in the perilipin-deficient animals continue to produce and secrete leptin. Mouse models for adipose tissue deficiency are leptin-deficient; administration of leptin either by infusion or transgenic overexpression reverses some of the pathology observed. Therefore, it would appear that normal leptin secretion in perilipin-deficient mice protects against excess lipid accumulation and the concomitant metabolic complications. Molecular biologic studies of perilipin-deficient mice based on microarrays and hyperinsulinemic-euglycemic clamps provide further explanation for the observed resistance to weight gain in perilipin-deficient mice. More in-depth microarray studies were conducted on white adipose tissue, kidney, liver, heart, and skeletal muscle. It is not surprising that white adipose tissue had the greatest number of transcripts affected by perilipin gene deletion, given the nature of the tissue distribution of perilipin in mice. Higher transcript levels for enzymes involved in beta-oxidation, the Krebs cycle, and the electron transport chain were all observed. Furthermore, expression of other genes involved in lipid biosynthesis was decreased. Increased expression of mRNAs for the uncoupling proteins UCP2 and UCP3 in both white and brown adipose tissue was identified. Such alterations in gene expression manifested themselves as changes in overall physiology. Insulin clamp studies revealed that perilipin-deficient mice had peripheral insulin resistance, decreased hepatic glucose production, and increased beta-oxidation.

Adipocyte Differentiation Related Protein (ADRP)

Adipocyte differentiation related protein was originally detected as an RNA transcript that was significantly induced during the course of cultured adipocyte differentiation; the subsequent determination that ADRP was structurally and functionally homologous with perilipin led to the discovery that ADRP coats the surface of lipid storage droplets in a wide variety of cell lines.

ADRP and Adipogenesis

In contrast to ADRP mRNA expression, ADRP protein levels decrease during the course of adipocyte differentiation, whereas perilipin mRNA and protein levels increase. Ontogenetically, young adipocytes have smaller ADRP-coated lipid droplets, whereas mature adipocytes have larger, perilipin-coated lipid droplets. As has been shown for perilipin, ADRP unbound by lipid droplets is degraded by the ubiquitin/proteasome pathway. Unlike the selective tissue distribution of perilipin (adipocytes, steroidogenic cells), ADRP distribution is more generalized to include many other tissues that accumulate lipids, with the major exception being mature human adipocytes.

ADRP Function in Cellular Lipid Metabolism

Studies elucidating the functional roles of ADRP have included gain- and loss-of-function experiments in cells, and gene targeting and silencing in mice. The bulk of such data supports a role for ADRP in regulating the interaction of lipases with the neutral lipid within the droplets, facilitating neutral lipid accumulation. How ADRP inhibits lipase homing to lipid droplets has not been determined empirically. Increased ADRP expression correlates with the expansion of lipid droplets and increased cellular triacylglycerol. ADRP stimulates intracellular triacylglycerol due to reduced triacylglycerol breakdown. These studies utilized triacsin C to inhibit triacylglycerol synthesis by inhibiting acyl CoA synthases. When cells are lipid-loaded and subsequently treated with triacsin C, the changes in cellular triacylglycerol levels over time hypothetically reflect the rate of triacylglycerol hydrolysis. In these experiments, lipid-loaded cells that express exogenous ADRP maintain greater triacylglycerol levels over time.

One way in which ADRP reduces lipolysis is by inhibiting the mobilization of lipase to lipid droplets; ADRP levels also seem to inhibit lipid droplet binding to TIP47. One proposed mechanism for ADRP excluding such proteins from lipid droplets is that in binding via hydrophobic interactions it coalesces the phospholipids that form the monolayer around each droplet. As a result, other such proteins that depend on hydrophobic interactions with the lipid droplet core to associate with lipid droplets may also be blocked, particularly if they bind with lower affinity.

ADRP Deficiency: Mouse Models

The physiologic relevance of ADRP in liver lipid metabolism has been determined using gene deletion, causing global deficiency for ADRP. ADRP-deficient mice exhibit normal adipose tissue differentiation and function, as assessed by histology, adipocyte marker expression, adipocyte triacylglycerol levels, white and brown adipose tissue amounts, basal lipolysis, and catecholamine-stimulated lipolysis. These mice also exhibit normal plasma lipids, glucose, and insulin levels. These data suggest that ADRP is unnecessary for the normal development and function of adipocytes. In contrast to this, ADRP deficiency is associated with decreased liver triacylglycerol content and attenuated hepatic steatosis caused by a high-fat diet. This reduction in liver triacylglycerol is not due to decreased fatty acid uptake, increased beta-oxidation, or reduced lipogenesis or lipogenic enzyme levels in mouse liver. There is a moderate increase in triacylglycerol secretion by the liver without any change in systemic triacylglycerol clearance. Thus the mechanism for reduced hepatic triacylglycerol in ADRP-deficient mice is unclear. One scenario that has not been considered is that by homology with the lung alveolus, ADRP-mediated neutral lipid trafficking between stellate cells (i.e. fibroblasts) and hepatocytes is disrupted.

Regulation of ADRP Expression and Function

ADRP mediates lipid accumulation in both the liver and vascular wall *in vivo*. Given the pathologic significance of these phenotypes in human disease, it is important to understand the factors that regulate the expression and function of ADRP. The ADRP gene is regulated by peroxisome proliferator activated receptors (PPARs), including PPARα in hepatocytes and PPARβ/δ in keratinocytes; PPAR agonists can induce ADRP, but the effect varies as a function of cell-type and species. ADRP stimulation does not increase in the subcutaneous adipose tissue of humans with impaired glucose tolerance after prolonged PPARγ agonist treatment; by contrast, stimulation of ADRP by fatty acids may provide a feed-forward mechanism whereby increased ADRP provides a reservoir to pool the final reaction products of neutral lipid synthesis.

The pharmacologic agents that affect the expression of ADRP also affect the expression of many other functionally related genes: while ADRP can be post-translationally modified by phosphorylation, acylation, or ubiquitination, phosphorylation of other lipid droplet proteins also influences the function and/or localization to lipid droplets by these proteins. Yet the functional significance of ADRP phosphorylation is unknown.

ADRP unbound to lipid droplets is modified by ubiquitin and targeted for proteasomal breakdown. Inhibitors of ubiquitinated protein targeting to proteasomes lead to increased ADRP and triacylglycerol levels in Chinese hamster ovary cells, but lysosomal and other protease inhibitors are unaffected. Moreover, loading of Chinese hamster ovary cells with lipid is associated with decreased levels of ubiquitinated ADRP, consistent with a model in which increased lipid accumulation protects ADRP from the effect of a ubiquitin ligase.

Shortened ADRP resists degradation, showing that exogenous fatty acids may increase ADRP levels both by stimulating transcription of the ADRP gene and by promoting triacylglycerol synthesis, thus providing additional lipid droplet surface area to protect ADRP protein from degradation. Triacylglycerol stabilization of ADRP within lipid droplets and, conversely, destabilization by delipidation, would account for why increased fatty acid dynamics correlates with increased ADRP protein and is associated with lipid droplets.

Study of ADRP binding to lipid droplets has utilized truncation, deletion, or mutation of the protein to determine its mechanism of action, only to find that no single domain or motif determines lipid droplet binding. Both the amino- and carboxy-terminal regions of the full-length protein appear to contribute to such binding, and although ADRP associates with lipid droplets via hydrophobic interactions, its association with lipid droplets may be regulated by protein complexes – ADP-ribosylation factor 1 is an ADRP-interacting protein that selectively associates with it. And ADRP-SNARE protein complexes have also been implicated in lipid droplet biology. In addition, ADRP complexes with dynein, the microtubule motor protein; disruption of dynein reduces lipid droplet formation.

ADRP Expressed in Non-Adipocytes

In non-adipocytes, ADRP expression correlates with intracellular levels of neutral lipid, and is associated with lipid droplet increases during states of increased lipid flux as additional lipid droplet surfaces form. Though ADRP's definitive function remains to be defined, manipulation of ADRP expression in cells and organisms alike is associated with marked changes in lipid metabolism. In some tissues, ADRP may promote lipid accumulation at the expense of decreased lipid secretion via lipoprotein secretion or reverse transport, but in other tissues ADRP may loculate lipid and attenuate lipotoxic damage. Despite the rising interest in ADRP, exploiting ADRP therapeutically in lipid-associated human diseases remains speculative.

TIP47

Tail-interacting protein 47 (TIP47) is homologous with perilipin and ADRP structurally, yet functionally perilipin and ADRP are both regulated by PPARs, whereas TIP47 is not. And like ADRP, TIP47 is expressed in virtually all tissues examined thus far. The structural ADRP-TIP47 sequence homology is functionally consistent with the effect of oleate on TIP47-associated lipid droplets. By contrast to perilipin and ADRP, TIP47 is stable in HeLa cells propagated in lipid-poor medium, and translocates to lipid droplets in response to lipid exposure; TIP47 functions similarly in 3T3-L1 adipocytes and in mouse liver.

Other Functional Roles for TIP47

TIP47 was first described as binding to the cytoplasmic tail of the mannose 6-phosphate receptor, mediating its movement from the endosomal compartment to the trans-Golgi network; it is also found in lipid droplets in association with decreased ADRP expression in a hepatoma cell line. Moreover, TIP47 has been implicated in HIV viral assembly, as a mechanism for targeting a protein-tyrosine phosphatase to secretory vesicles, and as an inhibitor of retinyl ester hydrolysis and hormone sensitive lipase in keratinocytes.

S3-12 and OXPAT, Tandem Genes with Reciprocal Expression

To round out this survey of lipid regulatory proteins, S3-12 and OXPAT are the newest additions to the PAT family. S3-12 is primarily found in specialized fat storage tissues, namely white adipose tissue, and in skeletal muscle and heart to a lesser degree. It is expressed in brown adipose fat to a negligible degree; conversely, OXPAT is largely expressed in tissues that have a high capacity for fatty acid oxidation such as the heart, brown fat, the liver, and skeletal muscle. The fact that S3-12 and OXPAT are differentially expressed in white and brown fat suggests that these tissues may logistically store triacylglycerol for different metabolic purposes – S3-12 may facilitate long-term neutral lipid storage, whereas OXPAT may mediate lipid storage for nearer-term utilization through oxidative pathways. These proteins appear to be functionally inert in the cytoplasmic compartment of cultured cells, though exposure to long-chain fatty acids reveals their regulatory roles, stimulating their translocation to the surface of lipid droplets; S3-12 translocates to lipid droplets when adipocyte lipolysis is stimulated.

Non-Mammalian PAT Proteins

PAT proteins are not unique to mammals. These are proteins with significant homology in many other animal species, ranging from sea anemones to insects, snails, frogs, and sea urchins. Functional data indicate that PAT proteins are fundamental to lipid homeostasis across a wide phylogenetic range – both fruit flies and fungi exhibit PAT proteins that localize to lipid droplets and promote storage of neutral lipids; such widely conserved structure-function relationships allow for an in-depth understanding of physiologic evolution, using model organisms to dissect the molecular and cellular roles of PAT family members. Access to such physiologically relevant molecular probes in such a wide array of organisms makes such evolutionary hypotheses testable and refutable.

In this chapter we have described the utility of understanding the nature of neutral lipid trafficking in the lung alveolus and its evolutionary exploitation for broadening our knowledge of other vertebrate physiologic traits. Chapter 5, entitled "Evolutionary ontology and epistemology," provides the rationale for reconfiguring the logic of biology and evolution.

5

Evolutionary Ontology and Epistemology

Contemplating Evolution as a Manifestation of Free Will

We shall not cease from exploration,
and the end of all our exploring will be
to arrive where we started
and know the place for the first time.
T.S. Eliot

Like Ernest Rutherford, who thought that all science was either physics or stamp collecting, the high priests of physics lord it over the biologists. Poring over the literature on physics and biology for many years now, after each foray we are bewildered by the lack of appreciation of the biologic imperative by the purist physicists and mathematicians. Of course in an abstract, ideal world we would simply apply such fundamental principles to biology and make sense of this complex emergent and contingent problem, but at its fundament biology is pseudophysics, starting with its origins as primitive cells reducing entropy, its internal energetic state far from equilibrium, in defiance of the second law of thermodynamics. Obviously, once you start with a premise that goes against such a fundamental principle of physics, you're in a different realm. As a corollary to that, biology is "perpetual motion," which we know is impossible in the world of physics, but is made feasible by biologic principles, which immortalize us through evolutionary adaptation and reproduction.

One of the papers in the realm of physics and biology is "The self-organizing fractal theory as a universal discovery method: the phenomenon of life" by Alexei Kurakin. In the article, Kurakin makes the case for the primacy of chemical reactions as the fundament of biology, yet when he discusses pathologic conditions, he merely invokes failed chemistry, never addressing the known mechanisms of homeostatic control for such chemistries in integrated biology and evolution – the physiologic context is lost on him. We assume that is because he starts with the given that all chemistry is one and the same, whether in the external physical environment, or the internal physiologic environment of the cell. One example he latches onto is the dynamic maintenance and breakdown of the matrix between cells, the material that cells generate to separate themselves from neighboring cells, noting the dynamic nature of this process. He characterizes these biologic "walls" between the internal

Evolution, the Logic of Biology, First Edition. John S. Torday and Virender K. Rehan.
© 2017 John Wiley & Sons, Inc. Published 2017 by John Wiley & Sons, Inc.

and external cellular environments as just that – walls – when in fact they are highly organically integrated with the phenotypes of the specific cells they sustain.

In our own research, for example, we see the direct relationship between the signaling mechanisms that determine connective tissue fibroblasts and the matrices they produce for homeostasis. Interestingly, it is the matrix that is one of the first structural elements to be affected by inflammatory diseases and cancer biology alike, suggesting that when the cell changes its phenotype during these disease processes, to maintain homeostasis it copes with the change in its environment by altering its chemistry, whereas Kurakin is saying that *the chemistry is the cause of the pathology*.

On further reflection, when you alter atomic structure by changing electrons, protons, and neutrons, the elements change their identities from one form to another. This is quantum mechanics. In contrast, in biology when homeostasis is altered, biology responds and adapts in reference to its ontogenetic and phylogenetic cell–cell signaling histories, which are its form of "quantum mechanics." So there is a fundamental difference in the ways physics and biology respond to change, the former obeying fixed stochastic rules, the latter making up its own rules as it goes along, pragmatic and existential – that is, free will – in contrast to the stochastic nature of chemistry and physics.

So this is an epistemologic problem of cause–effect relationships. If one starts from the chemistry of the Universe, and reduces the problem of biology to it, one gains one perspective. Conversely, if one starts from the spontaneous, self-referential formation of cells, which then provide an internal environment (*milieu intérieur*) that fosters the reduction in entropy, or negentropy, one gains a very different perspective. This is not merely a philosophical problem, it determines how we perceive ourselves in the Universe, either as having free will, or as being determined by the hard-and-fast rules of physics and chemistry. Obviously, without free will we could not be contemplating the nature of evolution! And of course, that realization also carries with it our burden of responsibility to Nature as its stewards.

Complementarity, or the Value Added by the Cellular Approach to Evolution

On 15 August 1932 Niels Bohr delivered a lecture entitled "Light and Life" to the International Congress on Light Therapy in Copenhagen, Denmark. The lecture provided Bohr an opportunity to reflect on the philosophical significance of recent developments in quantum theory for the life sciences. He had previously introduced the concept of complementarity in a 1927 lecture at Como, Switzerland, to address specific problems arising from quantum mechanics as a way of defining the conditions under which particular phenomena appear. But he was also interested in whether complementarity could be extrapolated to other scientific problems.

Complementarity entails two descriptions of the same phenomenon that require mutually exclusive experimental arrangements, but are both necessary for us to be able to understand it. This construct was specifically devised for the "dialectic" interrelationship between the particle and wave characteristics of light. By the same token, this analysis can be applied to the seeming genotype-phenotype dialectic in biology, characteristics that are likewise measured by different methods yet describe the same phenomenon. If the physicists can countenance the duality of light as an integrated whole, why can't biologists see things in the same holistic way? Instead, we continue looking at biology as the sum of its parts. No one has done that more deliberately than Ernst Mayr, who defined evolution as being either proximate or ultimate. He used the example of bird migration, the proximate process of reproduction being totally dissociated from the ultimate process of migration. This perspective discourages any thoughts about integrating the proximate and ultimate aspects of this phenomenon, reinforcing the mutation-selection dialectic that pervades evolution theory. Seen from its end results, evolution comprises two distinct entities, hindering any attempt to integrate the biologic processes of reproduction and migration. In contrast to this top-down approach, once the complex physiologic mechanisms involved in reproduction are reduced to the cellular level, they can hypothetically be interfaced with the migratory mechanism – how, perhaps, the seasonal production of neuroendocrine hormones is influenced by the seasonal changes in the wavelengths of light, affecting the pineal regulation of the reproductive hormones that determine seasonal breeding habits. Once the internal and external "environments" are reduced to compatible cellular components, they can be seen as a continuum rather than as loose associations, offering the opportunity to experimentally trace their deep evolutionary origins in the ontogeny and phylogeny of the organism.

Heliocentrism, the Age of Enlightenment, and the Recrudescence of Physiology

During the Renaissance, it was acknowledged that the Sun, not the Earth, was the center of the solar system, causing a sea change in the way humanity perceives its place in the Universe. The advent of the first principles of physiology would, at a minimum, impact the human psyche in much the same fundamental way. The knowledge that we are related to all of the other biota could be used to leverage how we treat the biosphere, moving forward. No longer would such behavior be hypothetical; rather it would be a biologic mandate to protect and respect our "relatives." The loss of any given species would effectively constitute the loss of all species.

Evolution as a Prism, not a Kaleidoscope (Fractals form Patterns, not at Random, but Because there are Underlying Principles that Generate those Patterns!)

The evolutionary biologic literature is largely represented by either pure theory, such as developmental systems theory – a collection of models of biologic development and evolution that argue that the emphasis that modern evolutionary synthesis places on genes and natural selection as the explanation for living structures and processes is inadequate; or by descriptive biology, including molecular mechanisms, but without their evolutionary origins. That is because we have not yet devised an effective way of thinking about evolution in "real time." The core problem seems to be that evolution is conventionally thought of from the present to the past, whereas the experimental reality proceeds from the present to the future. We suggest the following resolution for this conceptual problem (Figure 5.1). The classic representation for vertebrate evolution is as phyla, from fish to mammals – and, of course, we think of the adult stage in each phylum as being representative of each group, yet evolution engenders the entire life cycle, from embryo to adult, including reproduction, inevitably returning to the unicellular state (the fertilized egg in vertebrates). The description of phyletic evolution on the left of the schematic in Figure 5.1 allows us to organize the known biology in a first approximation of evolution, which does not lend itself to experimentation because it is based on chronological time and phenotypic homology, whereas experimentation must occur in the present. In order to make this conceptual transition to a paradigm in which evolutionary mechanisms can be tested, we must first focus on the embryonic stage within each phylum, comparing the cellular and molecular processes that give rise to specific structures and functions, both ontogenetically and phylogenetically.

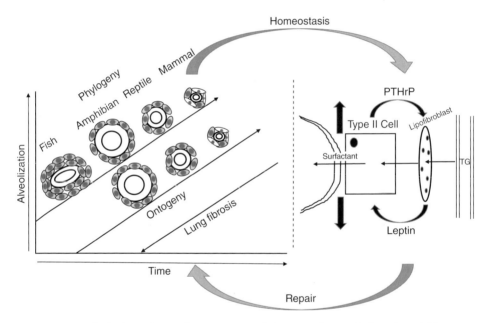

Figure 5.1 Lung biologic continuum from ontogeny-phylogeny to homeostasis and repair. The schematic compares the cellular-molecular progression of lung evolution from the fish swim bladder to the mammalian lung (*lefthand side*) with the development of the mammalian lung, or evo-devo, as the alveoli become progressively smaller (see legend in upper left corner), increasing the surface area-to-blood volume ratio. This is facilitated by the decrease in alveolar myofibroblasts, and the increase in lipofibroblasts, due to the decrease in Wnt signaling, and increase in parathyroid hormone-related protein (PTHrP) signaling, respectively. Lung fibrosis progresses in the reverse direction (lower left corner). Lung homeostasis (*righthand side*) is characterized by PTHrP/leptin signaling between the type II cell and lipofibroblast, facilitating lipofibroblast uptake and trafficking of triglyceride (TG) from the alveolar capillary circulation to the alveolar type II cell, coordinately regulating the stretch regulation of surfactant production and alveolar capillary perfusion. Failure of PTHrP signaling causes increased Wnt signaling, decreased peroxisome proliferator activated receptor gamma (PPARγ) expression by lipofibroblasts, and transdifferentiation to myofibroblasts, causing lung fibrosis. Repair (arrow from homeostasis back to ontogeny-phylogeny) is the recapitulation of ontogeny-phylogeny, resulting in increased PPARγ expression.

We use the lung as an example because we have begun using this organ for experimental physiologic evolution in our laboratory. We have found that the basic mechanism of lung morphogenesis changes progressively from the swim bladder of fish to the lung of mammals – parathyroid hormone-related protein (PTHrP) signaling intensity was amplified by a gene duplication for the PTHrP receptor during the phylogenetic transition from fish to amphibians. That gave rise to a progressive amplification of PTHrP-PTHrP receptor signaling from the swim bladder of fish to the lungs of frogs, alligators, birds, and mammals. The stepwise amplification of PTHrP signaling increases the efficiency of physiologic surfactant synthesis, which accommodates the phylogenetic increase in alveolar surface area-to-blood volume ratio, thus allowing for the increase in gas exchange during vertebrate evolution. This phenomenon was first described by John Clements *et al.*, demonstrating that when the amount of lung surfactant per unit area of the alveolus was regressed against alveolar surface area, the phylogenetic sequence of species formed a straight line from frogs to cows, via all the intermediate land vertebrates, indicating that surfactant production was an independent variable. The evolution of the surfactant has been well documented in a series of publications by Christopher Daniels and Sandra Orgeig. Yet the underlying mechanisms accounting for this pattern remain indeterminate.

PTHrP is also expressed in the mammalian anterior pituitary, where it regulates ACTH, and in the adrenal cortex, where it regulates corticosteroid production. That cascade will then stimulate adrenaline production by the adrenal medulla, mediating the "fight or flight" mechanism. The amplification of adrenaline during the water–land transition may have also caused the evolution of the heart, since experimental deletion of the adrenaline signaling mechanism inhibits heart development in mice. That mechanism is consistent with the coevolution of the lung and heart from fish to amphibians, reptiles, birds, and mammals.

The conventional assumption is that proximate and ultimate causation in evolution is exclusively due to random mutation and selection. Yet the mutations seem to occur within their physiologic context since it is observed that evolutionary changes are "preadapted" more often than not. How to reconcile randomness with ordered change? Perhaps the mutations occur within the constraints of the proximate mechanisms, and the selection is the consequence of selection pressure on the proximate mechanism. Development culminates in homeostasis, and both are generated by cell–cell interactions mediated by soluble growth factors. This oneness between the mechanisms of morphogenesis and physiology offers the opportunity for selection pressure acting through both the reproductive strategy of existing organisms (eco-devo), which then "translates" into genetic change during the developmental process. The same can occur through selection pressure on the developing conceptus, affecting the reproductive strategy (Barker hypothesis). Such mechanisms eliminate the need for invoking "ultimate causation" since both mechanisms are one and the same.

Upon Rereading Richard Strohman's Paper "Ancient Genomes, Wise Bodies, Unhealthy People: Limits of a Genetic Paradigm In Biology and Medicine"

Richard Strohman has pointed out the "deficiencies in the genetic paradigm of biology," and makes the case for the study of gene networks, bypassing Walter B. Canon and homeostasis. He doesn't seem to convey any functional knowledge of contemporary cell-molecular developmental biology – AMAZING! It's like we're all in silos, ignorant of one another's knowledge bases. This is not the fault of the individual, but of the system, which has become overly reductionist over the course of the last century. The advent of the Human Genome Project should have been met with novel out-of-the-box mechanistic ways of thinking about biology, but instead has resulted in "in-the-box" description at higher resolution. In a 1998 essay John Maddox stated that "Part of the trouble is that the excitement of the chase (of molecular causation) leaves little time for reflection. And there are grants for producing data, but hardly any for standing back in contemplation."

There have been a few "eureka moments" in science since Archimedes discovered the principle of buoyancy while in the bath, such as Copernicus's realization that the Earth is not the center of the solar system, the microbial origin of gastric ulcers, and M.E. Avery's discovery of surfactant deficiency as the cause of hyaline membrane disease. All of these instances resulted in a "paradigm shift." Herein, we propose a novel paradigm for evolution that is a departure from the top-down and bottom-up approaches; instead, we suggest a "middle-out" approach based on the ligand-receptor mechanisms that are the basis for development, homeostasis, and regeneration. We have previously suggested that by thinking of the process of evolution beginning with unicellular organisms evolving into metazoans, the mechanism of evolution could be discerned. Carl Woese and Thomas Cavalier-Smith have emphasized the importance of unicellular evolution, and Nicole King, Chris Todd Hittenger, and Sean Carroll have provided scientific evidence for the metazoan toolkit being present in unicellular organisms.

Physiology is Equivalent to Physics

Cells have entrained entropy through chemiosmosis and catalysis, so they are "derived from," but are not equal with, physics and chemistry. They employ physical principles, but they are not "machines" that comply directly with the laws of physics. They imitate physical principles, and in so doing they have been able to perpetuate themselves spatially by adapting to their environment temporally – it is this spatio-temporal nature of life that has baffled evolutionists, yet it is the fundament of development and phylogeny. By looking at the processes of ontogeny and phylogeny beginning with unicellular organisms, we have been able to deconvolute the process of evolution.

Physicists first became actively involved in biology after the Second World War. Prior to that, beginning with the Greek philosophers, followed much later by Emmanuel Kant, thinkers understood the unity of life intellectually, but had no scientific evidence. However, beginning with quantum mechanics, physicists felt empowered to comment on the meaning of life, ennobled by having discovered the operating principle behind the atom. Bohr was the first modern physicist to address the question of "what is life?" by applying the conceptual principle of the duality of light to biology, in his Como lecture in 1927. Yet this was metaphoric, which poets such as T.S. Eliot and Robert Frost have used to better effect, in my opinion. Take the comment in Frost's published notebook – "Life is that which can mix oil and water" – or the epigraph by Eliot that opens this chapter. Erwin Schrödinger later wrote a monograph, entitled *What is Life?*, in which he similarly tried to apply physical principles to the puzzle of the vital force. Others followed, such as the Nobelist Ilya Prigogine, and Michael Polanyi, who expressed their frustration at the realization that biology was seemingly "irreducible." More recently, in his Nobel Prize acceptance speech, Sydney Brenner stated that the problem of biology is soluble, citing his CELL project to map all of its intracellular pathways. Of course, the greatest of all physicists, Albert Einstein, kept the problem of "life" at arm's length, yet it was his intuitive insight that led him to $E = mc^2$, transcending the stigma of descriptive physics. He explains that he saw the "forest for the trees" as a 16-year-old, dreaming that he was traveling in tandem with a light beam (see Walter Isaacson's book *Einstein*). Like the concept of feng shui, Einstein was able to conceive of the fundamentals of the physical world – Brownian movement, the photoelectric effect, and relativity theory – all in his Wunderjahr of 1905. Of course he famously said that "G_d does not play dice with the Universe," so the conventional stochastic approach to evolution wouldn't have resonated with him.

But perhaps the solution to the evolution puzzle is not based on chance mutation and selection. In biology and medicine, we think that we are on the threshold of a breakthrough on a par with relativity theory. By subordinating descriptive biology to cell-molecular signaling as the essence of evolution, as the fundamental mechanism of life, we will be able to understand the "inner Universe" of physiology by starting from its origins.

The Historic, Systematic Exclusion of Cell Biology from Evolution Theory

The separation of cell biology from evolution theory has been recounted by several authors – Arnold de Loof, Gilbert Gottlieb, Betty Smocovitis, and Rudy Raff – though why this occurred is unclear. In a 1996 paper Scott Gilbert *et al.* clarified the rift. For one thing, the concept of the morphogenetic field was viewed as a threat to the concept of the gene as the unit of ontogeny and phylogeny. Yet it was the foundation of embryology at the turn of the twentieth century. Postulated by Theodor Boveri in 1910, and defined by Alexander Gurwitsch, the concept was popularized by Ross Harrison's limb transplantation experiments. Hans Spemann subsequently referred to the morphogenetic field as a field of organization, the cells within it being defined by their respective positions within it.

The Morphogenetic Field as the Mechanistic Basis for Going from Cells to Systems

From the 1920s through the mid-1930s embryology experienced a renaissance. This was the age of Spemann's laboratory and the foundations of his eponymous "organizer"; Harrison's demonstration of limb polarity; Hamburger's and Weiss's studies of neuron growth and specificity; Sven Horstadius's and Charles Manning Child's gradients; Benjamin Willier's and Mary Rawles' demonstration of the neural crest cell migrations; and Emil Witschi's observations of sex determination and gonadal differentiation. Joseph Needham, Conrad H. Waddington, and Jean Brachet were constructing a biochemical embryology, and it appeared as if the basis of morphogenesis was going to be discovered imminently. The discipline referred to as Gestaltungsgesetze attempted to discover the laws of ordered form. The basic paradigm of embryology, the idea that gave it structure and coherence, was the morphogenetic field. Yet the mechanism for this phenomenon would not emerge for 50 years, leaving a vacuum for other diversions to creep into evolution theory.

Proximate and Ultimate Causation in Biology: Artifact of the Absence of Cell Biology?

Fifty years ago, Ernst Mayr published a highly influential paper on the nature of causation in evolution, in which he made a distinction between proximate and ultimate causes. Mayr equated proximate causation with immediate factors (e.g., physiology), and ultimate causation with evolutionary explanations (e.g., natural selection). He argued that proximate and ultimate causes addressed different questions and were mutually exclusive. Mayr's account of causation remains widely accepted today, with both positive and negative ramifications. Several current debates in biology (e.g., over evolution and development, niche construction, cooperation, and the evolution of language) are linked by a common axis of acceptance-rejection of Mayr's model of causation. We argue that Mayr's formulations have, on the one hand, acted to stabilize the dominant evolutionary paradigm of mutation and selection against change, but on the other hand are now hampering progress in the biologic sciences.

Historic Dissociation of Cell Biology from Evolutionary Biology

Evolutionary biology is shrouded in metaphoric language and a deep history of internal conflict, resolved by language (synthesis) rather than scientific evidence. What are the mechanisms for Waddington's landscapes, Darwinian natural selection and descent with modification, epistasis, exaptation, punctuated equilibrium, cryptic genes, evo-devo, and so forth? How can one test such concepts without knowing what their mechanistic bases are?

This is particularly perplexing because in the wake of the recognition of the significance of developmental biology in evolution theory, or evo-devo, there is literally no cell biology, which is the fundament of contemporary developmental embryology – the only known mechanism for morphogenesis. According to Betty Smocovitis' *Unifying Biology*, this omission is due to the rift between the embryologists and evolutionists back in the nineteenth century, a rift that has never been reconciled. This is akin to studying physics without atomic theory. After all, Julian Huxley was able to achieve the evolutionary synthesis by reminding the community that Darwin was not aware of Mendel's gene theory of inheritance; similarly, Darwin was unaware of Schleiden and Schwann's cell theory.

In order to solve the puzzle of evolution, one must have a biologic model that accounts for the spatial and temporal changes that have occurred during the history of the organism. That's difficult to do in phylogeny, since all the pieces of the puzzle may not be available. But in developmental biology that is precisely what is being done – determining how an embryo develops from a zygote into a fully formed offspring, step by step.

Until the latter part of the twentieth century, such studies were descriptive, for lack of a mechanism to account for morphogenetic fields. But beginning in the 1970s, developmental biologists discovered that cells actively secreted soluble growth factors, such as epidermal growth factor (EGF), and that these growth factors affected neighboring cells by binding to their specific cognate receptors, causing the cells to grow and differentiate in ways that determined their developmental properties. Perhaps equally importantly, the differentiating cells produced growth factors that acted in a retrograde manner on the initiating cell-type to affect its growth and differentiation, accounting for the observed patterns of morphogenesis causally linked to homeostasis. Such insights were originally motivated by observations of hormones having profound effects on embryologic development, such as thyroid hormone, glucocorticoids, estrogens, and androgens.

Such hormonally induced patterns of development, leading to the mature form of the tissues and organs, was most dramatically seen in the case of glucocorticoid (GC) effects on fetal lung maturation. Graham Liggins was studying the effect of GCs on the birth process when he discovered that they could accelerate lung maturation in otherwise pre-viable lambs, provoking hundreds of studies on the effect of GCs on the cell-molecular development of the lung. In early studies of this kind, Barry Smith had made the seminal observation that the GCs did not directly affect the epithelial cells that produce lung surfactant, as had been expected, but instead stimulated the differentiation of the underlying connective tissue fibroblasts, which produced fibroblast-pneumonocyte factor (FPF), a low molecular weight peptide that stimulated alveolar type II cell differentiation, including surfactant synthesis. These studies gave insight to the cellular-molecular basis for lung development, culminating in its full function at the time of birth, providing a continuum from development to homeostasis. More recently has come the realization that altered fetal development can give rise to adult diseases. The Barker hypothesis has made us aware that there is a mechanistic continuum from development to pathology that can affect reproductive success, forming a link from one generation to another, like natural selection. This is particularly true when considering epigenetic mechanisms of health and disease.

The cellular-molecular mechanism being put forward accounts for space and time, both with respect to development and phylogeny, that is, evolution.

Evolution Theory Integrated with Cell Biology

There are no studies in evolutionary biology other than our own that reduce the problem down to the cellular level per se. Yet the perennial problem of how genes determine phenotypes is reconciled by the cell, begging the question why cell biology has been excluded from evolutionary biology. Strict Darwinists think that evolution is just genetic mutation and selection, yet we know that there are examples of evolutionary adaptations that are the result of serial changes, as in the case of fish jaw bones forming the ossicles of the mammalian inner ear, or the cellular basis of lung evolution. Such examples would provide a mechanistic basis for punctuated equilibrium.

One of the major advances in evolution theory has been the inclusion of developmental biology. Yet contemporary developmental biology is predicated on cell biology, particularly how cells differentiate in response to cell–cell signaling molecules and their cognate receptors. By reducing ontogeny and phylogeny to cell-molecular mechanisms that generate form and function, one can identify the intermediary steps that provide the selection pressure for such changes, and specifically how internal and external selection brought such physiologic changes about. Moreover, by focusing on the functional molecular phenotypes, the deep homologies are revealed through comparative functional genomics, the validity of which is seen when such analyses reveal the basis for both physiologic adaptation and complex disease processes. In this regard, the reduction of evolutionary biology to the cell-molecular level makes evolution theory accessible to all the other biologic disciplines, fulfilling Theodosius Dobzhansky's prophesy that "nothing in biology makes sense except in the light of evolution."

The modern synthesis reintroduced genetics to evolutionary biology. And evo-devo re-embraced developmental biology. But there was no integration of genes and cell biology, as there has been in cell-molecular developmental physiology. The core precept in that discipline is that development culminates in homeostasis, which enables the organism to repair or regenerate using developmental motifs. That precept also lends itself to evolutionary strategies in that the organism, by being sensitive and effectively responsive to environmental change, can faithfully adapt (or become extinct).

Ever since the case of *Tammy Kitzmiller, et al.* v. *Dover Area School District, et al.* gave credence to intelligent design we have been pondering why the creationists/intelligent designers have any credence in this day and age. We think we've finally figured out why. Evolution theory has never embraced cell biology, which is the fundamental discipline in contemporary biology. This is the residual of the historic rift between the evolutionists and embryologists back in the nineteenth century, which persists to this day.

There is likely to be a central principle for evolution, yet it is exceedingly difficult to demonstrate because biology has become so fragmented and specialized due to its descriptive nature. Despite these inherent problems in trying to reintegrate biology to be able to understand the principles of evolution, and in turn those of physiology, perhaps there is a way of seeing the forest for the trees that has been passed up in our zeal as scientists to reduce the problem *ad absurdum*. It is always seemingly easy in retrospect once you have all the pieces, like a jigsaw puzzle, but they have to be seen as an integrated whole through some set of guiding principles – we think that by focusing on the cell as the smallest functional unit, or "module," of biology, and how it has orchestrated metazoan evolution, we can trace the evolution of form and function.

Informaticists maintain that if they are provided with enough data, they can figure out any problem, including that of evolution. Apparently that attitude came out of the use of informatics at NASA, but evolution is not the sum of its parts. It is emergent and contingent, making it much more difficult to resolve than a stochastic event.

Dobzhansky is revered for his aphorism that "evolutionary biology is all of biology," yet there is little or no cell biology in evolutionary theory. Selection at the cellular level is rarely if ever addressed, for a variety of reasons, yet we know that there is variation, both adaptive and maladaptive, which would give insights to the mechanisms of evolution. Of course it is counterintuitive that selection pressure might act at the cellular level, given its seemingly protected nature in metazoans. Stephen J. Gould disavowed the relevance of cell selection, stating that "We neglect this subject because positive selection now so rarely occurs at this level in complex metazoans," but in the interim we have discovered such concepts as developmental instability, epigenetic inheritance, and the fetal origins of adult disease, or the Barker hypothesis, all of which would suggest that such positive internal selection is not rare at all.

For example, we have recently shown that there is a reciprocating pattern of internal and external selection that may explain how evolutionary selection pressure may have directly affected cellular evolution. Moreover, cells in metazoans do not act unilaterally, but are part of networks that determine homeostasis, and as such may provide a functional explanation for a variety of phenomena that are usually described metaphorically, such as developmental plasticity, cis-regulation, cryptic genes, reaction norms, punctuated equilibrium, exaptation, and epistasis, to name but a few.

To our knowledge, evolution has never been viewed as being mediated by homeostatic mechanisms, and as a result, many features of the evolutionary process have been described, but their underlying mechanisms have remained unexplored. In an effort to bring this gap in our knowledge to light, we will define these phenomena in cellular terms, as follows:

Adaptation – life is determined by the ability to change in harmony with the external environment, or adapt.

Allostasis – is homeostasis at the organismal level. But in reality, it is the same process seen at a different level.

Antagonistic pleiotropy – is a rationalization of descriptive biology for the process of aging. Pleiotropic traits are associated with loss of function during the aging process. However, seen from the perspective of the unicellular state, this is the natural consequence of the inhomogeneous distribution of bioenergetics during the life cycle, that is, because energy is disproportionately skewed toward the earlier stage of life, it fails as the animal ages.

Bias – the term Wallace Arthur used as the mechanism underlying embryonic growth and differentiation.

Burden – what Rupert Riedl referred to as the responsibility carried by a feature or decision.

Canalization – term coined by Waddington as a measure of the ability of a population to produce the same phenotype regardless of variability of its environment or genotype. Mechanistically, this is due to the compartmentation of specific gene motifs within germlines that interact during embryogenesis to recursively form structure and function.

Cis regulation – in cellular developmental terms, this refers to the ways in which external agents affect cellular structure and function by signaling to the genome to affect the molecular "read-out" of the cell.

Co-adaptation – the seeming synchrony of biologic traits due to their common origins in unicellular life.

Contingence – refers to the unicellular state as the data operating system for metazoan evolution.

Coordinative conditions – what L.L. Whyte referred to as the clue to the relation of physical laws to organic processes and to the unity of the organism.

Cryptic genes – during the course of evolution the organism mounts a response to its environment in order to adapt. That motif is part of the "memory" of the organism. When stress or injury occurs, the cellular homeostatic mechanism may be damaged, eliciting "cryptic genes" that were expressed in some earlier phase of the evolution of that organism.

Developmental plasticity – the process of development is mediated by soluble growth factors and their cognate receptors. Growth factor–receptor signaling operates within the developing embryo through cell–cell interactions that are determined by the physical environment. The signaling mechanisms can be affected by mechanical, biochemical, and molecular factors that vary between individuals to generate homeostatic "norms of reaction."

Emergence – refers to the recombination of pre-existing traits that have evolved from the unicellular bauplan.

Entelechy – Aristotle referred to it as a factor that directs the individual regularities of organisms.

Epistasis – one trait "balanced" by another. Walter Cannon had said that for every trait there must be "balancing traits" in order to maintain homeostasis, beginning with unicellular organisms.

Evolvability – the empiric recognition that there is a central mechanism of evolution.

Exaptation – sometimes referred to as a pre-adaptation, the process by which a biologic trait has its origins in a previous ancestral form. This is an artifact of the unicellular origin of metazoans that makes it appear that organisms are pre-adapted, when in reality they are already adapted but must modify their phenotype to survive.

Free will – the result of biology opting for negentropy and internalization of "information," sustained and perpetuated by homeostasis.

Genetic assimilation – a process by which a phenotype originally produced in response to an environmental condition, such as exposure to a teratogen, later becomes genetically encoded via artificial selection or natural selection. At the cellular level, this may occur through physiologic stress causing local genetic changes that accommodate such stress structurally and/or functionally.

Gradualism – described as evolutionary change over long periods of time. Mechanistically, this may have occurred internally at the cellular-molecular level, culminating in punctuated equilibrium (see below).

Homeostasis – the conventional synchronic view is that it maintains equipoise through negative feedback. The diachronic evolutionary perspective sees homeostasis as the mechanism that forms, sustains, and perpetuates evolution.

Internal principle – what many (such as Georges Cuvier, Etienne Geoffroy Saint-Hillaire and Richard Goldschmidt) have invoked as the mechanism of evolution without providing a mechanism.

Modularity – the unit-like nature of living organisms that results from their origins in the ultimate unit of life, the cell.

Pleiotropy – in descriptive biology, this refers to the multiplicity of uses of the same trait. Mechanistically, this is the result of the reallocation of resources from the unicellular state as the organism evolves. The former is information-gathering, whereas the latter provides insight to the nature of the evolutionary process.

Punctuated equilibrium – the bulk of the evidence regarding evolution is based on the fossilized record of the organism. However, we know that there are molecular changes that occur over time that are not part of the "hard evidence." Such data are referred to as "ghost lineages" – phylogenetic lineages that are inferred to exist but have no fossil record – and are more often than not detected much earlier than the hard data are because of their molecular nature. For this reason the intermediate steps in the evolutionary process have been hidden from view until the relatively recent advent of molecular evolutionary biology, only emerging in the fossil record when a critical mass of molecular changes has occurred, hence the "punctuated" nature of the process.

Reaction norms – see Developmental plasticity.

Spandrel – the term used by Gould and Richard Lewontin to dissuade evolutionists from attributing all phenomena in biology to evolutionary processes. However, they failed to appreciate the fact that all of metazoan biology is spandrels of unicellular organisms.

Systems biology – an attempt to determine the operating principles behind biology based on descriptive data. As a result, it merely indexes the information, but does not give insight to causation.

Ultimate and proximate causation – the seeming differences between structure and function resulting from the descriptive approach to biology. By focusing on the cell as the origin of life, one can see that there is one continuous process of life.

The Predictive, Integrative Nature of a Cellular Approach to Evolution

Starting from unicellular organisms interacting with their environments, cell–cell communication forms the basis for metazoan evolution. The power of this forward-directed analysis is that it provides a mechanistic continuum from development to homeostasis and repair, to reproduction, and back to development, coming full circle. The plasticity of the mechanisms involved in the cell-level intermediates is what gives rise to evolution. At each level of selection there is progressive evidence for interactive mechanisms that affect the epigenetic determinants of the life cycle, providing variability upon which evolutionary selection pressure can be applied. Developmentally, cell–cell signaling exhibits variability for the genes involved in growth factor–receptor interactions, referred to as reaction norms. This is particularly the case under stressful conditions such as food deprivation, hormonal signaling, or infection. The ability to cope with such conditions affects the homeorhetic mechanisms, and ultimately the homeostatic mechanisms, which are the culmination of the developmental process. This, in turn, affects the reproductive success of the offspring, reinforcing the developmental and homeostatic adaptations, ultimately translating into genetic effects, or what is referred to as the Baldwin effect. It is through such an evolutionary cascade that selection for one biologic trait can affect biologic traits in other tissues and organs through the same genetic mechanism. In the early 1800s the embryologist Johann Meckel the Younger observed this phenomenon, which recurs in gene knockout experiments with some frequency. This mode of evolutionary change accounts for many of the unexplained phenomena seen in evolutionary biology, such as stability and novelty, canalization, cryptic genes, reaction norms, genetic assimilation, and even the phylogenetic "recapitulation" seen during development, as observed by Ernst Haeckel. Perhaps the recapitulation process has been evolutionarily conserved in order for the developmental mechanism to remain accessible to integrate novel genetic motifs in a spatio-temporally appropriate manner – for example, Valérie Besnard *et al.* have recently observed that when cholesterol synthesis in the lung, which refers all the way back to the fish swim bladder, is genetically knocked out, there is compensation through lipofibroblast hyperplasia: lipofibroblasts don't appear in lung phylogeny until mammals, indicating that when the deep homolog is inhibited, the more terminally added mechanism can compensate. Such compensatory mechanisms undoubtedly exist in human populations, but we are not aware of them by and large because such individuals don't get sick, just like those cholesterol knockout mice. A classic example is the protection against malaria in sickle-cell disease. Indeed, exploitation of such evolved compensatory mechanisms could be of great benefit to medicine in the future.

Evolution of Regulatory Genes

As physiologic systems evolve, they have a propensity to do so from constitutive to regulated mechanisms, particularly the paracrine mechanisms of cell–cell development and communication. Understanding how and why such mechanisms have evolved developmentally and phylogenetically is key to our knowledge of the first principles of physiology because these processes have resulted from the interactions between external environmental-phylogenetic and internal physiologic-ontogenetic selection pressures, so as to mask the cause-effect relationships.

Stress Causes 'Canalized' Mutations

Polymorphisms are associated with evolved traits, suggesting that such mutations are "caused" by selection pressure, yet a mechanism for such a process has been lacking. The concept of cell–cell communication as the source of adaptive change would provide such a mechanism by mediating such molecular stresses as shear stress, resulting in the generation of radical oxygen species (ROS) that damage DNA, causing DNA editing within the constraints of the stressed condition. The iterative damage-repair mechanism would ultimately result in a context-specific adaptive mutation. Alternatively, the recruitment and selection for adaptive stem cell phenocopies could also generate novelty. (Such an adaptive mechanism might also accommodate carcinogenesis.)

Another way to think about this is to invoke the concept of modules, which was first suggested by J.T. Bonner in 1981. The cell is a module, and unicellular organisms are free-living life forms that reproduce, develop, and possess homeostatic mechanisms. As they have evolved, initially as unicellular life forms and then as multicellular organisms, they have utilized mechanisms of cell–cell communication to signal between one another and generate functional modules. The genes of development became homeostatic genes, making them integral to the structural and functional nature of the tissues and organs that they formed. As such, they act as transducers of both the internal and external environments, acting acutely to adapt to changing conditions. Such adaptations determine reproductive strategies, which result in differential developmental strategies that determine the phenotypes of the offspring. Over eons, such interactive, adaptational strategies generate evolutionary strategies. We have recently shown how the lung has evolved through being mediated by such cell-molecular transducive processes.

Evolvability

There are biologic traits that refer all the way back to those of unicellular organisms, some of which are more directly derived, whereas others are secondary, tertiary, quaternary, and so forth. Perhaps the term evolvability is a descriptor for the resonance or dissonance for a given set of traits and how they are affected by intrinsic and extrinsic factors. Take, for example, the impact of lung evolution on the evolution of physiologic systems – starting with the coevolution of lipid metabolism and respiration in eukaryotes, all of which have cholesterol in their plasma membranes, facilitating endo- and exocytosis, as well as gas exchange due to the thinning of the membrane. This theme of the coevolution of preadapted structures/functions, such as that of respiration and lipid metabolism, is reiterated during vertebrate evolution, starting with role of cholesterol in the function of the swim bladder as a lubricant, facilitating feeding, increasing the ingestion of algae, which are as high as 73% in lipid content. This is particularly pertinent to physostomous fish, which have the equivalent of a trachea that connects the esophagus to the swim bladder. The progressive evolution of the surfactant system promotes the expansion of the surface area of the lung by permitting the decrease in alveolar diameter, necessitating a progressively more efficient surfactant system to prevent alveolar collapse (Laplace's law: surface tension is inversely proportional to the diameter of the sphere). And as the lung became more efficient for gas exchange as vertebrates evolved from amphibians to reptiles and mammals, the cross-talk between the mesoderm and endoderm enhanced cis-regulation, a hallmark for evolution. The expansion of the lung surface area required differential control of blood pressure in the lung and the periphery, mediated by β-adrenergic receptors; the coevolution of the adrenal and lung generated synergy for the adrenal cortex, which developmentally determines the rate-limiting step in adrenaline synthesis, catalyzed by catechol-*O*-methyltransferase (COMT). As a result, there was more efficient oxygenation in the peripheral circulation, promoting the formation of fat, which facilitated the transition from poikilothermy to homeothermy, in and of itself enhancing the surfactant activity, dipalmitoylphosphatidyl choline being 300% more bioactive in reducing surface tension at 37 °C than at 25 °C. Moreover, the coevolution of the lung and kidney is exemplified by the commonalities between the epithelial linings of the alveoli (alveolar epithelial type II cells, or ATIIs) and glomeruli (podocytes) producing PTHrP in response to stretch, in the case of the alveolus promoting surfactant production and alveolar capillary perfusion (i.e., V/Q matching); in the case of the glomerulus, promoting fluid and electrolyte balance, having common roots *in utero*.

Seen from the perspective of phylogeny, all of these changes would have facilitated the emergence of vertebrates from water onto land – the lung facilitating oxygenation, the kidney facilitating water/salt balance, the adrenocortical system maintaining homeostasis, and elevated body temperature affecting whole animal physiology.

In this Chapter we have delved into philosophical aspects of the cellular-molecular approach to evolution. The identification of homeostasis as the underlying, overriding mechanism of evolution fundamentally affects the way in which we think of the process. Chapter 6, entitled "Calcium-lipid epistasis: like ouroboros, the snake, catching its tail!," delineates the epistatic balancing selection for calcium and lipids as the operating principle for vertebrate biology and evolution.

6

Calcium-Lipid Epistasis: Like Ouroboros, the Snake, Catching its Tail!

Life on Earth began shortly after the planet cooled off, approximately 4.5 billion years ago. Ever since then, life has been in constant flux, interacting with the environment through the mechanism of evolution in order to sustain itself and to flourish, or go extinct; there is no middle ground. In this vein, it may be helpful to consider a *gedankenexperiment* – imagine an organism that was perfectly adapted to the Earth's environment. Such an organism would have become extinct due to the ever-changing environment that has been the omnipresent driver for evolution.

Direct knowledge of the ecological basis for life early in the Earth's history is scant, though doubtless life emerged within environmental niches. The conditions of the nascent Earth were quite different from those we see and experience around us now. Yet over eons life evolved in synchrony with the geologic evolution of the planet, as evidenced by the diversity of life forms, both extant and in the fossil record. For example, when one reads *On the Origin of Species*, first published in 1859, one gets the sense that Darwin appreciated the intimate interrelationship between the biodiversity of Patagonia and its topography based on his exquisitely detailed descriptions of the land formations. But he did not make the connections overtly because he had no scientific basis on which to predicate such ideas, so he only implied such relationships.

The cell membrane of unicellular organisms made life possible by establishing the boundary between the inner workings of the organism and its physical surroundings. The plasmalemma acts to mediate the flow of matter and information in and out of the intracellular environment, or *milieu intérieur* as Claude Bernard phrased it. The lipid-protein bilayer is a highly interactive structure that determines the metabolism, respiration, and locomotion of the cell – the three fundaments of vertebrate evolution. The physical and chemical environments may be biologically advantageous or disadvantageous, so life must always be in flux. The Earth is tilted on its axis at an angle of 23.4 degrees, which generates the seasonal changes in the atmosphere, a unique feature of our planet. And the effect of our moon on the tides similarly generates periodic environmental changes that have churned biology since its inception (the lunar effect), perhaps even instigating it since the lapping of water on the shore generated the primitive lipid bubbles, or micelles, that gave rise to life.

Historic increases in the mineral content of the oceans have enhanced the biologic productivity of the seas. However, an over-abundance of certain elements, even essential ones, subjects organisms to physiologic stress. As a result, the cellular systems that determine homeostasis have evolved counterbalancing epistatic mechanisms to survive the ongoing and ever-changing threats imposed by the environment. Calcium is of

Evolution, the Logic of Biology, First Edition. John S. Torday and Virender K. Rehan.
© 2017 John Wiley & Sons, Inc. Published 2017 by John Wiley & Sons, Inc.

particular interest in this regard because the cell has to maintain a critical level of this mineral at all times, and any changes in calcium content, either up or down, severely impair the cell and if left unchecked eventually kill it.

Calcium is a divalent cation (Ca^{2+}) that is omnipresent in lake, river, and ocean water. It has variable degrees of hydration that can cause it to penetrate biologic membranes very rapidly. Thus, calcium is the fastest binding agent of all the bioavailable divalent ions in the environment. By contrast, Mg^{2+} reacts 1000 times more slowly. Calcium in the ocean mostly derives from rivers and hydrothermal effluents derived from the Earth's oceanic crust. Calcium and oxygen, major components of the Earth's crust, are essential for sustaining the ecosystem. A functional link between calcium and oxygen in living systems is found in the biologic molecules that bind calcium, with oxygen often found to regulate calcium homeostasis. Yet O_2 or calcium alone is toxic to cells and whole organisms alike; needless to say, when life commenced more than 3.5 billion years ago the concentrations of free calcium and O_2 surrounding the first cells were much lower than they are today.

Throughout evolution, the physiology of cells has been closely linked to their pathology, because of the evolutionary strategy for allowing substances like calcium and oxygen to become biocompatible (as a consequence of the dynamic interactions between negentropy and homeostasis), along with other potentially damaging chemical, physical, and biologic agents. It has been known for nearly a century that a rise in cytoplasmic free calcium is responsible for initiating cellular events such as movement, secretion, transformation, and division. Yet a prolonged high level of intracellular free calcium irreversibly damages mitochondria, and can cause chromatin condensation, precipitation of phosphate and protein, and activation of degradative enzymes such as proteases, nucleases, and phospholipases.

It has been shown experimentally that elevated concentrations of calcium lead to the dissolution of cells, and that this process is controlled by the activity of calcium-sensitive proteolytic enzymes. Calcium also mediates programmed cell death, or apoptosis. The calcium signaling mechanism, which triggers life at the time of fertilization, and is then reallocated to regulate the developmental program, suddenly transforms from a life signal to a death signal – a fine balance.

Is it purely by chance that the same calcium ion acts during cell mitosis, homeostasis, and apoptosis, regulating a myriad of cellular processes? The ability to sense and respond appropriately, efficiently, and effectively to environmental conditions is a crucial condition for the survival of all cells. The complex role of calcium in living systems, and its closely controlled concentration in the cytoplasm, must be an early signal carried by the cell, probably starting at the inception of life. What happened in the early Earth environment to cause this ion to play such a universal role in living systems? What is the role of the calcium ion in the origin of life and the major epochs in its evolution? When was this role of calcium in living systems established? When and how did signaling pathways and networks originally emerge? What was the calcium concentration in the Earth's earliest oceans, and what can calcium tell us about the environment of the early Earth? It has been hypothesized that changes in the marine calcium concentrations of the Precambrian were the crucial force driving major innovations in life, such as multicellularity, photosynthesis, the origin of eukaryotes, the origin of metazoans, biomineralization, and skeletogenesis.

Calcium within the Cell

Calcium acts as a universal intracellular signal for controlling many cellular mechanisms. The utilization of ionic calcium during the early stages of cellular biogenesis was probably determined by the ion composition of the early ocean. Among the available cations, calcium dominated nascent cells. It was selected for because calcium ions interact with biologic molecules quite readily due to specific properties such as flexible coordination chemistry, a high attraction to carboxyl oxygen (a common motif in amino acids), and rapid binding.

However, there is one disadvantage to having calcium present within the cell. All cells maintain calcium concentrations in their cytoplasm at a very low level ($\sim 10^{-7}$ M) because only this level allows for proper cellular homeostasis. At higher intracellular calcium concentrations proteins, nucleic acids, and lipids will denature. Throughout phylogeny, from bacteria to specialized eukaryotic cells, uncontrolled calcium has a ubiquitous toxic effect on cells. The deleterious effects of calcium also result from the very low solubility product of calcium and phosphate, causing rapid precipitation of calcium phosphate within the cell, which is toxic to cell function. Therefore, widespread utilization of phosphate-based cell bioenergetics is probably of very ancient origin, and is incompatible with high concentrations of calcium in the cytoplasm.

Three factors are necessary for maintaining low levels of calcium in the cytoplasm: (i) low permeability of the cell membrane, controlled by influx mechanisms; (ii) high buffering capacity; and (iii) an effective means of calcium export. In both prokaryotic and eukaryotic cells, a number of mechanisms have evolved to control cytoplasmic calcium concentrations.

The formation of extreme calcium gradients across the plasma membrane allows for the generation of both electrical and chemical intracellular signals. A rise of the calcium level 10–100-fold activates a wide spectrum of cell processes, ranging from gene expression to mitosis, contraction, secretion, stimulation of neurotransmitters, and the formation of exocrine products.

Calcium signaling is necessary for life, and an increased concentration of the ion is needed for it to act as such, though prolonged increased concentrations of calcium may be lethal, hence cytoplasmic calcium must be closely regulated at all times. Calcium signaling depends on increased intracellular calcium derived from either extracellular or intracellular sources. Therefore, all life forms require calcium homeostatic regulatory systems to maintain intracellular calcium at relatively low concentrations compared to those within the extracellular environment. Calcium signaling largely depends on an increase in the intracellular calcium concentration. Calcium concentration is lowest when the cell is at rest. It rises when a stimulus occurs, and this is responsible for the changes in cellular activity. Although this mechanism seems simple, there are many variations maintained by an extensive calcium signaling "toolkit." All cells control the levels of calcium within their cytoplasm using calcium channels. Even the most primitive prokaryotes have plasmalemma calcium pumps, calcium/H^+ and Na^+/calcium exchange systems. Genomic studies are now revealing the prevalence of conserved calcium channel-types involved in calcium signaling.

Calcium and the Earliest Life

Eukaryogenesis

According to the most widely held theory, eukaryogenesis was achieved by proto-eukaryotic "host" cells through endocytosis of chloroplasts and mitochondrial symbionts. Experimentally, endocytosis (both pinocytosis and phagocytosis) is calcium-dependent, with a maximum uptake at 0.1 millimolar external calcium concentrations. Considering this calcium range to be within that of the early Precambrian ocean when calcium levels were low, the first endosymbiotic cellular system was not feasible until the extracellular calcium level exceeded 0.001 millimolar. The existence of eukaryote-like microfossils in sediments 2 billion years old provides evidence that this level could have been attained quite early in evolution. It is unclear when the genetic material was shifted to the cell's interior, and became enclosed by a membrane to form the nucleus. Since nuclear calcium and cytoplasmic calcium are regulated independently, the transit of genetic material to the interior of the cell, surrounded by a membrane, may have been an adaptive response to protect the genetic material against deterioration by an excessive influx of calcium due to its build-up in the environment. Calcium is involved in cell cycle initiation of DNA synthesis, mitosis, and cell division, and the maintenance of chromosomal configuration. The transformation to a well-organized internal structure containing genetic material could have happened as early as the Archeae given their critical mass and diameter. A critical phase in the evolution of eukaryotes was the development of the cytoskeleton. There are indications that both synthesis and disassembly of the main cytoskeletal proteins tubulin and actin, comprising the microtubules and microfilaments, respectively, are highly calcium-dependent. Interestingly, changes in microtubule organization caused by changing external calcium concentrations occur very rapidly in some organisms, and are associated with dramatic morphogenetic effects.

Calcium and Early Unicellular Eukaryotes

Like prokaryotes, most unicellular algae and fungi have minimal calcium requirements. From an evolutionary perspective, yeasts are a good example, because they are known for their extremely low calcium requirements. Yeasts are characterized by enormous calcium-sequestering and calcium-storing capacities that can be interpreted as a physiologically atavistic trait, conserved from the distant past, when yeasts living in calcium-poor primordial habitats had to entrain calcium for cell homeostasis. Habitats with such labile calcium levels probably characterized areas in the early ocean where acidic, calcium-rich hydrothermal water mixed with highly basic, calcium-poor ocean water.

During the Proterozoic eon, just prior to the flourishing of complex life on Earth, unicellular eukaryotes became highly diverse. Near the Precambrian-Cambrian era some unicellular organisms reached millimeter sizes, while others evolved a great variety of shapes. It has been surmised that this trend in unicellular evolution is linked to a continuous increase in calcium concentration in the organisms' habitat, associated with the synergistic action of other metal ions and high phosphate levels. This notion is supported by experimental evidence that environmental calcium has profoundly impacted both the size and shape of unicellular organisms. For example, when various microalga species are propagated in calcium-depleted water they are much smaller than algae grown in calcium-rich water.

Given these observations, it would seem that three main strategies have been used by unicells to cope with progressive calcium accumulation in the environment to attain optimal calcium excretion (i.e., calcium detoxification) rates: (i) increasing the cell diameter; (ii) developing various secondary structures (spines, hairs, etc.), thereby increasing the cell excretion surface area; or (iii) a combination of these two strategies. There are empiric data for all three phenotypes.

Moreover, it has also been conjectured that near the end of the Precambrian, the calcium stress experienced by the planktonic algae was worsened by high levels of phosphate in seawater, evidenced by worldwide phosphorite deposits. With the presence of elevated levels of extracellular inorganic PO_4^{3-}, calcium entry into cells is greatly facilitated. The synergy between these ions can produce lethally toxic effects associated with the copious release of extracellular organic substances and increased cell size.

The Role of Calcium in Initiating Multicellular Life

Evidence suggests that the evolution of metazoans from protozoans happened several times independently in different branches of eukaryotes. Early on, eukaryotic organisms evolved cell polarity and cell contacts, forming colonies. This initial step toward multicellularity and cell differentiation necessitated the development of more complex and effective signaling systems. The transition to multicellular organization caused programmed cell death, which was evolutionarily adaptive, some cells dying selectively in order for others to develop and specialize within pre-existing niches. This required ever more complex pathways for calcium signaling. The necessity for precise control over cytoplasmic calcium concentrations forced cells to select for a calcium regulatory system, which evolved into several control pathways. Such a system must have appeared quite early in evolution, probably more than 3.5 billion years ago, and was maintained in subsequent lineages. Study of choanoflagellates – the organisms that preceded metazoans – has shown that they exhibited advanced functional regulation of the calcium signaling "toolkit," which means that this toolkit was important for the transition from the single-celled to the colonial state. Experimental manipulation of extracellular concentrations of calcium and other divalent cations has revealed that calcium plays a crucial role in inducing cell fusion. It has been demonstrated that calcium triggers cell aggregation, membrane lysis, and cell fusion processes. Such experiments have demonstrated that cell fusion can readily be enhanced by elevated pH (>10) at a temperature of 37 °C, and by the subsequent addition of calcium. These observations could help to clarify the notion that the earliest multicellular life forms appeared in the early alkaline Precambrian oceans. An abrupt rise in calcium concentrations at sites of hydrothermal activity or river estuaries could have provoked cell fusion and multinuclear organizational events.

The signaling role of calcium is well recognized in mammalian reproduction. Life begins with the fertilization of the egg by the sperm, causing calcium oscillations that last for hours, triggering the developmental program and cell division, in which calcium activates binary fission to form daughter cells. Later in development, calcium signals determine the differentiation of specific cell-types. And although we don't know what determines the process of parturition, the calcium-regulatory signal parathyroid hormone-related protein (PTHrP) increases in the uterus during the course of pregnancy, and is sensitive to the physical presence of the conceptus, implying that calcium signaling is intimately involved in parturition.

The role that calcium plays in simple multicellular organisms is exemplified by *Volvox*, which produces colonies composed of 4, 8, or 16 cells. Experiments have shown that the calcium level required for optimal growth of a four-celled colony is about two orders of magnitude lower than that required for growth of a 16-celled colony. At a very low level of calcium, *Gonium* cells fail to adhere to one another and produce colonies. Others have demonstrated how calcium controls the growth of colonies of the green alga *Coelastrum*. Medium having a calcium concentration of about 0.2 millimolar can only support the growth of unicells. In medium containing 30 times more calcium the number of colonies in the culture increases. A calcium concentration of 10 millimolar or higher has a toxic effect on *Coelastrum*. Increased abundance of algae during the late Proterozoic is especially visible in Mesoproterozoic and Neoproterozoic fossils. The intensive process of cell aggregation during the last part of the Proterozoic was probably enhanced by simultaneous excesses of both calcium and phosphate in seawater.

In metazoans, the integrity of cells is also controlled by extracellular calcium. Embryos and tissues do not adhere when extracellular calcium is below 0.1 millimolar. The most convincing example of the role of extracellular calcium in maintaining the integrity of cell aggregates is seen in sponges, in which the cells dissociate when the calcium concentration is less than 10^{-5} M, but reaggregate when the calcium concentration is raised above that level. It appears as though external calcium concentrations ranging between 10^{-5} and 10^{-4} M are necessary for multicellularity. Such concentrations could have been achieved quite early during the transition from the Archean to the Proterozoic, as a result of decreased alkalinity and a greater effluvial influx of calcium due to rapid crater formation in the lithosphere at that time. The advances in calcium-regulation and calcium-secretion systems in eukaryotes were probably the result of these ionic calcium shifts, and the evolution of a host of calcium-regulated proteins used subsequently by metazoan organisms.

Biocalcification and Skeletogenesis

The calcium-detoxification hypothesis considers the onset of biocalcification during the Precambrian-Cambrian transition as a common reaction of marine macrobiota to rapid exposure to sublethal calcium concentrations after a long period of exposure to relatively calcium-poor environments. This hypothesis is based on the universal function of calcium in cell physiology, and ubiquitous calcium regulation and signaling systems functioning in all eukaryotes, which likely evolved to maintain optimal cytoplasmic calcium concentrations at very low levels. At some point during the Proterozoic, the calcium levels in seawater began to rise, and solubilized carbonate chemistry determined calcium and Mg^{2+} balance. Such a rise could have been triggered by a rapid increase in dissolved sulfate in the ocean caused by the increase in oxygen in the atmosphere, again forming a causal link between calcium and oxygen. It should be noted that the end of the Proterozoic (the Ediacaran period) was a time of great flux in the marine realm associated with post-glacial oceanic geochemical upheavals documented by carbon, calcium, and boron isotope ratios.

Life forms had to effectively react to such conditions, or become extinct. The biocalcification that prevailed during the Precambrian-Cambrian transition was probably driven by the rise in calcium concentrations to levels that were either morbid or lethal to the biota. This event, and the fluctuating Ca^{2+} levels in the Phanerozoic seas, are thought to have forced a variety of protists and invertebrates to respond by depositing calcareous skeletons, and cyanobacterial mats to calcify *in vivo*. The biocalcification event at the Precambrian–Cambrian boundary finally changed carbonate deposition from a chemical to a predominantly enzymatic process, in which extraction of calcium from the Phanerozoic oceans became habitable by organisms. It should be noted, however, that many organisms also responded to calcium and other metal ion stresses by secreting complex molecules such as polysaccharides or glycoproteins. These molecules interact chemically with calcium and other metals to form organo-metallic complexes with high buffering capacities.

Calcium accumulation in the Precambrian oceans was one of the major factors driving the origin and evolution of complex life forms, consistent with geologic, chemical, and biologic data. Given how extensively life's functions are controlled by calcium, it is reasonable to assume that many evolutionary novelties are attributable to the rising concentrations of calcium in the oceans. Patterns of change in alkalinity and calcium concentrations provide a well-founded mechanism for catalyzing the evolution of living systems on Earth. Calcium regulation and signaling mechanisms common to all organisms appear to have evolved from the first cells, whose physiology is consistent with the prevailing conditions of the early Earth. An early Earth with an alkaline environment and very low calcium concentrations is a reasonable geochemical scenario for the inception and subsequent major innovations in living systems, offering a plausible explanation for many aspects of the evolution of life, from the first protocells to the rising complexity of multicellular organisms during the Phanerozoic.

Cholesterol and the Eukaryotic Cell Membrane

Given the primacy of calcium during the course of evolution, cholesterol and other related sterols were essential to the evolution of eukaryotic cell membranes as they related to multicellular organisms. The plasma membranes of eukaryotic cells contain large quantities of cholesterol, whereas prokaryotes are universally devoid of cholesterol. Konrad Bloch discovered the cholesterol biosynthetic pathway during the 1950s, hypothesizing that he could determine the sterol biosynthetic pathway, including the sequence for the intermediate steps. He hypothesized that each successive step in the chain of events from precursor to product would form a functionally superior molecule. Bloch showed that cholesterol evolved under the influence of rising levels of oxygen in the atmosphere, mediated by P450 cytochrome enzymes needed for cholesterol synthesis. According to Bloch, the evolutionary selective advantage conferred by cholesterol resulted from the increased cell membrane fluidity of the phospholipid bilayer.

Myer Bloom and Ole Mouritsen thought that the appearance of cholesterol in a more oxygen-rich environment removed a bottleneck in the evolution of eukaryotic cells. Such a proposal for the role played by the fluidity of the cell membrane in eukaryotic evolution correlates with Thomas Cavalier-Smith's characterization of eukaryotic cell evolution, enumerating "twenty-two characters universally present in eukaryotes that are totally absent from prokaryotes." He provides step-by-step accounts for the appearance of exocytosis and endocytosis underpinning the evolution of eukaryotic cells over the course of vertebrate evolution.

In addition, Bloom and Mouritsen hypothesized that cholesterol relieved the evolutionary constraint on membrane thickness. The presence of cholesterol in the phospholipid bilayer increased the orientational order without increasing its microviscosity, allowing relatively larger membrane curvature without increased permeability. The physicochemical effects of cholesterol on membrane thickness, endosymbiosis, and exocytosis is reprised by the evolution of pulmonary surfactant – cholesterol being the most primitive form – the increased oxygenation mediated by the lung, and the facilitation of feeding efficiency by the fish swim bladder, the phylogenetic antecedent of the vertebrate lung. Such a recapitulation of a trait that served one purpose during evolution, subsequently serving a molecularly and functionally homologous purpose later in evolution, is referred to as an exaptation. And if the reprised trait is found in other tissues, it is described as pleiotropic. Cholesterol is an archetypical molecular phenotypic trait that has been positively selected for, from the cell membranes of unicellular eukaryotes to the complex physiologic properties of lung surfactant, cell–cell signaling via G-protein-coupled receptors, and the endocrine system regulation of physiology, all being catalyzed by cytochrome P450 enzymes.

From their origin, eukaryotes have had endosymbiotic relationships with rickettsiae, which eventually evolved into mitochondria. And because rickettsiae possessed calcium channels and a Na^+/calcium exchanger, they naturally were able to regulate and store calcium (Figure 6.1). And regulation of calcium uptake by mitochondria regulated oxidative enzymes and ATP synthesis, so that calcium influx into the cell now interconnected cellular activity and energy production. The evolutionary origin of the endoplasmic reticulum (ER), is not as clear. Such tubular structures may have evolved endosymbiotically from proto-organisms, or due to invaginations of the plasmalemma, or even from

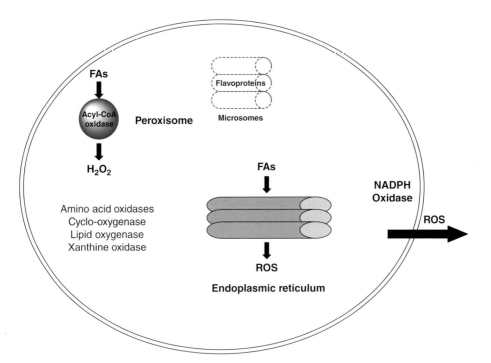

Figure 6.1 Ca^{2+}/lipid epistasis. Calcium dyshomeostasis was epistatically balanced by lipid. FAs, fatty acids; ROS, reactive oxygen species.

primordial vacuoles. The ER appeared very early in eukaryotic development, the endo-membrane providing an organelle for protein synthesis, post-translational protein modification, and calcium transport. Significantly, the ER evolved its own internal system of calcium homeostasis. In contrast to the maintenance of low cytoplasmic calcium concentrations, ER calcium concentrations are comparable with those in the extracellular space. Excessive ER calcium levels necessitated a second level of calcium signaling control, since calcium ions could now enter the cytoplasm from either the external space or from this regulated intracellular source. ER calcium stores are controlled by several intracellular channel types. The phylogeny of the intracellular channels is poorly delineated, though we know that primitive eukaryotes had G-protein-coupled receptors, and inositol trisphosphate ($InsP_3$)-associated second messenger systems triggered by intracellular calcium mobilization. With the advent of intracellular ER calcium stores, a new calcium influx pathway controlled by the calcium stores became necessary, and likely evolved very rapidly. Notably, the first examples of store-regulated calcium are detected in paramecia.

From Polarized Cells to Multicellular Organisms: The Success of Calcium Signaling

Polarized cells called for more articulated pathways of calcium signaling. Hence, the evolution of EF-hand proteins had both adaptive and maladaptive consequences. The calcium-binding proteins are extremely effective calcium buffers, many having dissociation constants in the nanomolar range. Consequently, they rapidly and efficiently buffer calcium ions entering through the plasma membrane, impeding calcium diffusion into the cytoplasm. This strong calcium buffering capacity constrains cytoplasmic calcium signals, ensuring spatial precision, exemplified by calcium microdomains. However, cytoplasmic calcium binding constrains distal cytoplasmic calcium diffusion in polarized cells. To achieve such distal calcium signaling, ER calcium mobilization was exploited. The ER harbors high concentrations of free calcium using calcium-binding proteins with dissociation constants in the millimolar range. Consequently, calcium rapidly diffuses across the ER lumen. This configuration facilitates long-distance calcium transport through intra-ER calcium tunnels. ER calcium generates increased endomembrane reactivity, so that any degenerative recruitment of ryanodine receptors (RyRs)/inositol trisphosphate receptors ($InsP_3$Rs) generates calcium waves.

The advent of primitive colonial multicellular organisms gave rise to cellular altruism, necessitating some cells dying off in order to allow other cells to survive and thrive. Again, calcium plays a central role here since rapid calcium dyshomeostasis is the fastest and most reliable way to biologically kill cells. Thus, calcium-dependent programmed cell death began functioning in its evolved role in tissue homeostasis and development. Cell polarization was an initial attempt at cellular specialization in multicellular organisms, and the subsequent need for intercellular signaling mechanisms. Cell communication occurs either directly via gap junctions or tight junctions, or indirectly, mediated by soluble growth factors, hormones, and neurotransmitters. Chemical signaling is the oldest form, with even proto-cells detecting molecular fluctuations in the environment. Therefore, chemical signaling was commonplace among ancient multicellular organisms. Utilization of such chemical signals required regulated release of packaged hormones by regulated exocytosis. Since release of vesicles involved calcium sensitivity, exocytosis regulation heralded another key function of calcium ions.

However, diffusion of hormones is challenging for complex multicellular organisms, since although their global actions may be beneficial for overall regulation, they are ineffectual for local information transfer. Thus, there was accruing selection pressure for release mechanisms for small amounts of chemicals that could exert their action in a highly specific, localized manner. Again, this feat was accomplished through calcium-regulated exocytosis within spatially and temporally limited release sites (as in neurons) – brief bursts of calcium entering the cell through plasmalemmal channels generate short-lived, high-amplitude calcium microdomains, controlling local and rapid release of neurotransmitters. Local, high-level expression of calcium signaling complexes and their receptors also generates such high-amplitude cellular microdomains. Living organisms initially devised relatively limited numbers of calcium-mediated systems, including membrane calcium channels, cytoplasmic calcium buffers, and membrane calcium transporters such as calcium pumps and exchangers. Such systems appeared early in evolution, subsequently becoming conserved phylogenetically. Bear in mind that various combinations of the components of calcium homeostatic systems provide almost endless possibilities for calcium regulation, determining the versatility and individuality of calcium signaling events in various cell-types under various conditions.

Calcium Dyshomeostasis and Neurodegeneration

The balance between degenerative and regenerative processes determines the plasticity of the nervous system. Given that calcium is a fundamental signaling mechanism involving nearly all cellular physiologic functions, small changes in its homeostasis can cause profound functional changes. For example, there is ample evidence that calcium dyshomeostasis affects normal brain aging. The decline in cognitive function associated with aging is not associated with significant neuronal loss – it results from the change in synaptic connectivity. Changes in calcium cellular machinery during the aging process have been observed to correlate with cognitive loss. For example, neurons in the hippocampus exhibit synaptic plasticity associated with increased calcium influx and L-type voltage-dependent calcium channel density during aging. And aging neurons also display increased injury due to ER stress and calcium leakage, decreased calcium extrusion, decreased cellular calcium buffering capacity, and decreased mitochondrial calcium sink capacity, activation of calcineurin, and calpains. In the aggregate, calcium dysregulation damages neuronal calcium-dependent potassium channels, ultimately compromising learning and memory. Calcium dyshomeostasis as a key factor in brain aging is further supported by pharmacologic interventions to inhibit the age-related calcium signaling increase. BAPTA-AM, which is a membrane-permeant calcium chelator, reduces the effects of impaired calcium dynamics in hippocampal synapses of senescent rats, enhancing spatial learning. Similarly, nimodipine, the L-type calcium channel blocker, blocks age-related learning impairment. Aging is the greatest risk factor for neurodegenerative diseases, characterized by the gradual neuronal loss of motor, sensory, or cognitive systems. Amelioration of age-dependent aberrations in calcium signaling, contributing to the initiation or progression of the neurodegenerative process, can possibly protect against neuronal vulnerability to metabolic and functional stressors. Multifactorial dysregulation of calcium and mitochondrial homeostasis is the final common pathway underlying the molecular mechanisms of neuronal loss in Alzheimer's, Parkinson's, and Huntington's diseases, amyotrophic lateral sclerosis, and other neurodegenerative disorders. The decreased mitochondrial functional capacity and resultant ATP production, accompanied by increased reactive oxygen species generation, reduced mitochondrial calcium buffering capacity, and enhanced calcium responses are all key elements that precipitate the cell death characteristic of most neurodegenerative disorders.

Barrier Function in Evolution

The evolution of a barrier function of membrane bilayers was critical for cellular evolution and the rise of complex life on Earth because it allowed cells and subcellular organelles to form their internal environments (*milieu intérieur*), which differ to one extent or another from the external environment. Other relevant and important properties are the selective permeability to molecules in directional or non-directional flow, which in most cases is performed by proteins; and signal transduction through the membrane, which is mediated either by proteins or lipids. Amphipathic lipids, in which the cross-section of their hydrophobic region is similar to the cross-section of their polar head group, form the matrix of biologic membranes due to their spontaneous capacity to form intact lipid bilayers. Such bilayers are the basic elements of biologic membranes.

Over the course of 3 billion years of evolution, biologic membranes have maintained their lipid bilayer phenotype, with proteins embedded. The lipid composition of these membranes has changed dramatically from those of the archaebacteria, in which the membrane is composed of a single layer of molecules having two head groups (bola lipids), and two very long hydrocarbon chains, to those of eubacteria, in which the matrix of the membrane is already a phospholipid bilayer. Whereas in eubacteria the number of lipid species is small and non-versatile, the membranes of mammalian cells include more than 2000 species of lipid molecules, including sphingolipids and sterols.

The appearance of sterols in biologic membranes is undoubtedly one of the important milestones in membrane evolution. Large changes in the environment caused eukaryotes to modify the lipid composition of their cell membranes by one or more mechanisms, in order to survive and/or optimize their adaptation to the environment. This led to the formation of large lipid membrane arrays, differing in their head groups and hydrocarbon side-chains. Sterols are a major means by which eukaryotic cells modulate and refine membrane properties. The barrier properties of membranes, as well as endocytosis, exocytosis, and cytoplasmic streaming, are strongly affected by the sterol content of the membrane.

Sterols with such properties are classified as membrane-active. The structural requirements for a sterol to be membrane-active include a flat fused-ring system, a beta-hydroxyl (or other very small polar group) at position 3, a cholesterol-like tail, and a relatively small minimal area at the air/water interface. Such minimal requirements correlate with the membrane-active sterols found throughout the evolutionary continuum, having specific effects on membrane order and organization. Cholesterol is the principal mammalian membrane-active sterol. The effect of such membrane-active sterols on membrane permeability is well known, being used routinely by those who study liposome permeability as a model for membrane structure-function relationships.

The physicochemical effects of cholesterol on membrane permeability have been studied extensively using virtually all the physical techniques available for membrane research. Such studies indicate that membrane-active sterols affect intramembrane short-range order and long-range lateral organization, typically stiffening chain ends, but lowering the chain order near the lipid head group. This happens preferentially in proximity to saturated hydrocarbon chains, which have higher affinity for cholesterol than unsaturated lipids, whereas polyunsaturated chains have no affinity for cholesterol.

In summary, membrane-active sterols have played an existential role in eukaryotic evolution, facilitating metabolism, respiration, and locomotion, the principal properties of vertebrate biology.

In this Chapter we have gone into detail with regard to calcium/lipid epistatic balancing selection as the founding principle of eukaryotic physiology and evolution. Chapter 7, entitled "The lung alveolar lipofibroblast: an evolutionary strategy against neonatal hyperoxic lung injury," describes the evolution and role of the lipofibroblast in lung development, homeostasis, repair, and evolution. The lipofibroblast is the epitome of lipid homeostatic balance in eukaryotic physiologic regulation.

7

The Lung Alveolar Lipofibroblast: An Evolutionary Strategy Against Neonatal Hyperoxic Lung Injury

In order to understand the evolution of the lipofibroblast, it must be seen in the context of cellular-molecular lung development.

Mechanism of Mammalian Lung Development

The development of the lung is mediated by coordinately integrated, mutually regulated networks of transcription factors, growth factors, matrix components, and physical forces, which all play important roles in determining lung structure and function. Lung development is divided into two major phases – the branching of the airways, followed by the formation of the alveoli. The lung begins as an out-pouching of the primitive foregut at 4 to 6 weeks of gestation (term gestation = 40 weeks) in humans – the proximal portion generating the larynx and trachea, while cells located at the distal end of the trachea give rise to the left and right main stem bronchi. Branching morphogenesis of the left and right bronchi forms specific lobar, segmental, and lobular branches. This process extends through the canalicular stage of lung development up to mid-gestation. Saccularization starts at mid-gestation, leading to alveolarization, which continues up to 8–10 years of age to generate the 300 million alveoli of the mature lung, providing an enormous gas-exchange surface, paired with an equally large and efficient alveolar capillary network. This sequence of biologic events has been positively selected for evolutionarily over biologic time and space, resulting in optimal gas exchange mediated by alveolar homeostasis. We have previously suggested that chronic lung disease (CLD) causes simplification of the lung alveoli in a manner that suggests reversal of the evolutionary process. Theorizing that by identifying those mechanisms that have evolved under selection pressure for optimal gas exchange, for example, the evolution of lipofibroblasts, we can effectively reverse the deleterious effects of CLD by mimicking the evolutionarily adaptive mechanism, rather than by superficially treating the symptoms.

Evolution, the Logic of Biology, First Edition. John S. Torday and Virender K. Rehan.
© 2017 John Wiley & Sons, Inc. Published 2017 by John Wiley & Sons, Inc.

Epithelial-Mesenchymal Paracrine Model of Alveolar Development

Under the influence of Sonic Hedgehog, the developing endoderm expresses parathyroid hormone-related protein (PTHrP) and its cognate receptor on the adjoining mesenchyme. PTHrP binding to its receptor on the mesenchyme activates the protein kinase A second messenger pathway, which actively downregulates the default Wingless/int (Wnt) pathway, and upregulates the adipogenic pathway through a key nuclear transcription factor, peroxisome proliferator activator receptor gamma (PPARγ), and its downstream target genes, such as adipocyte differentiation related protein (ADRP) and leptin. ADRP is necessary for the transit of neutral lipid from the lipofibroblast to the alveolar type II (ATII) cell for surfactant phospholipid synthesis. Leptin secreted by lipofibroblasts acts on its receptor on ATII cells, stimulating both surfactant phospholipid and protein synthesis. Therefore, epithelial (PTHrP)-mesenchymal (PPARγ) signaling provides a complete paracrine loop for the synthesis of pulmonary surfactant, maintaining alveolar homeostasis. Overall, PTHrP signaling, by inhibiting Wnt signaling, prevents the default myogenic phenotype, and by stimulating PPARγ signaling, induces the lipogenic phenotype, which is necessary for maintaining alveolar homeostasis through its paracrine effects on interstitial fibroblasts and ATII cells. Specifically, the interstitial lipofibroblast phenotype provides protection against oxygen free radicals (i.e., protection against oxotrauma), traffics neutral lipid substrate to ATII cells for surfactant phospholipid synthesis (i.e., protection against atelectrauma) or causes ATII cell proliferation (i.e., protection against any insult causing epithelial injury), thereby promoting alveolar growth, development, and injury repair. Homeostatically, this stabilizes the alveolus, preventing its collapse, maintaining adequate gas-exchange, and reducing energy expenditure by decreasing the work of breathing. On the other hand, although myofibroblasts (MYFs) may also be important for normal mammalian lung development, these cells are the hallmark of all Chronic Lung Diseases (CLDs) in both the neonate and adult. In the developing lung, MYFs are fewer in number and localize to the periphery of the alveolar septa, where they participate in the formation of new septa. However, in CLD MYFs not only increase in number but are also abnormally located in the center of the alveolar septum in great abundance.

Evolutionary Origin of Lipofibroblasts in the Mammalian Lung

There is strong circumstantial and molecular evidence suggesting that the increasing atmospheric oxygen tension over evolutionary time might have led to the formation of lipofibroblasts in the evolving lung: the lung is the first anatomic site where increased atmospheric oxygen would have exerted selection pressure for an evolutionary change. In this regard, the physiologic significance of oxygen in the atmosphere has long been recognized as the selection pressure behind vertebrate evolution. The role of oxygen in vertebrate adaptation has more recently been reprised by Robert A. Berner, who has found that oxygen tensions during the last 500 million years did not rise gradually, but instead fluctuated between 15 and 35%. This begs the question as to how vertebrates adapted to such variation in atmospheric oxygen. On the one hand, hypoxia is the most potent affector of vertebrate physiology; on the other hand, mammals have evolved to gestate under hypoxic conditions, which begs the question as to what the evolutionary strategy constitutes. Therefore, how animals might have adapted to such episodes of hyperoxia followed by hypoxia, or hypoxia followed by hyperoxia, is highly relevant, and has been readdressed recently.

Evolutionarily, the lipofibroblast is absent from the vertebrate lung until shortly before the appearance of land mammals, suggesting that these cells facilitated the adaptation to atmospheric oxygen. Lipofibroblasts are homologous with adipocytes, which differentiate from MYFs through the activation of the PPARγ gene, which determines adipogenesis. Direct evidence for oxygen sensing affecting the expression of this gene has shown that hypoxia inducible factor 1 (HIF-1) signals through DEC1/Stra13 to inhibit PPARγ expression; conversely, hyperoxia upregulates PPARγ. Csete and co-workers have shown that muscle satellite cells in culture will spontaneously become adipocytes in room air (=21% O_2), but not in 6% oxygen, suggesting that the episodic rises and falls in atmospheric oxygen over the last 500 million years have caused the evolution of fat cells both in the lung (lipofibroblast) and in the periphery. Such a mechanism provides a selection advantage since the lipofibroblast protects the alveolus against oxidant injury, and its production of leptin may have fostered modern-day stretch-regulation of alveolar surfactant, mediating ventilation-perfusion matching, the physiologic principle for alveolar gas exchange, thus having facilitated the evolution and homeostasis of the lung. The concomitant production of oxygen free radicals, lipid peroxides, and other oxidative products likely generated eicosanoids as a balancing selection for endogenous PPAR ligands. The improved alveolar gas exchange, with the resultant increased reactive oxygen generation, also likely led to the emergence of an increasing number of NADPH oxidase (NOX) homologs, the oxygen-reducing enzymes dedicated to reactive oxygen species in more complex metazoans. It is also interesting to note that NOX4 is required for the differentiation and activation of MYFs, the key cellular mediators of alveolar septation and lung injury-repair, perhaps representing "antagonistic pleiotropy," that is, the paradoxical selection of genes that are beneficial during early/reproductive life, but that may also mediate deleterious effects in later life.

The Evolution of Peroxisome Biology

Peroxisomes were discovered by Christian de Duve, whose laboratory was the first to isolate peroxisomes from rat liver and determine their biochemical properties. The basic mechanisms involved in peroxisome biology are shared by a variety of organisms, suggesting a common evolutionary origin. Speculation regarding peroxisome evolution began almost immediately after their discovery. Photomicrographic images suggested that there might be interactions between peroxisomes and the endoplasmic reticulum (ER), leading to speculation that peroxisomes were derived from the endomembrane system. Alternatively, the view that peroxisomes were independent organelles originating by endosymbiosis was subsequently proposed upon the observation that peroxisomes formed from the division of existing peroxisomes, and that they imported proteins, both aspects resembling bacterially derived organelles such as mitochondria and chloroplasts. The most flamboyant hypothesis regarding the evolutionary origin of the peroxisome was from de Duve himself, proposing a metabolic scenario for the endosymbiosis mechanism entailing the role of peroxisome enzymes in the detoxification of highly reactive oxygen species. Bolstered by the popularity of the serial endosymbiosis theory, this view has been the most widely accepted among biologists. Based on this scenario, the proto-peroxisome was acquired at a time when the atmospheric oxygen levels were increasing, representing a toxic compound for most living organisms. This concept is consistent with the evolution of the lung lipofibroblast, an example of the way in which vertebrates entrain otherwise highly toxic substances from the environment and adopt them as physiologic mechanisms.

More recently, the endosymbiosis theory for the origin of the peroxisome has been challenged. Experimentally, there is an interrelationship between peroxisome formation and the ER – specific peroxisomal membrane proteins must first be targeted to the ER prior to reaching the peroxisome, and peroxisome-less mutant yeast can form new peroxisomes from the ER upon introduction of the wild-type peroxisome gene. Phylogenetic studies have substantiated an evolutionary link between peroxisomes and the ER, showing homologous relationships between components of the peroxisome import machinery and those of the ER-decay (ERAD) pathway. Such data have led to the conclusion that the peroxisome originates from the ER, but they have not obviated the possibility that the peroxisome originated as an endosymbiont.

Based on the sequence homology with previously identified members of the nuclear hormone receptor superfamily, discovery of peroxisomes was followed by the identification of three PPAR isotypes (PPARα, β/δ, and γ), first in frogs and mice, and later in human, rat, fish, hamster, and chicken. These isoforms were initially shown to be activated by substances able to induce peroxisome proliferation. Various endogenous and exogenous PPAR ligands were later identified, namely fatty acids and eicosanoids, as well as synthetic hypolipidemic and antidiabetic agents; however, the physiologic relevance of each of the endogenous PPAR ligands can be questioned. Nitrated fatty acids (NFAs), produced by non-enzymatic reaction of NO and its products with unsaturated fatty acids, have been suggested as the newest endogenous PPARγ ligands. Though the total amount of NFAs in the bloodstream significantly exceeds their EC_{50} for PPARγ activation, whether concentrations of free endogenous NFAs are sufficient for efficacious PPARγ activation remains unknown. This is particularly relevant since critical roles for reactive nitrogen species such as NO and peroxynitrites have recently been suggested in both physiologic intracellular signaling as well as in mediating oxygen toxicity. Overall, though, PPARs are involved in rodent development; most importantly they are involved in lipid metabolism and energy homeostasis, PPARγ playing a role in adipogenesis and lipid storage, and PPARα playing a role in fatty acid catabolism, with the liver being the best characterized.

PPARγ Mediates the Evolutionary History of the Lipofibroblast: When Homologies Run Deep

During the Phanerozoic phase of vertebrate evolution (the last 550 million years) atmospheric oxygen rose to its current level of 21%, but it did not increase linearly. Rather, it increased and decreased several times, reaching concentrations as high as

35%, and falling to as low as 15% over this period. As mentioned previously, the oxygen increase may have induced the differentiation of muscle cells into lung lipofibroblasts, since the first place where increased atmospheric oxygen would have affected selection pressure for evolutionary change would be in the alveolar wall, as the lipids stored in these lipofibroblasts protect the lung against oxidant injury, consistent with this hypothesized adaptive response to the rising oxygen tension in the atmosphere. PPARγ must be upregulated for lipofibroblast differentiation to occur. Subsequently, leptin is secreted by the lipofibroblasts, binding to its receptor on the alveolar epithelial cells lining the alveoli, stimulating surfactant synthesis, and reducing alveolar surface tension, resulting in a more deformable and efficient gas-exchange surface. Such positive selection pressure could have led to the stretch-regulated co-regulation of surfactant and microvascular perfusion by PTHrP, recognized physiologically as the mechanism of alveolar ventilation-perfusion matching. The evolution of these molecular mechanisms could ultimately have given rise to the definitive mammalian lung alveolus, with maximal gas exchange resulting from coordinate stretch-regulated surfactant production and alveolar capillary perfusion, thinner alveolar walls due to PTHrP's apoptotic or "programmed cell death" effect on fibroblasts, and a blood–gas barrier buttressed by type IV collagen. We speculate that this last feature may have contributed generally to the molecular bauplan for the peripheral microvasculature of evolving vertebrates, given its effect on angiogenesis. One physiologic consequence of the increased oxygenation may have been the concomitant induction of fat cells in the peripheral circulation, which led to endothermy, or warm-bloodedness. The increase in body temperature synergized increased lung oxygenation because lung surfactant phospholipid is 300% more active at 37 °C than at ambient atmospheric temperature (i.e., the body temperature for cold-blooded organisms). For example, map turtles (*Graptemys geographica*) show different surfactant compositions depending on the ambient atmospheric temperature. Therefore, the advent of thermogenesis would have facilitated the physical increase in lung surfactant surface tension-lowering activity. These synergistic selection pressures would have been further functionally enhanced by the coordinate physiologic effects of adrenaline on the heart, lung, and fat depots, underpinned structurally by the increased production of leptin by fat cells, which is known to promote the formation of blood vessels and bone, accommodating the infrastructural changes necessitated by the evolution of complex physiologic traits.

Evolutionary Knowledge Explains the Benefits of Continuous Positive Airway Pressure

Parenthetically, the argument outlined above is not a "Just So Story" – the cited cell/molecular events that evolutionarily determine alveolar homeostasis follow a sequence that is consistent with the phylogeny and ontogeny of the vertebrate lung in both the forward and pathologically reverse directions, allowing us to suggest an approach to lung biology and pathophysiology consistent with evolutionary medicine. This is abundantly exemplified by the failure to explain the reduction in chronic lung disease by surfactant replacement in the surfactant-deficient premature infant, when traditional wisdom would predict amelioration of the disorder due to improvement in oxygenation and ventilation following provision of the deficient substance, namely, pulmonary surfactant.

This is because chronic lung disease is not due simply to the lack of surfactant in the alveoli, but more fundamentally it is due to the lack of fully established epithelial-lipofibroblast communications in the alveolar wall, which leads to surfactant insufficiency. Therefore, unless these homeostatic communications are established, regardless of what treatment is provided, it will not prevent or reverse chronic lung disease. This principle is likely the basis for the success of continuous positive airway pressure (CPAP). Specifically, it provides just the right amount of alveolar distension, capitalizing on billions of years of lung evolutionary phylogeny and development, stimulating the epithelial-lipofibroblast cross-talk induced by PTHrP, leading to a more physiologic milieu; that is why premature infants supported on CPAP are less likely to develop chronic lung disease.

Hyperoxia, Peroxisomes, and ROS

As outlined above, it is likely that to cope with the rising levels of oxygen in the atmosphere during the Phanerozoic eon, the generation of ROS emerged as a by-product of various metabolic pathways. For example, during the fatty acid oxidation that occurs in peroxisomes, high-potential electrons are transferred to O_2, which yields H_2O_2, the initial reaction being catalyzed by acyl-CoA oxidase. Microsomal ω-oxidation of fatty acids is catalyzed by cytochrome P450 enzymes, which form ROS through flavoprotein-mediated donation of electrons to molecular oxygen. Sulfhydryl oxidases in the ER catalyze oxidative protein folding, with the generation of disulfides and the reduction of oxygen to H_2O_2. Several enzymatic systems in the cytosol also generate H_2O_2, such as amino acid oxidases, cyclo-oxygenase, lipid oxygenase, and xanthine oxidase. H_2O_2 generated in the cytoplasm has the potential to perform signaling functions, as it may diffuse to various organelles, including the nucleus. Notable sources of ROS at the plasma membrane are the NADPH oxidases, which are associated with cell signaling rather than with a metabolic pathway.

Lipofibroblasts in the Human Lung

Given the evolutionary significance of lipofibroblasts in lung biology, it is rather surprising that their presence in the human lung was unequivocally demonstrated only recently, even though this cell-type has been extensively documented and studied in many other species for decades. Based on their adepithelial localization, and morphologic (lipid-staining), molecular (presence of characteristic lipogenic and absence of myogenic markers), and functional (triglyceride uptake) characteristics, which are the hallmarks of the rodent lung lipofibroblast, the presence of lipofibroblasts in both neonatal and adult human lung autopsy specimens has now been confirmed.

Mother Nature Opts for Lipofibroblasts to Maintain Homeostasis Too

Experimentally, Valérie Besnard *et al.* found that when they deleted a gene necessary for the synthesis of cholesterol specifically in ATII cells, the lungs appeared to function normally, even though cholesterol is necessary for effective surfactant surface activity. On further examination, it was found that the lung developmentally "compensated" for this deficiency by over-expressing the lipofibroblast population in the alveoli, suggesting that by "sensing alveolar dyshomeostasis" due to cholesterol-less surfactant, which is of poorer surface-active quality, the alveoli invoked an evolutionary strategy to facilitate surfactant production, both ontogenetically and phylogenetically, namely, by increasing alveolar lipofibroblast population.

PPARγ Agonists Turn on a "Master Switch" for Normal Lung Development, Universally Preventing Neonatal Lung Injury

It is clear from the work reviewed above that lipofibroblast PPARγ signaling plays a central role in epithelial–mesenchymal interactions, maintaining alveolar homeostasis and aiding lung injury repair. The lipofibroblast expresses PPARγ in response to PTHrP signaling from the ATII cell, resulting in both the direct protection of the mesoderm against oxidant injury, and protection against atelectasis by augmenting surfactant protein and phospholipid synthesis. Molecular injury to either the ATII cell or the lipofibroblast downregulates this molecular signaling pathway, causing MYF transdifferentiation – MYFs cannot promote ATII cell proliferation and differentiation, leading to the failed alveolarization characteristic of bronchopulmonary dysplasia (BPD). In contrast, maintaining the alveolar interstitial fibroblast's lipofibroblastic phenotype supports ATII cell proliferation and differentiation even under the influence of factors implicated in the pathogenesis of BPD. This scenario is validated by a plethora of *in vitro* and *in vivo* studies. Importantly, these studies show that exogenously administered PPARγ agonists can prevent or reverse MFY transdifferentiation, potentially preventing the inhibition of alveolarization in the developing lung, the hallmark of CLD of the newborn.

In summary, by identifying deep homologous mechanisms that have determined both the phylogeny and ontogeny of the lung, we have experimentally used exogenously administered PPARγ agonists to exploit the lung's evolved cellular strategy to combat hyperoxia and prevent neonatal lung injury leading to the chronic lung disease of prematurity. We rationalize that a diagnostic and therapeutic approach predicated on mechanisms that have resulted in the evolution of human lung under the selection pressure of increased atmospheric oxygen can be exploited to understand homeostasis, representing health, and dyshomeostasis, representing disease. However, on a cautionary note, though the above outlined approach appears to be robust, effective, and safe in promoting lung maturity and injury repair under experimental conditions, its clinical translation awaits further detailed pharmacokinetic and pharmacodynamic studies with specific PPARγ agonists for their safe and effective use in human neonates.

> We must trust to nothing but facts: these are presented to us by nature and cannot deceive. We ought, in every instance, to submit our reasoning to the test of experiment, and never to search for truth but by the natural road of experiment and observation.
>
> *Antoine Lavoisier*

In this Chapter we have exemplified the vertical integration from the molecular to the physiologic/pathophysiologic using the lung alveolus as a model. Chapter 8, entitled "Bio-logic," is a further exposition on the cellular approach to understanding the logical basis of physiology as a continuum from uni- to multicellular organisms.

8

Bio-Logic

We shall not cease from exploration
And the end of all our exploring
Will be to arrive at what we started
And know the place for the first time.
　　　　　　T.S. Eliot, Four Quartets

A hen is only an egg's way of making another egg
　　　　　　Samuel Butler

Introduction

The current state of the biologic sciences is reminiscent of the Tower of Babel – they have become so highly specialized that one discipline cannot effectively communicate with the other due to the hermeneutic, self-serving languages we all employ. In contrast to that, there is a strong sense that biology is the product of the process of evolution, and that there is an underlying driving mechanism which we haven't quite been able to figure out yet, sometimes referred to as teleonomy. That sense was perhaps best and most famously expressed by Theodosius Dobzhansky's dictum that "Nothing in biology makes sense except as evolution." He set the bar, and we've been trying to hurdle it ever since.

The fact that metazoans begin life as single-celled zygotes, and reproduce from that single cell, is indisputable. The mechanism involved in generating a whole organism from a fertilized egg involves cell–cell interactions mediated by soluble growth factors and their receptors, which mediate cell signaling through pathways that determine morphogenesis. Recently, experimental evidence has been put forth to indicate that single-celled organisms possess the complete genomic toolkit for multicellular organisms, documenting that single-celled organisms are the basis for the process of evolution. That provides the empiric rationale for examining evolution from its unicellular origin.

Statement of the Problem

When we think of evolution in terms of contemporary biologic phenotypes, we make the systematic error of reasoning backwards from the present to the past. Yet reasoning after the fact, by definition, is illogical. All of biology is formed from and by cells, which

Evolution, the Logic of Biology, First Edition. John S. Torday and Virender K. Rehan.
© 2017 John Wiley & Sons, Inc. Published 2017 by John Wiley & Sons, Inc.

emerged from the primordium 3–4 billion years ago, likely as primitive micelles formed from lipids. Such structures are semi-permeable, generating intracellular chemical gradients, a process referred to as chemiosmosis, ultimately allowing for the reduction of entropy within the cell, transiently circumventing the second law of thermodynamics. It was under these conditions that life began on Earth, initiated by prokaryotes, and perpetuated by the perennial competition between prokaryotes and eukaryotes, a battle that rages on to this day.

It is because of the emergent and contingent nature of life that Michael Polanyi failed to reduce it to physical principles, and Ilya Prigogine similarly failed. Back in the 1980s a group of physiologists – Ewald Weibel, C. Richard Taylor and Hans Hoppeler – attempted to determine if physiologic mechanisms were consistent with physical principles. They referred to their hypothesis as "symmorphosis." In the end, they concluded that physiology could not be predicted by the laws of physics. So by default, life merely imitates the physical world, sparked by the reduction in entropy within unicellular organisms.

The Solution to the Problem

We have identified a mechanism that integrates development and physiologic homeostasis: cell–cell interactions. Why not apply that mechanism to evolutionary biology as the long-term basis for phylogenetic change? Using that approach at the cell-molecular level offers the opportunity to determine how cellular composition has accommodated adaptation. In a recently-published book, entitled *Evolutionary Biology, Cell-Cell Communication and Complex Disease* (Torday & Rehan, 2012, Wiley-Blackwell), we exploited this approach to understand how the lung evolved to accommodate metabolic drive, based on the role of surfactant in facilitating both the developmental and phylogenetic increases in lung alveolar surface-area for gas exchange. By reducing this process to ligand–receptor interactions and their intermediate downstream signaling partners, we were able, for example, to envision the functional homologies between such seemingly disparate structures and functions as the lung alveolus and kidney glomerulus, the skin and brain, and the skin and lung.

Using such a reductionist approach to functional genomics has led to a mechanistic understanding of how internal selection pressure, brought on by physiologic stress within Claude Bernard's *milieu intérieur*, may have given rise to such lung diseases as Goodpasture syndrome and asthma. By linking together the cell-molecular pathways for basic physiologic mechanisms independently of their overt structural and functional appearances, particularly as they relate to extrinsic ecologic selection pressures, one can discern the "how and why" of evolution. By starting from the "middle" of the mechanism, tracing the signaling pathways linking genes to phenotypes, one can see how such pathways evolved across the space and time of biology as ontogeny and phylogeny.

The classic dissociation of proximate and ultimate causation in biology was elaborated by Ernst Mayr. Proximate causes deal with the mechanisms responsible for the makeup and functioning of the individual phenotype. Ultimate causes refer to the past conditions that led to the information encoded in DNA. According to Mayr, proximate causation takes place once the encoded genetic program is actualized in the individual, whereas ultimate causation determines the shaping of the program itself. This dichotomous scheme may be viewed as a logical consequence of the Weismannian separation of the soma from the germline. It assumes that we need different means to understand the phenotype and the genotype. Biologists studying proximate causes ask "how" questions about mechanisms, whereas those studying ultimate causes ask "why" questions about evolutionary epistemology. The phenomena involved at these different levels of causation occur on different timescales, and are referred to as diachronic. For example, by showing the continuum of the lung phenotype for gas exchange at the cell-molecular level, being selected for increased surface area by augmenting lung surfactant production and function in lowering surface tension, we have determined an unprecedented structural-functional continuum from proximate to ultimate causation in evolution. Beginning with cholesterol facilitating gas exchange through the unicell's plasmalemma, culminating in the alveoli of the mammalian lung, we have traced the cell–cell interactions that have facilitated surfactant production both ontogenetically and phylogenetically. By analogy, we can do the same using the example Mayr himself used to dissociate proximate and ultimate causation – that of migratory birds. At the time, being unable to reduce physiology to the cellular-molecular level, it was impossible to discern such a continuum for this complex physiologic trait, yet nowadays that behavior can be broken down to seasonal changes in the wavelength of ambient light, and its effect on the pineal gland in controlling neuroendocrine hormones, which ultimately determine the feeding patterns and reproductive strategies for bird migratory habits. By reducing this complex process to its cellular-molecular constituents, its causal nature can be hypothesized and experimentally tested, obviating the artificial siloing of biology as proximate and ultimate.

One fundamental insight from such molecular analyses is that the time dimension for evolutionary processes is a quantitative artifact of descriptive biology; once the underlying mechanisms are identified, the time dimension falls out of the analysis, other than to provide the sequence of events. Once achieved, the vertical integration literally and figuratively eliminates time. And the space occupied by the myriad forms of multicellular organisms is also eliminated once it is acknowledged that multicellular organisms evolved from unicellular organisms.

This ultra-reductionist point of view yields a very different perspective on life being simple, rather than complex. Moreover, it begs the question as to whether metazoans are merely a further extrapolation of such prokaryotic pseudo-metazoan traits as lateral inheritance, biofilm formation, and quorum sensing. Perhaps protozoans evolved such metazoan phenotypes as a way of monitoring the environment over multiple timeframes. After all, H.G. Wells wanted to teach us humility by having bacteria save humanity in *War of the Worlds*.

Putting Humpty Dumpty Back Together Again Based on Epigenetic Principles

Darwin initiated the search for the origin of species in 1859, using the metaphor of natural selection for its mechanistic basis. Ever since, those interested in pursuing the evolutionary process have been prone to using metaphors instead of mechanisms that could elucidate how and why evolution has occurred. Ironically, biology had a 50-year head-start on cosmology in its reductionist approach, yet the physicists have long since determined how the Universe "evolved," having determined that quantum mechanics and $E = mc^2$ enable us to see how the Cosmos was generated by the Big Bang. In contrast to that, biology lacks a central dogma to unify it. In any endeavor to formalize knowledge, the first phase involves collecting, describing, and organizing the information. Eventually, the scientific method is applied to the data to determine causation. Evolutionary biology has been in the descriptive mode for more than 150 years, whereas in the interim physicists have been able to devise theories and methods to determine the origins and composition of the Universe.

A systematic error in the reductionist approach to evolutionary biology is our failure to recognize that it is a mechanism, not a "thing" (namely, DNA). In order to understand how and why evolution works, one must first reduce it to its smallest functional unit of activity – the cell. In contrast, evolutionists describe the process dichotomously at the genetic and population biologic levels, neither of which is the smallest functional unit. Perhaps that is why cell biology is not part of the conventional analysis – it is not considered to be necessary – yet it is the fundamental mechanism of ontogeny; only in the recent past have we been able to determine the mechanisms underlying morphogenesis based on cell-specific production of soluble growth factors and their cognate receptor signaling partners on the surfaces of neighboring cell-types. These developmental mechanisms culminate in homeostatic control, providing a unified functional basis for physiology, repair, and regeneration. And since such processes are amenable to modification under selection pressure, they are also the mechanisms for phylogeny. Such cellular signaling mechanisms common to both ontogeny and phylogeny provide insights to the mechanisms of evolution, complying with the "emergent and contingent" nature of the evolutionary process.

It is high time that evolution moved on to the mechanistic phase. In order to do so, it must re-embrace cell biology, from which it isolated itself back at the turn of the twentieth century because embryologists such as Ernst Haeckel and Hans Spemann were unable experimentally to explain ontogeny recapitulating phylogeny (the "biogenetic law"), or identify the organizing principle, respectively. Instead, the evolutionists turned their attention to the burgeoning field of genetics, concluding that mutation (as variation) and natural selection were the only mechanisms necessary for descent with modification. As a consequence, evolutionists merely show associations between randomly occurring gene mutations and phenotypes, rather than how genes determine phenotypes. On the other hand, cell biology functionally integrates genes and phenotypes, and nowhere else is that more evident than in the case of developmental biology, particularly as it relates to physiology.

We humans have succeeded as a species because of our highly evolved brains. We have an obligation to both our ancestors and offspring to use our minds effectively so that we don't destroy ourselves, the biota, and the planet in the process. If we understood where we evolved from, and therefore where we are evolving to as a species, perhaps we would act in more socially responsible and humane ways. The key is to deconvolute evolutionary biology, which has become so complicated as to be useless in utilizing the human genome for the prediction and prevention of disease.

The solution to the puzzle of evolution is right under our noses, but instead we generate more and more neologisms and metaphors that allow us to circumlocute and evade the solution.

Conrad Hal Waddington actually foresaw that cell biology would reveal the workings of evolution, as expressed in his book *The Strategy of the Genes*, in which he stated that "somewhere hidden among the deepest secrets of the physiology of the cell, there must be the process by which the hereditary factors undergo those sudden mutations which are the basis for the long time-scale evolution." Those secrets were first revealed in the late 1970s, when it was discovered that cells secreted soluble growth factors that bind to their cognate receptors on nearby target cells, communicating to determine their mutual growth and differentiation during embryogenesis. We have used this approach to deconvolute the evolutionary process, which Waddington described as three time-scales – evolution, development, and physiology.

He contrasted biology with physics:

> Perhaps the main respect in which the biological picture is more complex than the physical one, is the way in which time is involved in it. In the Newtonian system, time was one of the elements in the physical world, quite separate from any of the others; a material body given mass just existed, unchanging and, indeed, quite indifferent to the passage of time. But time and change is part of the essence of life. Not only so; to provide anything like an adequate picture of a living thing, one has to consider it as affected by at least three different types of temporal change, all going on simultaneously and continuously.
>
> These three time-elements in the biologic picture differ in scale. On the largest scale is evolution; any living thing must be thought of as the product of a long line of ancestors and itself the potential ancestor of a line of descendants. On the medium scale, an animal or plant must be thought of as something which has a life history. It is not enough to see the horse pulling a cart past the window as the good working horse it is today; the picture must also include the minute fertilized egg, the embryo in its mother's womb, and the broken-down old nag it will eventually become. Finally, on the shortest time-scale, a living thing keeps itself going only by a rapid turnover of energy or chemical change; it takes in and digests food, it breathes, and so on.
>
> *C.H Waddington, The Strategy of the Genes*

Waddington recommended that this was the way to think of the process of evolution, but cautioned that it was still difficult to envision. Indeed, it would be another 20 years before growth factor signaling for embryogenesis would be discovered, providing the wherewithal to do as Waddington had suggested.

And with all due respect to Waddington, he was misguided by the seeming complexity of life, when in fact it may be the opposite. As I will discuss below, if you start from the premise that it's the unicellular state that is actually being selected for, then time and space can be factored out of the analysis.

Paracrine Growth Factors: From Morphogenesis to Homeostasis

Up until the late 1970s there was no known mechanism that explained how genes generated phenotypes. Up until then, biology was solely descriptive, attracting those who were skillful at limning biologic phenomena, like Richard Goldschmidt, Waddington, and Stephen Jay Gould. Then the soluble growth factors that mediate morphogenesis during development were discovered, beginning with experiments performed by Clifford Grobstein, who demonstrated that organs could autonomously develop in a totally defined culture medium in tissue culture, and that if he separated the endodermal and mesodermal layers of the developing kidney or lung, the isolated tissues would ball up and fail to develop. But if the tissues were recombined in culture, with a semipermeable membrane interposed between the tissue layers, they would inexplicably continue to grow and differentiate.

In 1967, James V. Taderera subsequently showed that low molecular weight developmental "principles" produced by the mesenchyme could be transmitted across a semipermeable membrane, thereby implicating soluble molecules in the mesenchymal regulation of organ development. The later discovery that specific growth factor receptors, and their downstream second messenger signaling cascades, determine form and function developmentally opened up the field of cellular-molecular embryology. This fundamental mechanistic insight to well-defined spatio-temporal relationships in biology has been totally ignored by the evolutionists, who are satisfied with merely characterizing the superficial genetic or phenotypic changes that occur over the course of ontogeny or phylogeny, reflecting their descriptionist heritage, which is why such individuals have self-selected to enter the field of biology. Alternatively, there are theoretical biologists who derive information for the sake of mathematically modeling the process of evolution. And there are pure philosophers who try to devise scenarios for the Darwinian "tangled bank" de novo. We maintain that these descriptive activities are all the direct consequence of a culture that has rejected cell biology for historic reasons.

Contemporary molecular embryology is based on growth factors signaling via their cognate receptors, depending upon spatio-temporal relationships that determine morphogenetic patterns. As such, these mechanisms provide a predictive magnitude and direction for the formation of structure and function. In this sense, it is no different from what we expect of a mechanistic basis for evolutionary biology, which is also trying to comprehend the magnitude and direction of biologic change, though the timescales are (seemingly) very different. But perhaps that's just an artifact of the descriptive modality. Once we transition to a mechanistic approach, such time and space considerations are independent of the mechanisms of interest, other than providing the nominal sequence of events.

More recent experiments have further demonstrated that paracrine growth factors such as sonic hedgehog (SHH), wingless/int (WNT) proteins, bone morphogenetic protein 4 (BMP4), scatter factor, and fibroblast growth factor 10 (FGF10) all play important roles in the lateral branching of the mouse lung bud. Genes encoding BMP4, WNT2, and SHH are expressed at high levels in the bud-forming distal epithelium, while genes encoding FGF10 and the SHH receptor patched (PTC) are expressed in the distal mesenchyme.

In the embryonic mouse lung, FGF10 determines the position and expansion of the lung bud. Mice homozygous for loss-of-function mutations of FGF10 lack limbs and lungs, while endodermal expression of a dominant negative for the FGF receptor FGFR2IIIb causes mice to lack terminal buds in their lungs. Moreover, the addition of FGF10 to 11.5-day-old embryonic mouse lung rudiments cultured in Matrigel™ causes extensive budding. FGF10 is seen in the mesenchyme around both the terminal and lateral branches.

The regulation of FGF10 appears to be controlled, at least in part, by sonic hedgehog and BMP4. SHH is expressed throughout the respiratory epithelium, with the highest expression occurring within the terminal buds. In lung rudiments where SHH is over-expressed, FGF10 transcription is reduced significantly. During normal mouse lung development, the lateral buds become surrounded with SHH-expressing mesenchyme after they form. During bud outgrowth, SHH and WNT7b from the epithelium induce FGF10 and cell proliferation of both the epithelium and mesenchyme cells. As outgrowth progresses, the level of BMP rises in the distal tip, until it reaches a level where it can inhibit FGF10. FGF10 expression then appears more laterally, where it initiates the formation of new buds. At the most distal region, a cleft appears, and extracellular matrix molecules stabilize this cleft.

During the fetal period of lung development, immature mesodermal cells are dominated by the WNT/β-catenin pathway, which confers the myogenic fibroblast phenotype. The developing epithelium expresses SHH, which stimulates mesodermal WNT/β-catenin through its receptor-mediated downstream interactions with PTC and GLI, actively promoting the myogenic fibroblast phenotype. Descriptively, as the endoderm and mesoderm of the alveolar interstitium mature, endodermal SHH signaling through the mesodermal WNT/β-catenin pathway decreases as endodermal PTHrP signaling to the mesodermal PTHrP receptor signaling pathway is concomitantly upregulated. We have exploited the stretch-regulation of PTHrP to test the hypothesis that fetal lung fluid stretches the alveolar interstitium and stimulates PTHrP signaling, which downregulates the mesodermal WNT/β-catenin pathway through cAMP-dependent PKA inhibition of GLI, upregulating the PTHrP signaling pathway, inducing the lipofibroblast phenotype (Figure 8.1). The mature lipofibroblast produces leptin, which induces endodermal type II cell differentiation. The downregulation of endodermal SHH expression by the mature epithelial type II cells ensures constitutive downregulation of the SHH/WNT/β-catenin gene regulatory network (GRN), molecularly stabilizing these key alveolar interstitial phenotypes.

So what is the value added in using a cell-molecular mechanistic approach? Using such an approach, we have been able to envision this continuum, and how it has fostered the evolution of the lung, for example. Based on our working knowledge of how paracrine growth factor–receptor interactions have mediated the development of the mammalian lung, we considered the overall ontogeny and phylogeny of the lung phenotype, that is, its evolution, as an overall selection pressure for increased surface area, from fish to human, in service to the metabolic drive underpinning the water-to-land transition. This has been realized by a progressive decrease in the size of the gas-exchange units, which increase the gas-exchange surface area-to-blood volume ratio over phylogenetic and ontogenetic space-time (Figure 8.2). This process could not have occurred without an increase in the net production of lung surfactant, which must physico-chemically compensate for the increased surface tension resulting from the decrease in alveolar diameter (by Laplace's law, surface tension is inversely related to the diameter of a sphere). The cellular regulation of surfactant production, in turn, is orchestrated by interactions between the alveolar epithelial lung cells that synthesize the surfactant, known as alveolar type II cells, and the adepithelial connective tissue fibroblasts that underlie them within the alveolar wall. The cell–cell interactions that regulate surfactant production have evolved from the secretion of cholesterol (the simplest form of surfactant) into the lumen of the swim bladder of fish to prevent the walls from adhering to one another, to a progressively more efficient means of synthesizing and secreting an increasingly complex biochemical surfactant mix of lipids and proteins

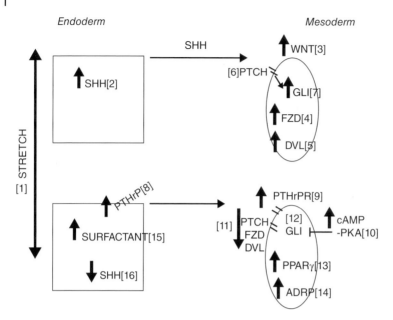

Figure 8.1 Schematic for maturation of the alveolar acinus. During early embryonic lung development (*top*) endodermal sonic hedgehog (SHH) [2] signals to the mesodermal wingless WNT/ PTC/GLI pathway [3–7]. Maturation of the interstitium (*bottom*) is driven by alveolar fluid distension [1], which upregulates the parathyroid hormone-related protein (PTHrP) signaling pathway between the endoderm and mesoderm [8–16], downregulating the WNT pathway by inhibiting GLI [12] and upregulating peroxisome proliferator activated receptor protein (PPARγ) [13] and adipocyte differentiation related protein (ADRP) [14]. Differentiation of the lipofibroblast stimulates differentiation of alveolar type II cell surfactant synthesis [15] and inhibition of SHH expression [16]. cAMP, cyclic adenosine monophosphate; FZD, frizzled; DVL, dishevelled; GLI, gliotactin-Kruppel family member; PTCH, patched; PKA, protein kinase A.

in order to accommodate the increase in surface area, as the lung has evolved phylogenetically. Along with the decrease in the diameter of the alveoli, the alveolar walls also became progressively thinner, further facilitating gas exchange between the alveolar space and the lung microcirculation. The "invention" of tubular myelin, an extracellular latticework of surfactant proteins and phospholipids generated from the lamellar bodies secreted by the alveolar type II cell, provides an extracellular homolog of the lipid barrier formed by the stratum corneum of the skin, including both the lipids and the antimicrobial peptides packaged within the lamellar bodies.

We maintain that tracing the changes in structure and function that have occurred over both the short-term history of the organism (as ontogeny), and the long-term history of the organism (as phylogeny), and how the mechanisms shared in common can account for both biologic stability and novelty, will provide the key to understanding the mechanisms of evolution. Like solving a mathematical fraction problem, the cellular-molecular approach determines the "lowest common denominator" for both ontogeny and phylogeny, eliminating the artifactual temporal-spatial differences between these processes.

It is important to bear in mind that there are certain gene-phenotype homologous relationships that are fairly readily apparent because of their position as "barriers" at the interface between the environment and the organism, such as the lung, skin, and gut,

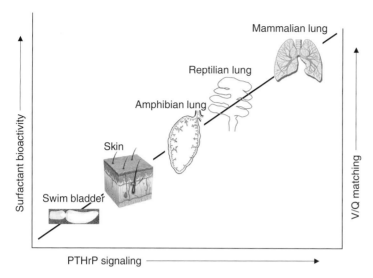

Figure 8.2 Structural evolution of the organ of gas exchange. During phylogeny from fish to mammals, the organ of gas exchange becomes increasingly complex. Starting with the swim bladder of fish, via the skin of amphibians, to the lung, there is an increase in surface area to accommodate the metabolic demand for oxygen. This is particularly true of the arboreal conducting airways and clustering of alveoli in the mammalian lung. Cellular changes in the interstitium of the lung from amphibians to reptiles and mammals are characterized by a decrease in myofibroblasts, and an increase in lipofibroblasts. There is a concomitant decrease in the diameter of the alveoli. We hypothesize that the structural changes are due to the progressive increase in the parathyroid hormone-related protein (PTHrP)/PTHrP receptor amplification signaling (*x* axis), which enhances surfactant production and ventilation/perfusion (*V*/*Q*) matching (*y* axes).

likely having originated from the cell membrane in unicellular organisms as their "common denominator." And then there are other homologies that are "derived" from those more readily apparent properties that must be deciphered based on their short- and long-term histories, particularly as they derive from those primary mechanisms. Instead of taking a "top-down" or "bottom-up" approach to understanding physiologic evolution based on superficial appearances, we have advocated for a "middle-out" approach based on the underlying cell–cell communication by which to determine the evolutionary origins of cell-molecular traits.

We have demonstrated the utility of a cell-molecular developmental physiologic approach in deconvoluting lung evolution, providing a cell-molecular mechanistic continuum from development to physiologic homeostasis and regeneration. Moreover, this tack allows for understanding the interrelationships between tissues and organs at a fundamental cell physiologic level, independent of their contemporary appearances and functions, effectively replacing the need for illogically reasoning after the fact. This approach has for the first time provided novel insights to the mechanisms of evolution for both the more directly evolved structures/functions of the lung, namely skin and bone, as well as for the deeper homologies of the kidney and brain, based on cell–cell signaling as the integrative mechanism.

We have learned from cell culture experiments that normal metazoan cells are not structurally or functionally autonomous; over time, differentiated cell-types lose their phenotypes. They exist within microenvironments created during development by cell–cell interactions between cells derived from different cell lines. The underlying mechanisms of development, physiologic homeostasis, and regeneration are mediated by soluble growth factors and their cognate receptors, which signal through second messengers to determine the metabolic and proliferative status of their surroundings. We maintain that these mechanisms are the basis for the evolution of complex biologic traits, and that by systematically analyzing these diachronic signaling mechanisms over time within and between species, the mechanistic basis for evolution can be discerned.

A Mechanistic Evolutionary Riddle: When is an Alveolus Like a Glomerulus?

As a prototypical working example of how to understand the evolution of a derivative structure, the lung and kidney appear to be distinctly different based on their overt structures and functions dedicated to gas- versus fluid/electrolyte-exchange, respectively. However, by starting with the developmental and physiologic commonalities between the alveolus and glomerulus as the functional units of the lung and kidney, one can find cell-molecular evolutionary homologies by ignoring the superficial differences. Both organs function to produce amniotic fluid during mammalian gestation, demonstrating developmental functional commonalities. But more importantly, these two seemingly disparate structures have common physiologic roots since both act as "professional" pressure transducers. Alveolar distension mediates gas exchange between the internal and external environments, whereas distension of the glomerulus mediates fluid and electrolyte balance to regulate the internal physiologic water and electrolyte milieus. Despite such functional differences, the physiologic distension of either the alveolus or the glomerulus is transduced by the same communicating cell-types (Figure 8.3): in the case of the alveolus (Figure 8.3, lefthand side of schematic), the distension of the alveolar wall stimulates the cross-talk between the alveolar epithelial type II cell and the interstitial lung fibroblast, causing coordinately increased production of parathyroid hormone-related protein (PTHrP) by the alveolar type II cell, increased production of leptin by the lipofibroblast, and increased prostaglandin E_2 production by the alveolar type II cell. As a result of the integrated upregulation of these molecules and their cognate receptors on their complementary epithelial and mesodermal cell-types, more surfactant is produced in response to the increase in alveolar surface area, maintaining reduced alveolar surface tension; alveolar capillary perfusion is also coordinately increased, since PTHrP is a potent vasodilator; calcium in the alveolar hypophase is regulated, since PTHrP is calciotropic, maximizing surface tension-reducing activity, allowing for efficient gas exchange in response to the expansion of the lung.

In the case of the kidney (Figure 8.3, righthand side of schematic), the podocytes that line the glomerulus also produce PTHrP when the glomerulus is distended with fluid, signaling to PTHrP receptors located on the surface of the mesangial fibroblasts; the mesangium monitors and controls fluid and electrolyte fluxes within the kidney in determining urinary output.

The functional relevance of these evolved mechanisms is reflected by the fact that in the case of both the alveolus and glomerulus, failure of the PTHrP homeostatic signaling mechanisms described above, due to a wide variety of insults (barotrauma, oxotrauma, infection, xenobiotics), causes increased WNT signaling in the fibroblasts

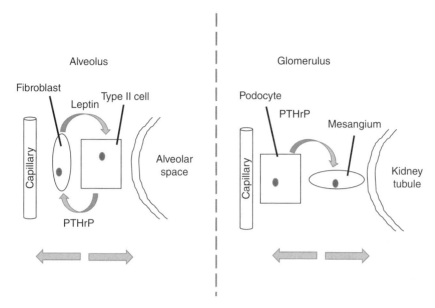

Figure 8.3 When is an alveolus like a glomerulus? The alveolus and glomerulus are stretch sensors. In the lung (*left panel*), the alveolar epithelium (square) and fibroblast (oval) respond to the stretching of the alveolar wall by increasing surfactant production. In the kidney (*right panel*), the mesangium (oval) senses fluid pressure and regulates blood flow in the glomeruli. PTHrP, parathyroid hormone-related protein.

of both organs (see Figure 8.3), resulting in either lung or kidney fibrosis and scarring due to the transdifferentiation of the resident homeostatic fibroblasts to myofibroblasts. This process of injury repair compromises both lung and kidney functions, yet it sustains organ function in an evolutionarily advantageous, quasi-homeostatic state, allowing the organism to survive and reproduce, passing its genetically adaptive cellular-molecular motifs on to its offspring. The ability to accommodate such vital injuries is a mechanistic expression of "survival of the fittest."

This counterintuitive, middle-out approach to understanding the cell-molecular origins of physiologic homologies is in contrast to the efforts of others to understand kidney evolution by a more superficial top-down molecular approach, as described by Rudolph Raff and Thomas Kaufman in *Embryos, Genes and Evolution*. Focusing on how the kidney handles nitrogen waste in the form of ammonia or urea, on the one hand, or hemoglobin synthesis on the other, does not recapitulate phylogeny – it is a "snapshot" of the consequences of the evolutionary mechanisms that have occurred over the course of the history of the organism. And to emphasize the difference between the top-down and middle-out approaches, unlike the evolutionary accommodation of gas or water through pressure transduction, there is no need to modify structure, so there is no demonstrable structural change. It is the determination of the "historic" functional cell-molecular homologies that reveals the evolutionary selection pressure and genotypic-phenotypic result. In the case of ammonia, urea, or hemoglobin the level of selection pressure is perceived to be only molecular, hence the lack of an integrated, structurally evolved trait. Another way to think about this is that biology cannot accommodate gas exchange by modifying oxygen, so instead it accommodates it by increasing the surface areas of the lung and kidney for exchange of gases, liquid, and electrolytes. Seen in this light, the kidney may have been exapted from those members of the species best able to upregulate PTHrP signaling for lung evolution, now facilitating kidney function during one of the reiterative water–land transitions in order to prevent desiccation.

The Water-to-Land Transition, PTHrP Amplification, and the Adaptation to Land

The evolution of PTHrP signaling known to have occurred during the water-to-land transition would provide a mechanistic explanation for the morphing of fish into land vertebrates, like Neil Shubin's *Tiktaalik*, the fossil remains of the transitional tetrapod discovered in 2004. All of the essential water–land adaptations – lung, skin, kidney, gut, and brain – would have been facilitated. At first glance, this event may seem like a "Just So Story" for vertebrate adaptation to land, yet we know that there were at least five separate attempts by vertebrates to breach land based on fossilized skeletal remains; this could not have occurred independently of the evolution of the visceral organs, particularly because many of the same genetic mechanisms are common to both bone and visceral organ development (PTHrP, WNT/β-catenin, TGFβ, PKA, PKC, SHH), so these events should also be viewed in the context of hypothetical internal selection mechanisms for cellular adaptation.

Mechanistically (Figure 8.4), the PTHrP receptor gene is known to have duplicated during the water–land transition, amplifying the PTHrP signaling pathways for the adaptive morphing of the lung, skin, and bone – all of these organs are dependent on the PTHrP signaling pathway for their development and homeostasis. Though the literature suggests that this occurred by chance, it could well have happened as a direct consequence of the generation of excess oxygen radicals and lipid peroxides due to vascular shear stress within the microcirculations of these very same tissues. On the one hand, these tissues and organs would have constrained land adaptation, but on the other, increased PTHrP signaling would have been advantaged by such gene duplication events. This process is formally known as the Baldwin effect.

In fact, if adaptation is thought of in the context of internal selection caused by vascular shear stress, the concept of plasticity becomes much more relevant, not to mention being experimentally testable; constitutive genes are the ones that were most vulnerable to mutation, since they were the genes being targeted by such selection mechanisms. And perhaps such unconventional internal selection was followed by classic Darwinian population selection for those members of the species that were best fitted to regulate those constitutive genes to survive, rendering the newly evolved homeostatic mechanisms regulatable. Theoretically, this may have been due to the fact that regulated mechanisms would be more resilient, and therefore less likely to generate mutagens than non-regulated constitutive genes. And this may also explain why humans have fewer than the predicted number of genes based on descriptive instead of mechanistic biology.

There have been numerous attempts to reconstruct biology from its component parts. Darwinian thought fostered the works of Haeckel, Waddington, Rupert Riedl, Adolf Seilacher, and Gould, to name only a few of those who have attempted to further our insights to evolution. And more recently, Harold Morowitz, and John B. West and colleagues have gained much notoriety by formulating comprehensive analyses of

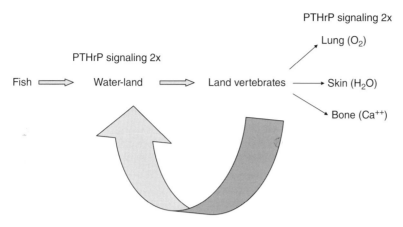

Figure 8.4 Role of the amplification of parathyroid hormone-related protein (PTHrP) in the water-to-land transition. Duplication of the PTHrP receptor occurred during the water–land transition, "amplifying" the PTHrP-PTHrP signaling pathway, fostering key adaptations for life on land. The lung, skin, and bone are all dependent on PTHrP signaling for their development; and since development is the mechanism of phylogenetic change, PTHrP signaling may also have facilitated the evolution of these structures.

physiology, but the problem with their approaches is that they reason backwards from existing structures and functions. They do not predict the changes that have occurred over the course of evolution, even given all the moving parts, and they thus leave biology as a loosely linked series of anecdotes, and medicine as virtually non-predictive and ultimately incomplete in its philosophic and functional scope.

Those members of the species best able to upregulate their PTHrP signaling in support of any one or all of the land adaptive traits – bone, skin, lung – would have had a higher likelihood of surviving on land. In turn, the other tissues and organs would also have been positively selected for their amplified PTHrP signaling capacity, making them more likely to survive. This is particularly relevant to the glomeruli of fish kidneys, which range from large (salt water), to small (fresh water), to being absent in some species; but glomeruli are ubiquitous in land vertebrates. Shear stress within the renal vasculature could have given rise to PTHrP signaling for glomerular function – PTHrP-mesangium signaling for water and electrolyte flux. Similarly, PTHrP is expressed in the pituitary and adrenal cortex of land vertebrates, making for a more robust physiologic "fight or flight" stress mechanism since the corticoids stimulate adrenaline (epinephrine) secretion as they course their way from the adrenal cortex through the adrenal medulla. But this amplified adrenaline response to stress is only applicable to amphibians and groups beyond them phylogenetically, since fish have an independent adrenal cortex and medulla. Such an evolved stress mechanism would have been advantageous for various physiologic adaptations to land, not the least of which would have been the positive selection for brain evolution – adrenaline inhibits flow through the blood–brain barrier, generating more neuronal interconnections within the central nervous system due to increased adrenaline and noradrenaline (norepinephrine) production within the brain.

Again, this is not merely a tautologic rationalization of the data. Developmentally, if you experimentally delete the PTHrP gene in the embryonic mouse, the bone, skin, and lung fail to develop the self-same characteristics for land adaptation. Phylogenetically, the PTHrP signaling pathway has been amplified through gene duplication, fostering stronger skeletal support, skin barrier function, and lung gas exchange.

In further support of the causal relationship between the water–land transition and the evolution of specific physiologic traits that actively accommodated the adaptation to life on land, there were two other gene duplications that occurred during the water–land transition: the β-adrenergic receptor (βAR), and the glucocorticoid receptor. The evolution of the βARs was necessitated by the demand for independent regulation of the systemic and pulmonary blood pressures to accommodate the expanding surface area of the evolving lung. The evolution of the glucocorticoid receptor from the mineralocorticoid receptor was necessitated by the increase in blood pressure due to the increased effect of gravity on land, causing increased blood pressure, generating further selection pressure for the βAR mechanism in alleviating the constraint on the expansion of the lung surface area; the effective stimulation of the βARs by glucocorticoids caused further positive selection pressure for the coevolution of both genes. Again, as in the case of the duplication of the PTHrP receptor, the specific effects of the physiologic stress due to land adaptation on shear stress in the lung and kidney may have specifically precipitated gene duplications in these capillary beds, functionally alleviating the physiologic constraints on these tissues and organs through internal selection, and further fostering these physiologic adaptations through external selection. For example (Figure 8.5), the episodic bouts with hypoxia due to the unmet physiologic needs of the organism as it attempted to adapt to land would have caused physiologic stress since hypoxia is the most potent stressor known, stimulating the pituitary-adrenal axis (PAA), with pituitary adrenocorticotropic hormone (ACTH) stimulating glucocorticoid (GC) production by the adrenal cortex, and subsequently amplifying adrenaline production by the adrenal medulla as the GC passed through it, stimulating catechol-O-methyltransferase, the rate-limiting step in adrenaline production. Acutely, adrenaline would have alleviated the hypoxic stress by stimulating surfactant secretion by the evolving alveoli, and the GCs would have increased βAR density, acting synergistically with adrenaline. As a result, the increased distension of the alveoli would have stimulated PTHrP production by the alveolar type II cells, promoting further alveolarization, alveolar capillary perfusion, and angiogenesis of both the capillaries and lymphatic vessels; those organisms that were most fit to upregulate this cascade would have been more likely to survive, providing a mechanism for its natural selection. Taken together, the evolution of alveolar PTHrP signaling coordinates the secretion and homeostasis of surfactant with gas exchange across the microvasculature at both the macro and micro level, since it functionally co-regulates calcium in the alveolar fluid hypophase with the regulation of surfactant removal from the alveolus via the lymphatic drainage. In the aggregate, this adaptive integration of the PAA and the pulmonary system would have fostered the phylogenetic adaptation of land vertebrates. And this cascade of physiologic adaptations may explain the evolution of PTHrP signaling for pituitary ACTH and adrenocortical glucocorticoid, since it would have further facilitated the positive selection for land adaptation by PTHrP receptor gene duplication.

Bear in mind that these events didn't occur all at once, but took place over a long stretch of land vertebrate evolution, both within and between species. Consistent with this scenario, elsewhere we have shown that in the course of lung evolution, there were alternating intrinsic and extrinsic selection pressures for the genes that facilitated the

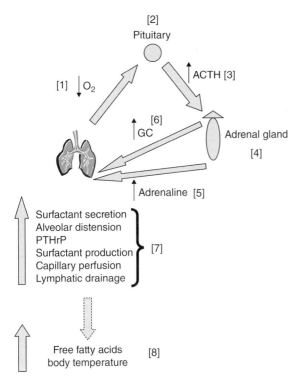

Figure 8.5 Parathyroid hormone-related protein (PTHrP) evolution. Periodic hypoxia [1] during evolutionary adaptation to land stimulated pituitary PTHrP [2], amplifying adrenocorticotropic hormone (ACTH) [3], stimulating adrenocortical PTHrP [4], amplifying adrenaline (epinephrine) [5] via glucocorticoid (GC) production [6]. GC enhances adrenaline activity in the lung, amplifying adrenaline-stimulated lung surfactant secretion [7], alveolar distension, increased PTHrP, and increasing surfactant production, alveolar capillary perfusion, and lymphatic drainage. Adrenaline also causes free fatty acid secretion from peripheral fat cells [8], further increasing metabolism, generating "body heat," increasing lung surfactant activity, and further increasing oxygenation.

increased surface area of the lung. This pattern may atavistically reflect the original mechanism by which the cell membrane of unicellular organisms facilitated the adaptation of the cell to the environment.

The predictive power of this cellular-molecular approach for understanding the evolution of complex physiology is underscored by the synergistic evolution of the lung and endothermy. There are a number of theories for the evolution of endothermy, but none that integrates it in a functionally relevant way to the ontogeny and phylogeny of vertebrates. In contrast to that, the following hypothesis for the origin of endothermy is based on the physiologic interactions between the respiratory, neuroendocrine, and metabolic systems that would have occurred under the episodic hypoxic conditions encountered during the water–land transition, and the subsequent fluctuations in ambient oxygen levels theorized by Robert A. Berner. The adrenaline effect on surfactant secretion in response to hypoxia, alluded to above, would have stimulated the secretion of fatty acids from peripheral fat cells, providing substrate for the tandem increases in respiration, metabolism, and the consequent increase in body heat. This would have caused further positive selection for lung evolution since surfactant phospholipid is 300% more active at 37 °C than at 25 °C, thus providing additional oxygen for metabolic drive. The consequent stress-induced increases in glucocorticoids would have further enhanced the adrenaline effect by amplifying βAR activity in fat cells. At the cellular level, these effects of adrenaline and glucocorticoids are consistent with their mechanisms of action on phospholipid composition in both the lung surfactant and in somatic cell membranes – during the processes of ontogeny and phylogeny there is an increase in *saturated phosphatidylcholine* in lung surfactant, caused by the effect of glucocorticoids on its synthesis in the alveolar type II epithelial cell. In the periphery, adrenaline has been found to increase the *unsaturated phosphatidylcholine* content of the cell membrane, similarly amplified by the effect of glucocorticoids on βARs. In the lung, the increased production of surfactant saturated phosphatidylcholine is physiologically advantageous because its phase transition temperature (the temperature at which it fluidizes) renders it more surface-active in reducing surface tension at higher body temperatures; in the periphery, the opposite occurs, since unsaturated phosphatidylcholine renders the cell membrane more fluid at lower body temperatures due to its lower phase transition temperature, making it more permeable to oxygen. Hence, the same adrenaline mechanism that facilitated lung evolution also facilitated gas exchange in the periphery – a synergistic "win-win" that put mammals at an advantage in adapting to terrestrial life.

Such mutual positive selection for both lung gas exchange and increased body temperature is consistent with the evolution of endothermy. And these interrelationships may have been exapted since both the lung lipofibroblast and the peripheral fat cell produce leptin; in the lung, leptin promotes surfactant synthesis, whereas in the periphery, leptin may affect body temperature – among its many physiologic effects, it has been shown to increase body temperature, perhaps due to its inflammatory interleukin homology. Interleukins have been implicated in the evolution of endothermy as a mechanism in support of host defense. Experimentally, treating ectothermic fence lizards with leptin increases their basal metabolic rate and body temperature. Thus, the integration of pulmonary physiology and host defense may have led to selection pressure for endothermy.

Also worth noting is that in hibernating animals hypoxia is associated with increased unsaturated cell membrane phospholipids, rendering the cell less permeable to oxygen at low temperature. This metabolic adaptation in heterothermic animals is a "reverse-evolutionary" strategy for conserving oxygen under hypoxic conditions.

Dinosaurs and birds are also warm-blooded, but because their lungs are attached to the thorax the above-cited adrenaline effect on surfactant does not apply. This may be why the bird adrenal is not compartmentalized into cortex and medulla, instead being composed of randomly associated corticoid and chromaffin cells that would not have amplified adrenaline production as in the case of mammals.

This is not surprising since there is a functional homology between host defense and the surfactant systems. There are four surfactant apoproteins – A, B, C, and D. Apoproteins A and D are collectins, which are members of the host defense system. In experiments designed to determine the role of leptin in *Xenopus* lung development, we treated frog tadpole lung tissue with leptin and found that it had the same effect on alveolar development that it does in mammals – increased surfactant synthesis in combination with the thinning of the gas exchange surface. Yet this was counterintuitive since frogs are buccal breathers – they actively force air into muscle-lined faveoli, which are gas exchange spaces each a thousand times larger than an alveolus, and so are unaffected by surface tension, obviating the need for surfactant to prevent atelectasis. However, in retrospect, the stimulation of surfactant proteins necessary for host defense makes sense since the lung evolved as an expansion of the foregut, creating a potential site for infection. Therefore, the impetus for surfactant production by the evolving lung may have been predicated on increased antimicrobial peptides, followed by surfactant phospholipids, known to be produced by the gut. Thus, the forward-directed approach to evolution provides a causal chain of events rather than a series of loose associations, at best.

The functional interrelationship between the neuroendocrine and respiratory systems and endothermy is an exaptation that refers all the way back to the origins of eukaryotic life itself. The advent of cholesterol fluidized the cell membranes of unicellular eukaryotes, facilitating gas exchange, metabolism, and locomotion, the three major traits in vertebrate evolution. This may have been the molecular evolutionary prototype for the coevolution of the neuroendocrine and surfactant systems that fostered endothermy. As a note added in support of this hypothesis, the epidermal growth factor (EGF) signal mediator neuregulin is fundamental to both lung development and myelinization, for example.

Moreover, the mutual positive selection for endothermy and gas-exchange efficiency was driven by an increasingly robust neuroendocrine system, marked by the progressive physical integration of the adrenal cortex and medulla during the water–land transition. The latter must have been due to Darwinian selection.

As added evidence for the interrelationship between key gene duplications that occurred during the water–land transition and physiologic stress causing internal selection, type IV collagen also evolved novel polymorphisms in the basement membranes of the lung and kidney phylogenetically from fish to humans during this period. The NC1 domain of type IV collagen forms a natural physicochemical barrier against fluid exudation from both the lung and kidney due to its molecular electrostatic and polar properties, preventing the loss of fluid across the alveolus and glomerulus that would otherwise have occurred due to the increased physiologic demand on these structures during the water–land transition.

Moreover, pathophysiologically, loss of any of these evolutionarily adaptive properties causes cellular-molecular malfunctions consistent with "reverse evolution." For example, loss of PTHrP expression by alveolar epithelial type II cells due to over-distension, infection, or oxidant injury causes transdifferentiation of lipofibroblasts to myofibroblasts, leading to increased alveolar diameter, which is a reversion to earlier phylogenetic forms of the lung seen in reptiles and amphibians. Compromised βAR function similarly leads to chronic lung disease, and glucocorticoid deficiency leads to bronchopulmonary dysplasia in the developing lung. And the abnormal molecular composition of the NC1 domain of type IV collagen in Goodpasture syndrome can cause physiologic failure of both the lung and kidney.

Contrast Evolutionary and Developmental Biology as Descriptive Versus Mechanistic

If the key to understanding evolution is as a mechanism for spatial-temporal relationships of genes as determinants of phenotypes, and these relationships are mediated by soluble growth factors and their cognate receptors, then by following the latter we can understand the former. After all, how can you generate an "arrow of time" without a mechanism for the magnitude and direction of its trajectory? Ironically, the evolutionary biology literature has virtually no orientation to growth factors as the mediators of evolution, or their signaling to cognate growth factor receptors, which are the determinants of the "arrow of time" described by evolutionists. As a result, evolutionary biology is purely descriptive, offering no biologic mechanism to explain natural selection.

On the other hand, as mentioned earlier, contemporary developmental biology is predicated on the functions of growth factors and their receptors as the determinants of morphogenesis. The big breakthrough in molecular embryology occurred in the late 1970s with the discovery that soluble growth factors and their receptors underlie and mediate the patterns of development. And developmental physiology as the outcome of embryonic development acknowledged that the denouement of development is integrated homeostasis. Recognition of such developmental and homeostatic mechanisms as a continuum provides deep insight into the mechanisms of evolution. By superimposing cell–cell signaling on conventional ways of thinking about descriptive evolution, one can begin to understand such otherwise nebulous terms and concepts as survival of the fittest, descent with modification, natural selection, the biogenetic law, Spemann organizers, canalization, genetic assimilation, exaptation, modularity, evolvability, systems biology, developmental systems theory, pleiotropy, and so forth.

Conrad Waddington invoked canalization, aka homeostasis, in the context of evolution. When a cell biologist looks at Waddington's adaptive landscapes, which resemble tents, supporting poles and all, they want to look under the canvas and see what has caused those hills and valleys (I know I do). In so doing, they have been able to determine the cellular/molecular basis for morphogenesis, which is where evolutionists began in the nineteenth century, but were unable to provide the mechanistic basis for Haeckel's biogenetic law or Spemann's organizer. So the geneticists wrested the subsequent inquiry into evolution from the embryologists, and have been reducing evolutionary biology to mutation and selection ever since. Cell biology has literally been eliminated from evolution theory for these historic reasons, yet it has revealed how single cells can create whole organisms, much the same as evolution has. And suffice it to say that evolutionists are not trained in cell biologic methods. Therefore, it would seem productive to let the cell biologists back into the tent. How would this advance our understanding of the mechanisms of evolution? Perhaps by addressing some of the major concepts in evolution theory in cellular terms (see above), we may see how developmental biology would facilitate our thinking in this field, which has the potential for being the basis for a unifying theory of biology in practice, as well as in principle – science is deductive, not inductive. We suffer from too many metaphors and too few experimentally refutable hypotheses.

As mentioned above, Ernst Mayr artificially (and in the present day and age, artifactually) separated evolutionary biology into proximate and ultimate causation in an effort to protect biology against the onslaught by reductionist physicists back in the 1950s and 60s. The advent of genomics has yet again threatened to reduce evolutionary biology to systems biology, but the reasons for the breakdown between these sub-disciplines have been resolved, potentially enabling a rapprochement between the "biologies."

Rudolf Raff's recounting of this era makes it clear that "a boundary discipline exists, and its investigations can yield important complementary insights not possible in either discipline alone," namely evolutionary developmental biology. The reintegration of developmental biology and evolutionary biology was a major step in advancing our understanding of both disciplines. But there is still a huge gap in this effort due to the strong presence of cell biology in developmental biology, and its virtual absence from evolutionary biology. The gap appears to be due to the long-standing rift between these two disciplines, yet Walter Garstang observed that because the morphology of animals arises anew in each generation, evolution of new animal forms had to be viewed as a problem in the evolution of development. In reformulating the modern synthesis, those advocating for the reintroduction of developmental biology into evolutionary biology failed to challenge the evolutionary community to use contemporary methods of cell-molecular embryology, which is dependent on the mediation of gene products by soluble growth factors and their receptors expressed on different cell-types that participate in morphogenesis. One can speculate as to why this lapse occurred, but for whatever reason, it seems to have left evolutionary biology without a way of integrating genes and phenotypes in the same way that developmental biology does. This is ironic, since these principles have resolved the problem of the Spemann–Mangold "organizer" by demonstrating how soluble growth factors and their cognate receptors mediate spatio-temporal signaling to generate form and function, providing the basis for developmental physiology. By determining the molecular basis for the development of physiologic principles, we now have a working model for a mechanistic continuum from development to homeostasis, repair, and aging. By focusing on the serial mechanisms that generate phenotypic change in adaptation to the environment, we eliminate the need for "time," other than as the sequence of events. And "space" is also eliminated, if indeed we are all derivatives of unicellular organisms. Therefore, the "evolution" of such biologic mechanisms should obviate the need for the artificial dissociation between the proximate and ultimate mechanisms of evolution. And yet such precepts persist, impeding the functional integration of genomics into evolution theory.

For example, John Bonner had introduced the concept of "modularity" into evolutionary biology, which was seen as a breakthrough idea that would advance thinking in the discipline. Had the evolutionists embraced cell biology, they would have avoided the need to introduce yet another metaphoric circumlocution into the discipline, alleviating the need to devise experiments to determine how developmental motifs form the basis for evolution at the cell-molecular level.

It is universally held that genes determine biologic structure and function. However, genes do not directly interact with other genes, and therefore they must be considered within their cellular contexts. Nowhere is this more apparent than in the process of development, in which genes determine morphogenesis by spatio-temporally regulating soluble growth factors and their receptors, dictating the growth and differentiation of other cells within their niche. This phenomenon was first described by Hans Driesch (in 1897), and later refined by Spemann and Dorothy Mangold as morphogenetic fields, but without having knowledge of ligand–receptor interactions; that mechanism only emerged in the late 1970s. Similarly, it is acknowledged that development mediates evolutionary change, yet evolutionists rarely if ever reduce the analysis to cells and their products. The reason for this is somewhat obscure; in her *book Unifying Biology: The Evolutionary Synthesis and Evolutionary Biology*, Betty Smocovitis has attributed the absence of cell biology from evolution theory to the rift between evolutionists and embryologists in the late nineteenth century. It is unfortunate that those who have been advocating for the rapprochement between evolutionary biology and developmental biology, or evo-devo, have overarched cell biology yet again in favor of random muta-tion and selection – biology is not stochastic, it is pragmatic and existential in nature.

One often reads of molecular biologists alluding to the highly conserved nature of genes of interest as validation for their relevance to some biologic process or structure, but what does that mean functionally? That it is expressed far back in the history of the organism, inferring that it has been present through much of the evolution of the spe-cies. But rarely if ever is this pursued mechanistically in order to determine how and why such a conserved gene was involved in the evolutionary mechanism. Other than the process of development, there is no system in which to test such mechanisms.

Although this is a simple concept, there was considerable difficulty in actually execut-ing studies based on the idea. Development and evolution certainly offer a facile sort of analogy to each other: both are processes of change. Although this analogy was compel-ling during the nineteenth century, it was sterile until the developmentalists discovered soluble growth factors and their cognate receptors, which were able to mediate the spatio-temporal aspects of the developmental process. Development is a programmed and reproducible process. If we accept Darwinian mutation and selection, evolution can be neither. Evolution can consist of internal and external selection, with internal stability being homeostasis, which can exhibit "reaction norms" that are heritable based on the Baldwin effect. The process of evolution is described as "emergent and contin-gent." Canalization can be seen in the context of homeostatic regulation, which, when it fails, can generate cryptic genes that represent the history of the organism, now reprised to provide a physiologic "safety net" that allows for the healing to occur; as such, it allows for reproduction even in the face of illness. The apparent inevitability of develop-ment was daunting. To connect it effectively with evolution, two major ideas had to be accepted. The first, pointed out by Garstang, is that the larval stages also face the rigors of life (reminiscent of the Barker hypothesis, that adult diseases originate *in utero*). Mendelian genetics allows new traits to appear at any developmental stage, and natural selection potentially operates upon them as it does upon traits expressed in adults. The second major point is that although ontogeny appears inevitable and inextricably orchestrated in its flow, it is not a single process. There are a large number of processes at work, some more or less coupled to others. It was Joseph Needham who, in 1933, using an engineering metaphor of shafts, gears, and wheels, suggested the idea of dis-sociability of elements of the developmental machinery. He pointed out that it is pos-sible to experimentally separate differentiation from growth or cell division, biochemical differentiation from morphogenesis, and some aspects of morphogenesis from one another. The implication of this idea is enormous: developmental processes could be dissociated in evolution to produce novel ontogenies out of existing processes, as long as an integrated developmental program and organismal function could be maintained.

Epistemology: Maybe we Got it Backwards?

The integrated mechanism for physiology has long been accepted to be a fait accompli, yet we know that there are processes of development, evolution, and regeneration-repair that comply with some unknown, underlying bauplan. The recent experimental evidence for the complete metazoan toolkit being present in the unicellular state of sponges provides the rationale for such an integration of structure and function, by definition. Mechanistically, the insertion of cholesterol in the plasma membrane of eukaryotes facilitated endocytosis, locomotion, and respiration, providing the impetus for their evolution. Moreover, it is striking that the cytoskeleton collectively mediates homeostasis, mitosis, and meiosis alike, suggesting the phenotypic autonomy of these unicellular organisms. The significance of this is evinced by subjecting yeast, the simplest of eukaryotes, to microgravity, causing both loss of polarity and failure to bud. Without polarity, there is no calcium flux or reason to locomote – where is up, down, sideways? – and budding is the reproductive strategy of yeast; loss of these fundamental traits by "disorienting" the cytoskeleton underscores the adaptation to the one element in the environment that is omnipresent, unidirectional, and perpetual from the inception of the planet. So perhaps multicellularity was merely the eukaryotic ploy used to combat lateral inheritance, biofilm formation, and quorum sensing in our age-old competitors, the prokaryotes.

Conclusion

The multicellular form may merely be a derivative of the unicellular state, acting as a matrix for it to monitor the oncoming environment so that the gene pool knows what epigenetic marks acquired during the multicellular phase of the life cycle to include or exclude in the next generation. For example, *Dictyostelium* exists in two forms, a free-swimming amoeboid form and a colonial fruiting body. Under conditions of abundant nutrients, the slime mold remains in its free-swimming amoeboid form; under conditions of low food abundance, the amoeboid free-swimming phenotype forms colonies. Logic would dictate that this organism evolved under high nutrient abundance conditions, and therefore its unicellular form is the primary phenotype, the colonial form being derivative.

We need better to understand evolution from its unicellular origins as the "Big Bang" of biology.

The current Chapter was intended to look at complex physiology as a "vertically synthetic," internally consistent, scale-free process, both developmentally and phylogenetically. Chapter 9, entitled "Cell signaling as the basis for all of biology," focuses on cell–cell signaling as the mechanistic basis for all of the principles of physiology.

Suggested Readings

MacDonald BA, Sund M, Grant MA, Pfaff KL, Holthaus K, Zon LI, Kalluri R. (2006) Zebrafish to humans: evolution of the alpha3-chain of type IV collagen and emergence of the autoimmune epitopes associated with Goodpasture syndrome. *Blood* **107**;1908–1915.

Smocovitis VB. (1996) *Unifying Biology: The Evolutionary Synthesis and Evolutionary Biology*. Princeton University Press, Princeton, NJ.

Torday JS, Rehan VK. (2012) *Evolutionary Biology, Cell-Cell Communication and Complex Disease*. John Wiley & Sons, Inc., Hoboken, NJ.

Waddington CH. (1957) *The Strategy of the Genes*. George Allen & Unwin, London, UK.

9

Cell Signaling as the Basis for all of Biology

We dance round in a ring and suppose,
But the Secret sits in the middle and knows.
              Robert Frost, The Secret Sits

Introduction

By reducing ontogeny and phylogeny to cellular/molecular biology, the "first principles of physiology" (FPPs) can be discerned for the first time. Such FPPs were exapted by vertebrates during the water-to-land transition, originating from the acquisition of cholesterol by eukaryotes more than 500 million years ago. The introduction of cholesterol into the plasmalemma of unicellular eukaryotes facilitated unicellular evolution over the course of the Earth's history in service to the reduction in intracellular entropy, far from equilibrium. That mechanism ultimately gave rise to the embryonic primordial germ layers that form the multicellular homologs of the gut, lung, kidney, skin, bone, and brain. The key concept is that the gene regulatory networks that give rise to these tissues and organs embryologically are highly conserved and specific to their germline origins, providing a frame of reference for both existing and novel structures and functions in subsequent generations (= evolution). The central principle is that the underlying homeostatic control mechanisms that sustain structure and function are flexible, allowing for the generation of a range of physiologic set-points instead of one genetically fixed state. This perspective is 180 degrees out of sync with the prevailing genetic dogma, in which evolution is considered to probabilistically result in random mutations, giving rise to novelty through natural selection. Yet such plasticity is totally in keeping with newer concepts like the Barker hypothesis, that diseases emanate from the fetus, and the role of Lamarckian epigenetic inheritance, and niche construction theory, namely, that species create their own environments.

Tiktaalik, the fish/tetrapod fossil first documented by Neil Shubin in 2004, provides a hypothetical way of thinking about the cellular-molecular ontogenetic and phylogenetic changes that occurred during the vertebrate water-to-land transition (WLT). To successfully survive that transition, *Tiktaalik* had to have been "preadapted" for air breathing as well as for kidney, skin, gut, bone, and brain function on land. Conventionally, in the phylogenetic context of fish physiology as the antecedent for such a critical transition, logic would dictate that the gill is the homolog of the lung. However, from the cellular-molecular perspective, the swim bladder has definitively been shown to be structurally, functionally (as a gas-exchanger), and genomically homologous with the tetrapod lung. Both the swim bladder and lung are developmental outpouchings of the foregut, mediating the uptake and release of oxygen and carbon dioxide, for buoyancy in the case of the swim bladder, and for metabolism in the case of the lung. Developmentally, parathyroid hormone-related protein (PTHrP) is among the most highly expressed genes in the swim bladder. Furthermore, the PTHrP receptor gene underwent duplication during the WLT, phylogenetically marking the transition from fish to amphibians. That event made feasible atmospheric gas-exchange on land since PTHrP–PTHrP receptor signaling is essential for lung alveoli to develop. Experimentally deleting the PTHrP gene in developing mice causes the death of the offspring at birth due to the absence of alveoli. PTHrP is expressed in both the epithelial cells that line the swim bladder of fish and in the alveoli of land vertebrates. PTHrP stimulates the production of lung surfactant in the alveoli, maintaining the structural and functional integrity of the alveoli by reducing surface tension; without surfactant, the alveoli collapse, rendering them dysfunctional.

PTHrP, the Evolutionary Basis for Lung Homeostasis

PTHrP is a paracrine hormone produced and secreted by alveolar type II cells in response to distension of the lung. PTHrP acts locally via cell surface PTHrP receptors on mesenchyme to induce specialized connective tissue fibroblasts that differentiate into lipofibroblasts (Figure 9.1). Lipofibroblasts are critical to the ontogeny, phylogeny, and ultimate evolution of the lung because: (i) they protect the alveolus against oxidant injury by actively recruiting and storing neutral lipids from the alveolar microcirculation, which act as antioxidants; and (ii) because the stored neutral lipids are actively mobilized from the lipofibroblasts to the alveolar type II cells for surfactant synthesis through the mechanically coordinated effects of PTHrP, leptin, and prostaglandin E_2,

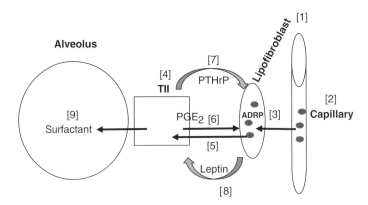

Figure 9.1 Neutral lipid trafficking from lipofibroblast to type II cell. Lipofibroblasts [1] actively take up and store neutral lipid from the circulation [2] by expressing adipocyte differentiation related protein (ADRP) [3]. Alveolar type II cells (TII) [4] recruit these neutral lipids [5] by secreting prostaglandin E_2 [6]. Parathyroid hormone-related protein (PTHrP) [7] and leptin [8] coordinate mechanical stretching and lung surfactant production [9] to integrate distension of the alveoi with breathing.

which act by means of their cognate receptors residing on the neighboring cell surfaces of epithelial type II cells and lipofibroblasts.

PTHrP is overtly responsible for alveolar calcium homeostasis, specifically that of the hypophase, providing a link to its deep evolutionary history: concentrations of calcium on the surface of the alveoli determine the conformation of tubular myelin, which in turn determines the effect of surfactant on alveolar surface tension; tubular myelin is a lipid-protein composite that is homologous with the lipid-β-defensin barrier formed by the stratum corneum in the skin that prevents the leaking of fluid and protects against bacterial infection. The skin is the most primitive organ of land vertebrate gas exchange, so it may have provided a cellular-molecular source for the homeostatic exaptation of the lung.

The evolution of the alveolar type II cell and lipofibroblast independently, culminating in the coordinate regulation of surfactant, initially in the skin and ultimately in the lung, would have required longer than the existence of the Earth. Conversely, positive selection pressure for PTHrP signaling between the endoderm and mesoderm in sequential support of buoyancy and air breathing could have evolved the fish swim bladder adaptation for buoyancy, which could have been exapted by land vertebrates for homeostatically regulated surfactant production.

Adipocyte Differentiation Related Protein (ADRP) as a Deep Homology that Interconnects Evolved Functional Homologies

In the lung alveolus, ADRP (see Figure 9.1) is physiologically upregulated by the distention of the alveolar type II cells, which produce PTHrP in response to being cyclically stretched; the extracellular PTHrP then binds to its cognate receptor on the surface of adepithelial lipofibroblasts, stimulating the nuclear receptor peroxisome proliferator activator receptor gamma (PPARγ), which then signals the upregulation of ADRP. This mechanism hypothetically evolved initially to protect the alveolar wall against hyperoxia, since rising atmospheric oxygen tension causes the differentiation of myofibroblasts into lipofibroblasts, and the neutral lipid content of these cells prevents oxidant injury. This mechanism could later in evolution have been co-opted to regulate surfactant synthesis during the vertebrate WLT, consistent with the phylogenetic adaptation of the alveolus from the aquatic swim bladder of fish to the highly air-adapted lungs of mammals and birds. This phenomenon is of particular interest in the context of exploiting such functional molecular homologies when one considers the homologies between the alveolar lipofibroblast and endocrine steroidogenesis. For example, oxygen in the atmosphere has not increased linearly over time from zero to 21%, but has undulated episodically, ranging between 15 and 35% over the last 500 million years of the Phanerozoic eon. Since hypoxia is the most stressful of all physiologic agonists, fluctuations in atmospheric oxygen would have placed significant constraints on both the evolving lung and endocrine system. Perhaps fortuitously, the vertebrate pulmonary and endocrine systems were preadapted for such an adaptation due to the presence of PAT (perilipin, adipocyte differentiation related protein, and TIP47) genes; the well-known effects of the adrenocortical system on lung development, homeostasis, and evolution can be seen as part of a logical sequence of alternating external and internal epistatic adaptations.

This is not a tautology or teleologic rationalization since the same morphogenetic mechanisms occur during both ontogeny and phylogeny, though on different time-scales, and the effects of geologic environmental changes are reflected therein. Moreover, we observe the reversal of this evolutionarily causal process in mechanisms of chronic lung diseases, in which there is "simplification" of the alveo-lar bed, resulting in a frog-like structure in mammals. Experimentally, for example, it has been found that when a gene necessary for the synthesis of alveolar choles-terol, the most primitive of lung surfactants, is deleted specifically from mouse lung alveolar type II cells, the lung developmentally "compensates" by over-expressing the lipofibroblast population in the alveoli, suggesting that these cells have an evolutionary capacity to facilitate surfactant production, both ontogenetically and phylogenetically.

This logical, forward-oriented approach to understanding how and why the lung evolved can be further exploited, since β-adrenergic receptor signaling was essential for the local regulation of blood pressure in the lung independently of the systemic circulation, permitting further increases in the surface area of the evolving lung. Vertebrate ancestors were organisms able to survive the otherwise deleterious effects of alternating bouts of hyperoxia and hypoxia due to fluctuations in oxygen in the atmosphere by adaptively modifying their pulmonary and endocrine systems (see below). Here again, as in the case of the PTHrP receptor, the β-adrenergic receptor also duplicated during the WLT, facilitating lung-specific control of blood pressure, allowing for a further increase in lung surface area to support metabolic demand. At this phase in vertebrate evolution, the glucocorticoid receptor is documented to have evolved from the mineralocorticoid receptor, perhaps as a counterbalancing mechanism for the agonal effect of mineralocorticoids on blood pressure. There was a substantial increase in the effect of gravity in the transition from water to land; that effect may have been functionally offset by shunting some mineralocorticoid signaling to glucocorticoid signaling – increased vascular shear force stress may have generated radical oxygen species, causing the addition of two amino acids to the mineralocor-ticoid receptor, rendering it glucocorticoid receptive. The emergence of the physiologic glucocorticoid mechanism may have been further facilitated by the presence of penta-cyclic triterpenoids, which are unique to the land environment, produced by rancidify-ing land vegetation. These compounds inhibit 11β-hydroxysteroid dehydrogenase type II (11βHSD2), which inactivates cortisol's blood pressure-stimulating activity, causing positive selection pressure for the tissue-specific expression of 11βHSD1 and 2 in glucocorticoid target organs, including the lung, thereby permitting local activation and inactivation of cortisol.

Reinforcing this hypothesis, when pituitary adrenocorticotropic hormone (ACTH) stimulates glucocorticoid production by the adrenal cortex, the hormone passes through the intra-adrenal portal vascular system of the adrenal medulla, known to exist in rodents and primates alike, providing it with uniquely high local concentrations of glucocorticoids. These high concentrations are needed to induce the medullary enzyme, phenylethanolamine-*N*-methyltransferase (PNMT), which controls the synthesis of catecholamines, thus coordinately upregulating both of the primary adrenal stress hor-mones for a maximally adapted "fight or flight" response.

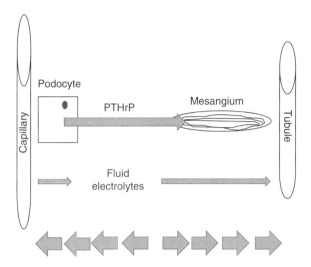

Figure 9.2 The glomerulus as a parathyroid hormone-related protein (PTHrP)-mediated stretch sensor. When fluid and electrolytes enter the glomerulus from the kidney capillary, the fluid stretches the podocytes lining the glomerular space, stimulating secretion of PTHrP. PTHrP binds to its receptor on the mesangium, which controls the secretion of the fluid and electrolytes into the kidney tubule for excretion.

PTHrP and the Evolution of the Kidney Glomerulus

Akin to its role as a stretch-regulated gene product that maintains alveolar homeostasis, PTHrP is also essential for renal physiology (Figure 9.2). PTHrP is produced by the epithelially derived podocytes lining the glomerulus, homologous with the alveolus. Kidney PTHrP is also functionally homologous with the alveolus in maintaining the homeostatic control of the mesangium, a stretch-sensitive fibroblastic structure that determines systemic fluid volume and electrolyte homeostasis. Parenthetically, this molecular homology between the lung and kidney should come as no surprise since both structures contribute to the formation of amniotic fluid during development. And it should be borne in mind that the glomerulus also makes its appearance during the phylogenetic transition from fish to amphibians, and subsequently to reptiles, mammals, and birds.

PTHrP and Skin Cell-Molecular Evolutionary Homeostasis

PTHrP is necessary for skin morphogenesis via the cell–cell paracrine interactions between melanocytes and keratinocytes. The net result is the formation of the stratum corneum, a bipurposed barrier against bacteria and water, essential for preventing water loss from the skin in terrestrial vertebrates. It is noteworthy that both the alveolar and skin epithelium exhibit functional homologies at the cell/molecular level, packaging lipids together with proteins and host defense peptides, and secreting them in the form of lamellar bodies to generate lipid-based barriers against water loss and host invasion.

The evolutionary significance of the functional-structural homology between lung and skin as barriers against environmental factors is further exemplified by the patho-physiology of asthma. Patients with asthma often also have the skin rash atopic dermatitis. Both of these pathologic phenotypes are seen in humans and dogs, and have been mechanistically linked at the molecular level through a common defect in β-defensins, which mediate innate host defense in both skin and lung. In dogs, β-defensins determine coat color, which serves a multitude of adaptive advantages, ranging from protective coat coloration to reproductive strategies, both of which are existential. The β-defensin CD103 has also been shown to cause both atopic dermatitis and asthma in dogs, since it is also found in dog airway epithelial cells. Therefore, hierarchically, host defense and reproduction take evolutionarily adaptive precedence over wheezing due to asthma. Interestingly, a similar interrelationship between asthma and atopy has been demonstrated in humans: Chinese children with a polymorphism for human beta defensin-1 exhibit both diseases.

The Goodpasture Syndrome as a Heuristic

Vertebrates transitioned from water to land roughly 300 million years ago, causing selection pressure for barrier function in the skin, lung, and kidney. Type IV collagen acts to physically maintain the integrity of the epithelium throughout the body. Among the internal organs, it sustains the barrier function of the alveoli and glomeruli, which were rapidly evolving in the WLT. Comparative molecular evolutionary studies of the Goodpasture's syndrome epitope have revealed that the isoform of type IV collagen that causes Goodpasture's syndrome evolved during the phylogenetic transition from fish to amphibians due to selection pressure for specific amino acid substitutions that rendered the molecule more hydrophobic and negatively charged.

Goodpasture's syndrome is an autoimmune disease that can cause both alveolar and glomerular dysfunction, resulting in catastrophic failure of both the kidney and lung epithelial barriers, caused by pathogenic circulating autoantibodies targeted to a set of discontinuous epitope sequences within the non-collagenous domain 1 (NC1) of the α3 chain of type IV collagen [α3 (IV) NC1], referred to as the *Goodpasture autoantigen*. *Caenorhabditis elegans*, *Drosophila melanogaster*, and *Danio rerio* do not react to Goodpasture autoantibodies. In contrast to this, frog, chicken, mouse, and human basement membrane preparations bind the Goodpasture autoantibodies. The α3 (IV) chain is absent in worms (*C. elegans*) and flies (*D. melanogaster*), and is first detected phylogenetically in fish (*D. rerio*). Interestingly, native *D. rerio* α3 (IV) NC1 does not bind Goodpasture autoantibodies. In contrast to the recombinant human α3 (IV) NC1 domain, there is a complete absence of autoantibody binding to recombinant *D. rerio* α3 (IV) NC1. Three-dimensional molecular modeling of the human NC1 domain suggests that evolutionary alteration of its electrostatic charge and polarity due to the emergence of critical serine, aspartic acid, and lysine amino acid residues, accompanied by the loss of asparagine and glutamine, contributes to the emergence of the two major Goodpasture epitopes on the human α3 (IV) NC1 domain, as it evolved from fish over the ensuing 450 million years. The evolved α3 (IV) NC1 domain forms a natural physicochemical barrier against the exudation of serum and proteins from the circulation into alveoli and glomeruli due to its hydrophobic and electrostatic properties, which likely provided the molecular selection pressure for the evolution of this protein, given the rising oncotic and physical pressures on the evolving barriers of both the lung and kidney during the WLT. Taken together, the lung, kidney, and skin evolved critical physiologic barriers against dessication in land-dwelling vertebrates.

Internal and External Selection, PTHrP, the β-Adrenergic Receptor, Glucocorticoids, and the Water-to-Land Transition

Regarding evolution as a linked series of interactions between the organism and the ever-evolving environment is an informative way to conceive of adaptive cellular-molecular changes in structure and function. By aligning major internal and external selection pressures with the cellular-molecular changes in structure and function of the alveolus, the processes involved in evolution can be visualized (Figure 9.3).

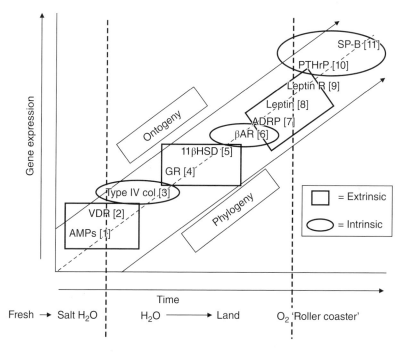

Figure 9.3 Alternating extrinsic and intrinsic selection pressures for the genes of lung phylogeny and ontogeny. The effects of the extrinsic factors (salinity, land nutrients, oxygen) on genes that determine the phylogeny and ontogeny of the mammalian lung alternate sequentially with the intrinsic genetic factors, highlighted by the rectangles and ovals, respectively. [1] AMPs = antimicrobial peptides; [2] VDR = vitamin D receptor; [3] type IV col = type IV collagen; [4] GR = glucocorticoid receptor; [5] 11βHSD = 11β- hydroxysteroid dehydrogenase; [6] βAR = β-adrenergic receptor; [7] ADRP = adipocyte differentiation related protein; [8] leptin; [9] leptin R = leptin receptor; [10] PTHrP = parathyroid hormone-related protein; [11] SP-B = surfactant protein-B. Along the horizontal axis major geochemical events that caused the cell-molecular changes in lung evolution are depicted: Fresh → salt H_2O = transition from fresh to salt water; H_2O → land = water-to-land transition; O_2 "roller coaster" = fluctuations in atmospheric oxygen tension over the last 500 million years. From Torday, 2013. Reproduced with permission of the American Physiological Society. (*See insert for color representation of the figure.*)

Such *external* environmental constraints to the transition from water to land as air breathing, gravitational orthostatic forces, and desiccation were all hypothetically adapted to through internal cell/molecular pathways for development and homeostasis expressed in common – PTHrP and its cognate G-protein-coupled receptor. Importantly, this model is also predictive, since PTHrP is a potent vasodilator and an angiogenic factor (promotes capillary formation), hypothetically explaining why glomeruli, as microvascular derivatives of the renal artery, may have evolved during the transition from fish to amphibians.

The role of PTHrP in facilitating the vertebrate WLT may have transpired as follows: terrestrially adaptive vertebrates would have been selected for their spontaneous overexpression of PTHrP–PTHrP receptor signaling, initially for lung evolution from the swim bladder, particularly in physostomous fish like zebrafish, which possess a trachealike pneumatic duct that connects the esophagus and swim bladder for gas filling and emptying. Such foregut plasticity was prognostic for the future adaptive plasticity in the descendants of physostomes, in contrast to physoclistous fish, in which the swim bladder is not derived from the gut. At the cell/molecular level, the smooth muscle that forms both the pneumatic duct and trachea is determined by fibroblast growth factor 10 (FGF-10) expression. In support of this hypothesis, Ewald Weibel and colleagues have found that the lung is "over engineered," that is, its physiologic capacity far exceeds its needs, consistent with positive selection pressure for the lung trait, and thus for PTHrP signaling over-expression. The PTHrP-mediated mechanisms in the kidney and skin may have been a consequence of such positive selection, since their evolution under the influence of amped-up PTHrP signaling would have protected land vertebrates against desiccation during the WLT. Adaptive increase in the calcification of weight-bearing bones in response to increased gravitational force on land would have further facilitated adaptation to land living. Wolff's law predicts that bone will adapt to the load under which it is placed. PTHrP is a mechanically sensitive paracrine hormone that regulates calcium deposition during bone development and homeostatic adaptation, determining bone calcium uptake and incorporation into cartilaginous structures, and facilitates the adaptation of terrestrial organisms to environmental gravitational forces.

These iterative processes for the acquisition of traits that facilitated the WLT are consistent with data documenting that vertebrates attempted the WLT on at least five separate occasions. Based on the principle of parsimony, one can hypothesize that these terrestrial properties were realized through the known duplication of the PTHrP receptor gene, beginning with bone in response to increased gravitational stress. Those organisms that had demonstrated their ability to upregulate their PTHrP–PTHrP receptor signaling for lung evolution evolved as the forebears of contemporary land vertebrates. It is only reasonable to assume that the lung adaptation was mirrored molecularly by the skeletal adaptation. The over-expression of PTHrP signaling would also have benefitted the skin, kidney, and brain since all of these organs are affected by the experimental deletion of PTHrP in mice. In contrast, lineages that were unable to accomplish this feat became extinct.

This perspective is supported by our demonstration of the correlation between the cell/molecular genetic motifs common to ontogeny and phylogeny of the lung and major environmental epochs (see Figure 9.3). Note the apparently seamless alternation between internal and external selection mechanisms in association with major ecologic stresses; we postulate that there are seemingly no gaps between these genetic adaptations because the data are derived from contemporary land vertebrates;

conversely, those lineages of the species that failed to adapt died off, and thus would not be represented in this analysis.

Tiktaalik

In 2004 the discovery of the fossilized remains of an organism representing the embodiment of the WLT was first published. But in the era of quantum mechanics, we have come to expect more than just descriptive evidence for the biologic "Big Bang." The ability to live on land was physiologically very challenging, so why did vertebrates emerge out of water onto land? According to the Romer hypothesis, it was necessitated by the increase in carbon dioxide in the primordial atmosphere, causing the Earth's lakes, ponds, rivers, and oceans to dry up, forcing our vertebrate ancestors to seek refuge on land or face extinction.

So from a physiologic evolutionary perspective, how could fish have evolved into tetrapods (Figure 9.4)? The biggest constraint was the inability to breathe air. It had long

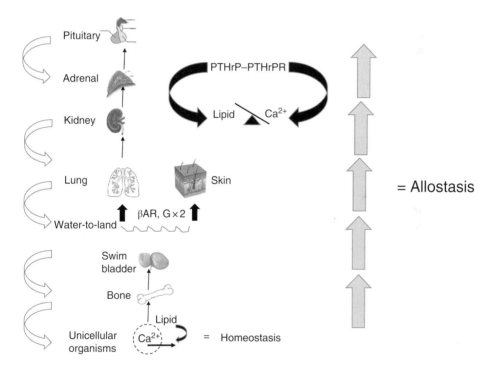

Figure 9.4 Vertical integration of physiologic land adaptation. The ontogenetic and phylogenetic integration (∫) of calcium-lipid homeostasis, from unicellular organism incorporation of lipid into the plasmalemma to multicellular organism calcium/lipid epistatic homeostasis fostered the evolution of metazoans. This figure focuses on the specific stress of the water-to-land transition on the evolution of a wide variety of organs – bone, lung, skin, kidney, adrenal – resulting from the duplication of the PTHrP receptor gene in fish, followed by the β-adrenergic receptor (βAR) gene, culminating in integrated physiology, or allostasis (*far right*). Internal selection was mediated through selection pressure on homeostatic mechanisms mediated by paracrine cell–cell interactions. As vertebrates adapted to a terrestrial environment, the PTHrP signaling mechanism iteratively allowed for physiologic adaptations to air breathing (skin, lung), prevention of (skin, kidney), and "fight or flight" (pituitary-adrenal). From Torday, 2015. Reproduced with permission of John Wiley and Sons.

been thought, though controversially, that the swim bladder of fish evolved into the lung of land vertebrates, since both are gas exchangers derived from the plastic foregut. The notion that evolution co-opted an organ for buoyancy into one that mediated oxygenation for metabolism is attractive, though there are certain anatomic constraints. That controversy has been put to rest by a recent study showing that at the molecular level, the swim bladder expresses all the homologous genes for lung development, including PTHrP, which is a gravity-sensing gene that is necessary for the formation of alveoli in mammals. Thus, there is a functional genomic link between the WLT and PTHrP signaling; the latter underwent a gene duplication sometime during the fish-to-amphibian transition, thereby helping to provide an explanation for the evolution of the lung from the swim bladder.

As also discussed above, equally important in considering the mechanisms for WLT is that the organs necessary for barrier function against desiccation are also PTHrP-dependent, both developmentally and physiologically. The skin, kidney, and gut all express PTHrP, and signaling between the mesoderm and epithelium of these organs is mediated by binding between PTHrP and the PTHrP receptor, which determines the structural and functional development of these organs to form homeostatically regulated physiologic barriers against water loss. Moreover, PTHrP is necessary for the calcification of cartilage, so it would have facilitated the evolution of the bony tetrapod limbs of *Tiktaalik* to accommodate the increased gravitational force on its skeleton on land, exhibiting the plasticity of the lung, skin, and bone. In support of this interrelationship, mice in which the PTHrP gene is deleted exhibit gross morphogenetic defects in lung, skin, and bone.

The angiogenic properties of PTHrP are another feature relevant to its utility in the WLT and organ adaptation. PTHrP promotes vascularization of bone and skin, particularly when the vascular endothelium is cyclically distended, as in conditions of increased physiologic stress such as those involved in the WLT. PTHrP receptors exist in the lymphatic microcirculation as well. Additionally, PTHrP is a vasodilator, ultimately epistatically relieving tension on the remodeled microvasculature, while simultaneously providing increased perfusion for remodeling of the adjacent parenchyma in further adaptation to internal physiologic stress. Such an adaptive mechanism is consistent with the progressive expansion of the surface area of the lung, and may help explain the evolution of the glomerulus, which is absent from the kidneys of fish, but is omnipresent in amphibians, reptiles, mammals, and birds.

The PTHrP receptor gene duplication during the water-to-land migration of *Tiktaalik* may not have been by chance. There could have been periodic increases in shear stress on the capillaries of the specific organs that were most affected physiologically in enabling the WLT – lungs, adrenals, skin, kidney, skeleton – generating radical oxygen species (ROS) and lipid peroxides that affected those vascular beds. ROS damage DNA, causing gene duplications. ROS are also normal embryonic signal transducers that can act to promote structural and functional remodeling during morphogenesis. As a corollary to this hypothetical mechanism of adaptation, the endothelium is known to be highly heterogeneous, each endothelial cell acting like an adaptive non-linear input/output device. The input comes from the extracellular environment, consisting of biomechanical and biochemical forces. The output consists of the heterogeneity of the endothelial cell population – cell shape, calcium flux, protein expression, mRNA expression, migration, proliferation, apoptosis, vasomotor tone, hemostatic balance, release of inflammatory mediators, and leukocyte adhesion/transmigration.

In that vein, the long-term consequences of the alternating hyperoxia and hypoxia during the Phanerozoic eon could have been facilitated by the PTHrP receptor gene duplication, fostering the coevolution of the lung, adrenal, and pituitary, since all three structures express homeostatic PTHrP (see Figure 9.3). Hypoxia due to evolutionary pulmonary insufficiency would have caused physiologic stress, causing PTHrP-mediated stimulation of pituitary ACTH, stimulating PTHrP-mediated adrenocortical corticoid secretion, in turn stimulating medullary adrenaline (epinephrine) production. This would ultimately have compensated for the underlying compromised alveolar gas exchange by stimulating surfactant secretion, increasing alveolar distensibility, and alleviating the short-term constraint on gas exchange by allowing for an adaptive increase in alveolar function. But over the long haul, such recurrent microvascular shear stress to all three of these structures – lung, adrenal, and pituitary – could have caused adaptive evolutionary remodeling. This fractal perspective on the evolution of physiology is significant because it exemplifies the scale-free nature of the process, vertically integrating the self-same principles from unicellular to multicellular organisms.

In support of this mechanism of land adaptation facilitated by the PTHrP receptor gene duplication, there was another gene duplication during this same period, that of the β-adrenergic receptor (βAR). This was of critical importance for the increase in alveolar surface area for gas exchange during the WLT: the increase in βAR density allowed for the independent regulation of alveolar capillary blood pressure, protecting the lung vasculature against the deleterious effect of increased blood pressure in the systemic circulation resulting from the stresses of land adaptation, including gravitational effects and the demand for increased metabolism in the evolving Phanerozoic atmosphere that was driving vertebrates onto land in the first place. The physiologic mechanisms driving the increase in βAR expression were both the increase in adrenaline caused by the PTHrP stimulation of the hypothalamic-pituitary axis (HPA) and the associated evolution of glucocorticoids from mineralocorticoids, which also occurred during the WLT. This may have been the result of epistatic balancing selection to decrease the environmental stimulation of blood pressure, which was reduced by biochemical modifications to the mineralocorticoid molecule, namely, 11- and 17-hydroxylation, which are necessary for glucocorticoid bioactivity. Such modifications of the mineralocorticoids may have been due to random selection, driven by the beneficial effect of glucocorticoids on βAR expression, relieving the hypertensive effects of both the external and internal environments recounted above. The evolution of the glucocorticoids may also have been a consequence of the stimulatory effect of PTHrP on adrenocortical glucocorticoid synthesis. Parenthetically, this close integration of pulmonary and endocrine evolution in adaptation to land may form the basis for the effects of glucocorticoids on both parturition and vertebrate embryonic development.

Such constrained mechanisms of development and mutation for evolutionary change would have fostered adaptive internal and external selection, depending on both the nature and magnitude of the agent and the physiologic constraint, suggesting a mechanistic way of thinking about the meaning of "evolvability." More importantly, organisms such as *Tiktaalik*, which were able successfully to surmount the WLT, exhibited the *plasticity* necessary for the remodeling of vital organs for adaptation to terrestrial life. *Tiktaalik*'s ancestors were thus "preadapted" for terrestrial life, having survived oxygen's challenges on numerous occasions during the continuum of vertebrate evolution from its unicellular origins.

Cellular Growth Factors, the Universal Language of Biology

The fundamental *a priori* understanding of vertebrate physiology, not as a top-down descriptive process but as a series of co-options originating from the cell membrane of unicellular organisms, will lead to understanding of the first principles of physiology based on its evolutionary origins. The actualization of such FPPs would have numerous ramifications, including a predictive model for physiology and medicine, as well as a functional merging of biology, chemistry, and physics into a common algorithm for the natural sciences. Such a perspective would allow us to de-emphasize the human signature from our anthropocentric view of our physical environment, on a scale akin to the recentering of the solar system on the Sun in the sixteenth century, which corrected our perception of our Universe, and later that of other universes.

Predictions of a Cellular Approach to Evolution

Since ontogeny is the only biologic process known that can generate new structures and functions, it should be exploited to understand evolution, since it generates physiologic traits throughout the phylogenetic history of an organism. By specifically focusing on cell–cell interactions mediated by soluble growth factors and their cognate receptors, one can determine the evolution of the lung, structurally and functionally tracking it back to the swim bladder of physostomous fish. And since the adaptation of fish to land was contingent on efficient atmospheric gas exchange, the lung can be seen as the cellular/molecular template for the evolution of other physiologic adaptations to land life. By systematically tracing the molecular homologies between the lung, adrenals, skin, kidney, gut, bone, and brain developmentally, phylogenetically, and pathophysiologically in tandem across space and time, the FPPs can be determined. Once such relationships are tracked back to their unicellular origins in the plasmalemma, the underlying principles can be used to replay the evolutionary tape, and predict and prevent homeostatic failure as disease. Ultimately, the reduction of biology to ones and zeros can be used to merge biology, chemistry, and physics into one common user-friendly algorithm as the "periodic table" of Nature.

The current Chapter was intended to look at complex physiology as a "vertically synthetic," internally consistent, scale-free process, both developmentally and phylogenetically. Chapter 10, entitled "Information + negentropy + homeostasis = evolution," explains the interrelationships between physical chemistry and biology, fueled by avoiding the second law of thermodynamics.

10

Information + Negentropy + Homeostasis = Evolution

Introduction

There is a consensus that information lies at the heart of all Nature as first stated by Claude Shannon, the "father of Information Theory." The Big Bang converted potential energy into kinetic energy, scattering the elements throughout the Universe as a function of their mass, thus creating a *de facto* hierarchy of information. Biology has mimicked the Big Bang, creating an "internal Universe" through negentropy and biocompartmentation, which also constitutes information. The "internal Universe" is energetically linked to the negentropy within the system as the functional basis for these interrelationships. But it is very difficult to then integrate the aforementioned principles with traditionally descriptive physiology. That is because when you base the process on the phenotypic traits involved you are reasoning after the fact. So although the precepts are correct, they do not fit together readily with the reality, and as a consequence, they are not predictive. In an effort to reconcile this mismatch of principles and applications, contemporary biology has either become overly reductionist, on the one hand, or overly descriptive, on the other. As a result, for example, this is the case for systems biology attributing causation to single genes; in the case of the latter, mathematical modeling of bioinformatics yields a compendium of genes that are associated with one another, but does not show how and why these interrelationships exist – it is a snapshot, not a continuum.

In contrast to this failure to understand physiology from the top-down or the bottom-up, we have suggested a "middle-out" approach (see Torday & Rehan in "Further reading") as a mechanism, guiding us all the way back from complex homeostatic regulation to our unicellular origins. If one starts from the spontaneous formation of primordial cells as semi-permeable membranes forming a protected space within themselves, over evolutionary time that space has filled up with endomembranes that reduce entropy through chemiosmosis. By compartmentalizing the bioenergetic mechanisms gleaned from the information in the external environment, internal information forms as a consequence. And the mechanism by which negentropy is sustained within and between generations is homeostasis, the process by which the cell continuously senses its environment and self-regulates accordingly, both within and between life cycles.

The difference between the descriptive and mechanistic approaches to physiology is what the linguistics theorist Noam Chomsky referred to as the difference between descriptive and explanatory adequacy – one concept of a phenomenon being superior

Evolution, the Logic of Biology, First Edition. John S. Torday and Virender K. Rehan.
© 2017 John Wiley & Sons, Inc. Published 2017 by John Wiley & Sons, Inc.

to another; that is, that one theory explains a certain phenomenon better than another, the latter continuing to leave the phenomenon a "mystery."

The information gained from this process of biosensing for both stability and change is then converted into knowledge through storage of the organism's history, that is, ontogeny and phylogeny, in its DNA. The organism thereby generates a "toolkit" that allows it to adapt to its ever-changing environment based on prior experience applied to current and future conditions, which we commonly refer to as the process of evolution. In his classic 1976 paper on evolution as "tinkering," François Jacob described the utility of the toolkit, but glossed over the underlying mechanism; the core of the current book is that it is the cell that forms the "logic" for the evolutionary process.

The cellular cooperativity that generates multicellular organisms can thus be seen as the sharing of knowledge as a strategy for survival, rather than merely an epiphenomenon for a way to improve metabolism as Geoffrey West and Harold Morowitz would have us think of it. And the processes of development, homeostasis, regeneration, and aging can all be viewed as a continuum, serving to create, sustain, recreate, and perpetuate biologically functional information, made useful as knowledge within a cellular context.

The premise of our first book on cellular evolution was that cell–cell communication was central to vertebrate evolution. Implicit in that book was the fact that biologic information was being communicated. Where did that information originate from? If one starts with the formation of micelles from lipids, it provides an "environment" in which chemiosmosis can function to generate negative entropy, or negentropy. That physically untenable state, violating the second law of thermodynamics, has been sustained by homeostatic mechanisms since the beginning of life. In that condition, cellular information is constantly being gleaned from the environment in the form of ions aligned on either side of the intracellular chemiosmotic membrane as information. Over time, the formation of endomembranes – endoplasmic reticulum, nuclear envelope, peroxisomes, Golgi apparatus – provided compartments in which bioactive substances such as heavy metals, lipid oxides, and gases could have been compartmentalized and appropriated to support and advance life. This is particularly the case in the competition between prokaryotes and unicellular eukaryotes, the former having a hard exoskeleton, whereas the latter have a soft exterior and a harder cytoskeleton. In the competition between these two groups, the prokaryotes evolved such "colonial" characteristics as quorum sensing and biofilm formation; eukaryotes competed and fought back by developing mechanisms of cellular cooperativity, ultimately leading to the formation of metazoans.

The advent of cholesterol in the plasmalemmas of eukaryotes was a critical step in their evolution. Cholesterol allowed for the thinning of the cell membrane, facilitating gas exchange, increased locomotion due to cytoplasmic streaming, and endocytosis, which greatly expedited "cell eating" since the cell could directly ingest nutrients from its environment, unlike its prokaryotic cousins, which excrete digestive enzymes extracellularly to break nutrients down so that they can then be absorbed molecularly. Cholesterol also provided for the formation of lipid rafts, the matrix in which cell-surface receptors reside and mediate cell–cell communication.

As a working example of such vertical integration from cells to tissues, organs, and organ systems, consider that the apolipoprotein ApoE4, which has been found to protect against childhood diarrhea, predisposes adults to Alzheimer's disease. This phenomenon is referred to as antagonistic pleiotropy, or the expression of the same gene in different structures, acting adaptively early in life, but maladaptively later in life. The evolutionist George Williams ascribed the aging process in general to antagonistic pleiotropy. But seen at the functional level, ApoE4 acts to facilitate the disposition of cholesterol in the cell membrane in support of "barrier" function. In the case of diarrheal insult, this would make sense since bacterial infection causes epithelial barrier failure and the loss of electrolytes through the enterocytes that form the epithelial lining of the gut wall. In the brain, ApoE4 also facilitates cholesterol, but in this case in the myelinization of the axons. But with age, it can cause calcium dyshomeostasis, probably precipitating the formation of amyloid plaque as a protective scarring mechanism. This injurious response ultimately causes failure of neuronal transmission of electrical activity. The common origins of both the gut and brain in the unicellular plasmalemma could explain such an event over the evolutionary history of the organism. Based on conventional top-down descriptive biology, these events are merely associations, but seen at the cellular-molecular level the causal mechanisms can be tested. One reason to think that these pathologies are interrelated is because peroxisome proliferator activated receptor gamma (PPARγ) agonists that stimulate lipid metabolism can prevent both the diarrhea and the Alzheimer's disease; PPARγ antagonism of calcium dyshomeostasis in the endoplasmic reticulum is an epistatic mechanism that goes all the way back to the origins of calcium regulation by peroxisomes in unicellular organisms, referred to as the de Duve hypothesis.

Richard Guerrant and his colleagues have exploited such knowledge of the role of ApoE4 in childhood diarrhea to treat the disease: They have used a dietary supplement composed of glutamine, zinc, and vitamin A to bolster the leaky epithelial barrier. Interestingly, the treatment was successful, but only in girls. Upon further reflection, based on the calcium dyshomeostasis hypothesis, estrogen also causes calcium disruption in cells, so it is possible that the formulation only affected the constitutive aspect of calcium flux, not the regulated alterations present in both boys and girls. It is hypothetically possible that a more fundamental effector of calcium metabolism, such as PPARγ stimulation of peroxisomes, might have a universal effect on boys and girls alike. Moreover, given the effect of PPARγ in preventing Alzheimer's disease, use of a PPARγ agonist might be a ubiquitous treatment for ApoE4 over-expression.

The Physico-Chemical Origins of Cellular Life

Cells are composed of membrane-bound compartments, creating an internal cellular environment in which genetic material can reside and metabolism can occur freely. Cell membranes are composed of complex mixtures of amphiphilic molecules, one end soluble in water, the other end soluble in lipid, such as phospholipids, sterols, and so forth, as well as proteins that transport molecules and perform enzymatic functions. Phospholipid membranes are stable under a wide variety of physiologic conditions, making them extremely good permeability barriers. As a result, cells have tight control over the uptake of nutrients and the export of wastes through the actions of specialized channel, pump, and pore proteins embedded in their membranes. A great deal of complex biochemical machinery is also required to mediate the growth and division of the cell membrane during the cell cycle. The question of how a structurally simple protocell could accomplish these essential membrane functions is critical to understanding the origins of cellular life.

Isolated vesicles spontaneously formed by fatty acids have long been used as models for protocell membranes. Fatty acids are attractive as fundamental building blocks for prebiotic membranes since they are chemically much simpler than phospholipids. Fatty acids with saturated acyl chains are extremely stable compounds, and therefore might have accumulated at biophysically significant levels, even given their relatively slow, conditional synthesis. Moreover, the bonding of fatty acids with glycerol to yield the corresponding glycerol esters provides a highly stable membrane component. Finally, phosphorylation and the addition of a second acyl chain yields phosphatidic acid, the simplest phospholipid, providing a conceptually simple pathway for the transition from primitive to more advanced biomembranes. But an even more compelling reason for considering fatty acids as fundamental to the nature of primitive cell membranes is their dynamic properties within membranes, which are essential for both membrane growth and permeability.

Fatty acids are single-chained amphiphiles with a less hydrophobic surface than phospholipids, so they assemble into membranes only at much higher concentrations. This equilibrium property is reflected by their kinetics: Fatty acids are not as firmly anchored within the membrane as phospholipids; they can enter or exit the membrane within seconds to minutes. Fatty acids can also exchange between the two leaflets of a membrane bilayer on a sub-second timescale. Such rapid transit is needed for membrane growth when new amphiphilic molecules are supplied by the environment. New molecules enter the membrane primarily from the outside leaflet, and the rapid movement allows the inner and outer leaflets to equilibrate, causing uniform growth.

Since protocells were mechanically simple, they had to rely on their intrinsic membrane permeability properties. Membranes composed of fatty acids are, in fact, reasonably permeable to small polar molecules, and even to charged ions and nucleotides. This appears to be a result largely of the ability of fatty acids to form transiently unstable structures and/or transient complexes with charged solutes, which facilitates transport across the membrane.

Prebiotic vesicles were probably composed of complex mixtures of amphiphiles. Amphiphilic molecules isolated from meteorites, as well as those synthesized under simulated prebiotic conditions, are highly heterogeneous, both in terms of acyl chain length and head group chemistry. Membranes composed of mixtures of amphiphiles often have superior properties to those composed of homogeneous species. For example, mixtures of fatty acids, their corresponding alcohols, and/or glycerol esters generate vesicles that are stable over a wide range of pH and ionic conditions, and are more permeable to ions, sugars, and nucleotides. This is in striking contrast to the apparent requirement for homogeneity in the nucleic acids, where even low levels of modified nucleotides can be destabilizing, and can block replication.

Fatty-acid vesicle enlargement has been shown to occur through at least two distinct pathways: through the donation of fatty acids by added micelles, or through fatty-acid exchange between vesicles. The enlargement of membrane vesicles from micelles has been observed following addition of micelles or fatty acid precursors to pre-existing vesicles. When alkaline fatty acid micelles are mixed with a buffered solution at a lower pH, the micelles become thermodynamically unstable. As a result, the fatty acid molecules are either incorporated into pre-existing membranes, leading to enlargement, or undergo self-assembly into nascent vesicles.

Another mechanism for the growth of vesicles entails fatty acid exchange between vesicles. This process may lead to proliferation of a subpopulation of vesicles that compromises surrounding vesicles. Isotonic vesicles do not manifest such exchanges resulting in changes in size over time, nor do populations of uniformly swollen vesicles; rather, such vesicles are in equilibrium with a lower concentration of fatty acids because the tension on the membrane of the swollen vesicles makes it energetically more favorable for fatty acid molecules to reside in the membrane. When ionically swollen vesicles are combined with isotonic vesicles, fatty acid exchange occurs rapidly, resulting in further swelling of the vesicles, and concomitant shrinking of the relaxed vesicles. Because vesicles can swell due to high concentrations of nucleic acids, this mechanism favors the growth of vesicles containing genetic polymers over empty vesicles. Since more rapid replication increases the vesicle nucleic acid concentration, this process of competitive vesicle enlargement potentiates a direct physico-chemical link between replication of nucleotides and the rate of protocell growth.

Given that ionically swollen vesicle division is normally distributed, protocells that develop some heritable means of faster replication and growth would have a shorter cell cycle, and would therefore gradually take over the population. Such a physico-chemical mechanism might represent Darwinian competition at the cellular level.

The current Chapter was intended to explain how physical chemistry, catalyzed by the reduction in intracellular entropy, has given rise to and perpetuated evolution. Chapter 11, entitled "Vertical integration of cytoskeletal function from yeast to humans," explores the physiologic stress of the water–land transition in order to determine how specific genetic changes facilitated vertebrate adaptation to terrestrial life.

Suggested Reading

de Duve C. (1969) Evolution of the peroxisome. *Ann N Y Acad Sci* 168:369–381.

Jacob F. (1977) Evolution and tinkering. *Science* 196:1161–1166.

Morowitz HJ. (1955) Some order-disorder considerations in living systems. *Bull Math Biophys* 17:81–86.

Orgel LE. (2004) Prebiotic chemistry and the origin of the RNA world. *Crit Rev Biochem Mol Biol* 39:99–123.

Szilard L. (1929) On the decrease of entropy in a thermodynamic state by the intervention of intelligent beings. *Z Phys* 53:840–856.

Torday JS, Rehan VK. (2012) *Evolutionary Biology, Cell-Cell Communication, and Complex Disease.* John Wiley & Sons, Inc., Hoboken, NJ.

West GB, Brown JH, Enquist BJ. (1997) A general model for the origin of allometric scaling laws in biology. *Science* 276:122–126.

11

Vertical Integration of Cytoskeletal Function from Yeast to Human

In our ongoing quest to decode the evolution of vertebrate physiology, we have heavily exploited parathyroid hormone-related protein (PTHrP) signaling because of its combined central importance in the water-to-land transition (WLT), and its linkage to the deep homologies in adaptation to gravity, the oldest, constant (magnitude and direction) of all environmental cuing mechanisms. We will review both of these aspects of PTHrP signaling, and outline the commonalities between them.

Calcium/Lipid Homeostasis: Lessons from the Alveolus, or PTHrP Signaling and the Water-to-Land-Transition, or "Adventures in Pleiotropy"

PTHrP is expressed in all vertebrate epithelial cells, acting as a paracrine differentiation and growth factor during development, homeostasis, and repair. Maturation of the PTHrP–PTHrP receptor signaling pathway culminates in homeostatic regulation of a wide variety of functions, ranging from respiration to cartilage mineralization, skin metabolism, glomerular filtration, brain homeostasis, and reproduction. The question that naturally arises is whether there is a unifying evolutionarily (= ontogeny + phylogeny) adaptive mechanism that accounts for this wide diversity of functions. One way to determine if such a mechanism exists is to trace PTHrP signaling phylogenetically, particularly during the severe constraints of the WLT. This "experiment of nature" allows us to see and hopefully understand how the expression of PTHrP signaling could have facilitated this physiologically challenging transition, knowing that it was amplified by the documented duplication of the PTHrP receptor – perhaps that event occurred deterministically, rather than merely by chance mutation and selection, as it is popularly expressed in the literature by molecular biologists and evolutionary biologists alike. If the former were the case, it would provide insights to how and why PTHrP mechanistically facilitated land adaptation in vertebrates physiologically.

The first anatomic structure challenged by the WLT may have been the skeleton, which had to be able to tolerate the increased gravitational force of living on land (relative to that in water), under duress due to the drying up of lakes, ponds, and rivers (the Romer hypothesis). The skeleton is well suited to such "plasticity" according to Wolff's law, formulated by the German anatomist and surgeon Julius Wolff (1836–1902) in the nineteenth century. It states that weight-bearing bone in a healthy person or animal will adapt to the load under which it is placed. The fossil record shows that bone must have

Evolution, the Logic of Biology, First Edition. John S. Torday and Virender K. Rehan.
© 2017 John Wiley & Sons, Inc. Published 2017 by John Wiley & Sons, Inc.

had great capacity for plasticity, since there were at least five independent skeletal adaptations that occurred during the WLT period. It is now widely accepted that the first tetrapods arose from advanced tetrapodomorph stock (the elpistostegalians) in the Late Devonian, probably in Euramerica. However, actual terrestrial forms did not emerge until much later in geographically far-flung regions during the Lower Carboniferous. The complete transition occurred over the course of about 25 million years; definitive emergences onto land took place during the most recent 5 million years. The sequence of character acquisitions during the transition can be seen as a five-step process, involving:

1) Higher osteichthyan (tetrapodomorph) diversification in the Middle Devonian (beginning about 380 million years ago [mya]).
2) The emergence of "prototetrapods" (e.g., *Elginerpeton*) in the Frasnian stage (about 372 mya).
3) The appearance of aquatic tetrapods (e.g., *Acanthostega*) sometime in the early to mid-Famennian (about 360 mya).
4) The appearance of "eutetrapods" (e.g., *Tulerpeton*) at the very end of the Devonian (about 358 mya).
5) The first truly terrestrial tetrapods (e.g., *Pederpes*) in the Lower Carboniferous (about 340 mya).

By inference, since PTHrP is a gravisensor that was expressed as far back in vertebrate phylogeny as fish, it could hypothetically have facilitated skeletal remodeling in the adaptation to land. PTHrP is expressed in bone, where it regulates calcium homeostasis under the influence of mechanical strain, according to Wolff's law (see earlier). PTHrP has also recently been found to be expressed during the development of the physostomous fish swim bladder, finally allowing it to be recognized as the functional molecular homolog of the mammalian lung. And in both structures, PTHrP expression is gravity-sensitive, suggesting that this homology goes very deep into vertebrate evolutionary history, since gravity is the oldest omnipresent (unlike the Sun), omnidirectional, and magnitude-constant environmental affector of adaptation. It is tempting to speculate that over the course of the several known attempts at the WLT, those species most capable of upregulating their PTHrP signaling expression would have been the most likely to succeed in the land assault and ultimate habitation. Since the swim bladder is the organ for buoyancy in adaptation to feeding and swimming, its PTHrP expression may also have been positively selected for, synergizing with bone PTHrP expression. The functional homology between the swim bladder and the lung emanates from the complementary effects of PTHrP in adapting to both gas-exchange and metabolism. In the case of the swim bladder, it acts to adapt for both buoyancy and feeding; in the case of the lung, it accommodates stretch-regulated surfactant production. That is to say, in the case of both the swim bladder and lung, oxidative metabolism is the selection pressure for gas-exchange, surfactant production being the final common pathway. Lung physiology is contingent on PTHrP expression, since the latter is necessary for alveolar formation, followed by the induction of the lipofibroblast, its expression of leptin, adipocyte differentiation related protein, and prostaglandin E_2 (PGE_2) by the alveolar type II cell, all coordinately regulating stretch-dependent surfactant production for vertebrate adaptation to air breathing (see Figure 1.3).

Independent experimental evidence for the evolutionary relevance of this signaling cascade comes from a study by Valérie Besnard and co-workers (Figure 11.1) in which they genetically disrupted the cholesterol-biosynthetic gene *Scap* in alveolar type II cells of the developing mouse lung. The loss of cholesterol expression by these cells – cholesterol being the most ancient form of surfactant – resulted in compensation for the loss of surface tension-reducing activity by over-expression of alveolar lipofibroblasts. These data suggest that the lipofibroblast was expressed during the evolution of the alveolus in mammals (since they are not present in earlier stages of phylogeny), allowing for the decrease in alveolar diameter in order to increase the surface area-to-blood volume ratio and gas-exchange.

As a demonstration of the predictive power of the cellular-molecular evolutionary approach, the PTHrP mechanism of lung evolution would have been constrained by the necessity for independent regulation of blood pressure within the microvasculature of the alveoli. Without such an adaptation, increases in systemic blood pressure would have damaged the lung during peak excursions of hypertension. What mitigates against this constraint in vertebrate evolution on land is the advent of increased β-adrenergic expression in the alveolar microvessels, facilitating alveolar capillary vasodilation. β-Adrenergic signaling activates the PTHrP receptor, synergizing the effects of β-adrenergic signaling and PTHrP signaling for vasodilation and surfactant production. Therefore, it is likely that in the phylogenetic transition from fish to amphibians the duplication of both the PTHrP receptor and β-adrenergic receptor was due to environmental stress causing internal selection pressure simultaneously for both of these genes within the endoderm, rather than merely being a chance occurrence, as Darwinian evolutionists would have us think. *Bear in mind that such an integrated view of physiology is readily derived from this evolutionary perspective.*

This scenario for PTHrP-facilitated WLT can be further exploited to gain insight to the evolution of other physiologic traits necessary for the WLT, and for integrated organismal physiology, or allostasis (Figure 11.2). There must have been protracted periods of relative hypoxia as the lung evolved when tetrapods emerged from water to colonize the land, particularly given the episodes of increased and decreased atmospheric oxygen during the Phanerozoic eon (the Berner hypothesis). Hypoxia is the most stressful of all physiologic insults, causing secretion of adrenocorticotropic hormone (ACTH) by the pituitary; it was recently discovered that PTHrP is expressed in the anterior pituitary, where it stimulates ACTH secretion. Moreover, PTHrP is also expressed in both the mammalian and avian adrenal cortices, where it mediates the ACTH effect by stimulating corticosteroid production. The corticosteroids subsequently pass through the adrenomedullary portal vasculature, where they stimulate phenylethanolamine-N-methyltransferase (PNMT), the rate-limiting step in adrenaline (epinephrine) synthesis from noradrenaline (norepinephrine). This specialized microvascular bed has only been reported in the mammalian adrenal medulla, suggesting an evolved trait for maximizing the "fight or flight" response. This mechanism may have been particularly robust under hypoxic conditions, since cortical PTHrP, a potent vasodilator, would also have caused microvascular dilation, further enhancing adrenaline production. This amplification of adrenaline production stimulates the secretion of surfactant by lung alveolar type II cells, further facilitating gas exchange by the episodically evolving lung.

Another organ system in which PTHrP is stretch-regulated is the urogenital tract. PTHrP is expressed in the uterus, and its expression increases with advancing gestation.

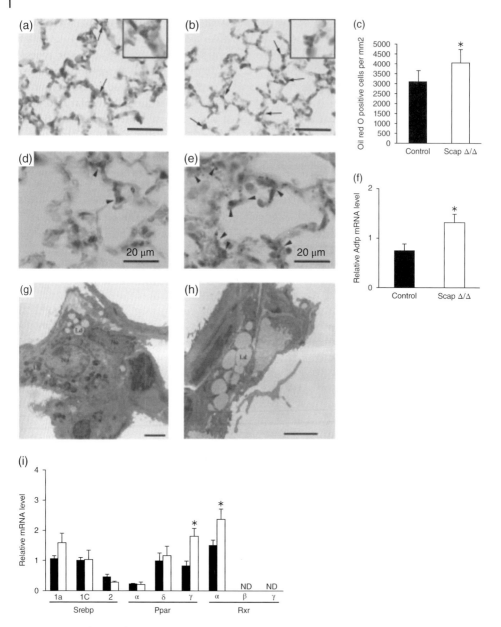

Figure 11.1 Deletion of *Scap* reveals the deep role of cholesterol in vertebrate evolution. B, E, and H are photomicrographs illustrating compensatory increases in lipofibroblasts in the alveoli of *Scap*-deleted mouse lung, including increased amounts of neutral lipid droplets (H). I shows increases in sterol regulatory element binding protein 1a (Srebp), peroxisome proliferator activated receptor gamma (Ppar), and retinoid X receptor a (Rxr). Reproduced from Besnard V, Wert SE, Stahlman MT, Postle AD, Xu Y, Ikegami M, Whitsett JA. Deletion of Scap in alveolar type II cells influences lung lipid homeostasis and identifies a compensatory role for pulmonary lipofibroblasts. *J Biol Chem* 2009 Feb 6; 284(6):4018–30. doi: 10.1074/jbc.M805388200.

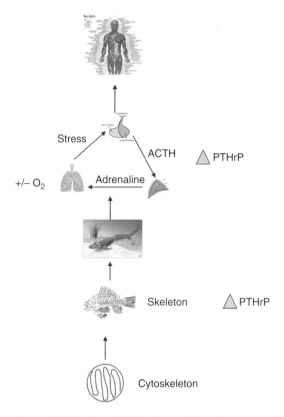

Figure 11.2 Positive selection for parathyroid hormone-related protein (PTHrP) signaling during vertebrate evolution. Starting with adaptation to the physical environment by the unicellular cytoskeleton, skeletal adaptation mediated by PTHrP was key to vertebrate water-to-land transition. Positive selection for PTHrP signaling was subsequently adaptive for hypoxia, stimulating the pituitary-adrenal axis for adrenaline production, relieving the constraint on the alveoli by stimulating surfactant secretion. ACTH, adrenocorticotropic hormone. (*See insert for color representation of the figure.*)

If the gravid uterus is experimentally emptied of its fetal contents, the uterine expression of PTHrP decreases to pre-pregnancy levels, suggesting that PTHrP is somehow involved in parturition. PTHrP is also expressed in the urinary bladder and kidney. In the glomerulus, the podocytes lining the glomerular space express PTHrP, which binds to the mesangium and determines water and electrolyte filtration rates.

The expression of stretch-regulated PTHrP signaling in such seemingly disparate organs as the skeleton, lung, and urogenital tract raises the question as to whether there was some common ancestral homolog for these biologic traits that was generated by internal selection? Or is this simply a consequence of Darwinian "external selection"? Because of the orthogonality of the PTHrP mechanism during the WLT, we have formulated the former as a working hypothesis.

Both mammalian lung and bone cells are gravisensors for PTHrP expression. When either lung epithelial cells or osteoblasts are attached to Sephadex beads and allowed to go into free fall in a rotating wall vessel bioreactor, simulating $0 \times g$ conditions, PTHrP mRNA expression decreases nine-fold over 8–12 hours. When these cells are returned to $1 \times g$ conditions, the PTHrP expression returns to pre- $0 \times g$ levels within 8 hours, indicating that these cells are gravity-sensitive. Physiologically, this phenomenon manifests itself as mechanotransduction, determining the rate of lung surfactant production in the alveolus, and calcium biofixation in bone. In either case, the mechanical adaptation to gravity affects the homeostasis of these tissues and organs.

In yeast, microgravity affects both cell polarity, and reproduction in the form of budding, offering insight to the homologous expression of PTHrP in the urogenital tract – both mammals and yeast are eukaryotes. During mitosis and meiosis, the cytoskeletal microfilaments become the spindles that organize the chromosomes before the cell divides into daughter cells or during chromosomal segregation. This may be the functional homolog for PTHrP expression as a gravisensor in both the urogenital tract of mammals, and during cell division in yeast. In multicellular organisms, cellular orientation translates into metabolic cooperativity for lipid trafficking in the alveolus, and for higher-order reproduction in facilitating uterine physiology during pregnancy. *This exercise in molecular homology demonstrates the value added in this evolutionary perspective for understanding the mechanistic origins of complex physiology.*

In contrast, the authors of the Besnard paper never took the long-haul evolutionary view. They cited studies demonstrating a direct relationship between altered cellular lipid composition and endoplasmic reticulum (ER) stress. Increase or depletion of the fatty acid or cholesterol content activated ER stress responses, indicating the requirement for precise regulation of cellular lipid composition to maintain normal cell function. This study and other models in which lipid homeostasis is perturbed share an increased expression of ER stress-induced transcription factors (e.g., ATF3, ATF5, and CHOP), as well as elevated expression of oxidative stress response genes, and those influencing amino acid biosynthesis and transport pathways, suggesting a common response to alterations in cellular lipid composition. ER stress can modulate the expression of sterol regulatory element-binding proteins (SREBPs), indicating potential reciprocal interactions between ER stress and the SCAP/SREBP pathway. It is highly likely that these deep homologies refer back to the balancing selection for the evolution of the peroxisome (the de Duve hypothesis), whereby ER stress leading to calcium dyshomeostasis was epistatically countered by lipid homeostasis. as proof of principle, peroxisome proliferator activator receptor gamma (PPARγ) agonists prevent disruption of surfactant homeostasis, including cholesterol biosynthesis.

The possibility that such balancing selection may have given rise to the lipofibroblast was not alluded to in the Besnard paper. Elsewhere, Barbara Wold's group has demonstrated that muscle cells spontaneously differentiate into adipocytes in 21% oxygen, but not 6%. This seminal observation suggests that the oxygen-induced ER stress on the alveolar type II cells during the evolutionary adaptation to the rising levels of oxygen in the Phanerozoic eon may have been epistatically "balanced" by the evolution of the lipofibroblast, which is dependent on PPARγ expression, resulting in increased surfactant production, ultimately alleviating such stress by increasing oxygen uptake, as illustrated below.

Like the alveolus, the glomerulus of the kidney is also a "pressure transducer" (see Figure 9.3), the mesangium determining the flux of water and electrolytes in the kidney. PTHrP is expressed by the podocytes lining the glomerular space, signaling to the PTHrP receptor on the mesangium to maintain its function. Selection pressure for increased PTHrP receptor expression may have facilitated the formation of the glomerulus in amphibians during the WLT, since the fish kidney doesn't have a glomerulus; rather it has a more primitive glomerulus-like glomus. PTHrP is angiogenic, so upregulation of PTHrP signaling in the microvasculature of the evolving kidney renal artery may have generated the glomerular microvascular capillary arcade due to shear stress.

This scenario of PTHrP expression in a wide variety of tissues and organs to alleviate physiologic stress in the WLT can be carried one step further in the hierarchical evolution of land-dwelling vertebrates. PTHrP is also expressed in the pituitary and the adrenal cortex, stimulating ACTH production in the former and corticosteroids in the latter (see Figure 1.3). Moreover, both structures exhibit a highly articulated microvasculature that may similarly have been induced by increased local expression of PTHrP promoting angiogenesis. Corticosteroids stimulate adrenaline in the adrenal medulla as they pass through the medullary vascular arcade, having a net acute adaptive effect of decreasing physiologic stress on the alveolus by increasing surfactant production, and on the kidney in adaptation to land by slowing renal blood flow, retarding the elimination of water and electrolytes.

Consider the functional homology between the β-adrenergic receptor effect on surfactant production and on fatty acid release from peripheral fat cells for thermogenesis. Warm-blooded mammals evolved subsequent to amphibians and reptiles. It was the advent of peripheral fat cells that gave rise to homeotherms, requiring higher metabolic rates, and therefore gas exchange. So which came first? Increased body temperature would have passively facilitated lung physiology since lung surfactant is 300% more active at 37 °C than at 25 °C. And experimentally, when fence lizards are treated with leptin their basal metabolic rate rises, as does their body temperature, making them more like warm-blooded animals. This suggests that it was systemic leptin produced by fat cells that gave rise to the mammalian lung, not the other way round. This is consistent with Markus Lambertz's recent paper suggesting that it was the alveolar mammalian lung that was foundational for the fixed lungs of reptiles and birds. Mechanistically we know that the alveolar lung can "simplify" under stressful conditions of development and disease, so this hypothesis is testable.

Evolutionary Lessons from the Role of PTHrP in Middle Ear Evolution

Duplication of the PTHrP receptor gene during vertebrate evolution and the WLT may also bear on the evolution of the middle ear ossicles from the fish jaw, since this is a classic example of Wolff's law relating to bone plasticity. There are effects of PTHrP gene deletion on Meckel's cartilage, which is thought to have been the structure affected during evolution to give rise to the ear ossicles. The evidence for this relationship is as follows.

PTHrP-null mutant mice exhibit skeletal abnormalities both in the craniofacial region and the limbs. In the growth plate cartilage of the null mutant, a diminished number of proliferating chondrocytes and accelerated chondrocytic differentiation are observed. In order to examine the effect of PTHrP deficiency on the craniofacial morphology and highlight the differential features of the constituent cartilages, the various cartilages in the craniofacial region of neonatal PTHrP-deficient mice were examined. The major part of the cartilaginous anterior cranial base appeared to be normal in the homozygous PTHrP-deficient mice. However, acceleration of chondrocytic differentiation and endochondral bone formation was observed in the posterior part of the anterior cranial base and in the cranial base synchondroses. Ectopic bone formation was observed in the soft tissue-running mid-portion of Meckel's cartilage, where the cartilage degenerates and converts to ligament in the course of normal development. The zonal structure of the mandibular condylar cartilage was scarcely affected, but the whole condyle was reduced in size. These results suggest the effect of PTHrP deficiency varies widely between the craniofacial cartilages, according to the differential features of each cartilage.

Compared with the cranial base cartilages, Meckel's cartilage and condylar cartilage showed characteristic changes in homozygous mice. Meckel's cartilage has been reported to play a supportive role in mandibular development, and there are three distinct regions, each having an apparently different fate:

1) The anterior region of Meckel's cartilage contributes to mandibular development and undergoes endochondral ossification.
2) The most posterior region also undergoes endochondral ossification and gives rise to the malleus and incus.
3) The mid-region, distal to the ossification center of the mandibular anlage, degenerates and gives rise to the sphenomandibular ligament.

The most dramatic difference observed in the Meckel's cartilage of PTHrP-deficient mice was in this soft tissue mid-region. During normal development, the degeneration of this region has been reported to start at around day 18 of gestation in mice lacking type X collagen. Histologic study of the extracellular matrix in rats lacking type X collagen showed that this process of degeneration starts in the perichondrium, where macrophage- and fibroblast-like cells appear to degrade the unmineralized cartilage matrix, and the chondrocytes are finally attacked by giant cells. In homozygous PTHrP-deficient mice, the cartilage in this area was degenerated, and was surrounded by the presumptive bone matrix. Similar ectopic bone formation has also been reported in the perichondrium of rib cartilage in the homozygous mice. These observations suggest that PTHrP deficiency might alter the mechanism of normal bone cell differentiation.

The presence of three ossicles in the middle ear is one of the definitively evolved features of mammals. All reptiles and birds have only one middle ear ossicle, the stapes or columella. How the two additional ossicles appeared in the middle ear of mammals has been studied for the last two centuries, representing one of the classic examples of how structures can change during evolution to function in novel ways. From the combined evidence of the fossil record, comparative anatomy, and developmental biology it is now apparent that the two newly acquired bones in the mammalian middle ear, the malleus and incus, are homologous to the quadrate and articular, which form the articulation for the upper and lower jaws in non-mammalian jawed vertebrates. Incorporation of the primary jaw joint into the mammalian middle ear was only possible due to the evolution of a new way of articulating the upper and lower jaws, with the formation of the dentary-squamosal or temporo-mandibular joint (TMJ) in humans. The evolution of the three-ossicled middle ear in mammals is thus intimately connected to the evolution of a novel jaw joint, the two structures evolving together to form the distinctive mammalian skull.

The middle ear ossicles of mammals reside in an air-filled cavity, straddling the gap between the external and inner ear. Vibration of the tympanic membrane (eardrum) is picked up by the manubrium of the malleus, and transferred to the incus and stapes, conducting the vibrations to the inner ear via the oval window. Defects lead to conductive hearing loss. In birds and reptiles, only one ossicle bridges the air-filled middle ear cavity, passing vibrations from the external to the inner ear. In birds, this ossicle is known as the columella auris, while in reptiles it is known as the stapes.

The middle ear ossicles are found in the auditory bulla, comprised of several bones, namely the tympanic ring, the bulla, and the malleus; the gonial bone lies between the tympanic ring and malleus, facilitating the function of the latter. The malleus derives from both endochondral and gonial bone sources.

Since both reptiles and birds possess only one middle ear ossicle, the origins of the malleus, incus, tympanic ring, and gonial have been controversial. Karl Reichert (in 1837) was the first to hypothesize that the malleus and incus were homologous to the articular and quadrate bones of the non-mammalian jaw joint, which has been supported by extensive interdisciplinary evidence from the fossil record, developmental biology, and molecular biology. Such studies have generated an integrated theory for the mechanisms involved in forming the mammalian ear and jaw.

Evidence from Developmental Biology

Meckel's cartilage is composed of two rods of cartilage that overarch the sides of the mandible; the proximal portion forms the jaw bone in all but mammalian vertebrates. In avian embryos, portions of Meckel's and the quadrate cartilage derive from the first pharyngeal arch, whereas the retroarticular process that develops proximal to the articular and the columella derive from the second pharyngeal arch. Separated by the jaw joint, cartilage generates the two skeletal elements, giving rise to the quadrate and articular bones of the jaw.

The malleus and incus are formed from a single cartilage that subdivides, whereas the stapes derives from a separate cartilage that extends toward the incus to form a joint. The malleus and incus derive from the posterior of Meckel's cartilage like the other two ear ossicles; the malleus remains attached to Meckel's cartilage throughout most of embryonic development, forming a conduit between the jaw bone and the middle ear.

In mice, the cartilaginous connection between the jaw and ear breaks down after birth, starting on or about day 2 of life, with the transformation of Meckel's cartilage next to the malleus into the sphenomandibular ligament. The dissolution of Meckel's cartilage functionally separates the ear from the jaw. Meckel's cartilage supports the bones that ossify along its length.

Genetic data are also consistent with Reichert's theory. The homeodomain transcription factor Bapx1 is found in the jaw joints of birds, fish, and reptiles, whereas it localizes to the middle ear in developing mammals, pointing to the common origin and homology of the ear ossicles.

The homologies between the ear ossicles suggested by comparative anatomy nearly two centuries ago have been confirmed by molecular and developmental biology. Additional fossil evidence has facilitated further documentation of the transition from the fish jaw to the mammalian middle ear bones.

PTHrP Vertically Integrates the Evolution of Vertebrate Physiology

This compilation of evolved PTHrP-dependent traits is based on mechanistic relationships between genes and phenotypes that have evolved over the course of ontogeny and phylogeny. As such they are not "Just So Stories" but are experimentally testable and refutable. For example, our laboratory has shown that alveolar type II cell PTHrP regulates leptin expression by lipofibroblasts in the mammalian alveolar interstitium. Leptin shows the same stimulatory effect on both mammalian and amphibian lung development. Yet the resulting stimulation of the surfactant system in both species was counterintuitive based on traditional physiology, since the frog lung does not require surfactant for surface tension reduction. However, the surfactant serves dual roles of surface tension reduction and host defense due to the antimicrobial function of the surfactant proteins. Leptin stimulated the expression of surfactant proteins, and promoted the thinning of the alveolar wall to promote gas exchange, which would otherwise have put the organism at risk of infection absent the surfactant proteins. It is these latter ontogenetic and phylogenetic properties that were probably the basis for the selection pressure for leptin signaling since the lung evolved initially through the increased surface area of the gut, necessitating increased host defense. Both the thinning of the alveolar wall and the expression of the surfactant proteins occur prior to the onset of alveolar epithelial surfactant phospholipid expression. It is the former traits that were preadapted for lung evolution.

Therefore, by looking at the process of evolution in the forward or prograde direction, both phylogenetically and ontogenetically, the nature of the evolutionary process becomes clear. In contrast, by reasoning after the fact, the evolutionary strategy is obscured.

The current Chapter has examined the roles of the mechanical "superstructure," the cytoskeleton and skeleton, as organizing principles for integrated physiology as an existing form that relates all the way back to the origins of life itself. In Chapter 12, entitled "Yet another bite of the 'evolutionary' apple," we go through the reverse logic of physiology emanating from the unicellular state. It may sound redundant, but because of the counterintuitive nature of the approach it is helpful to see this viewpoint in multiple ways.

Suggested Readings

Besnard V, Wert SE, Stahlman MT, Postle AD, Xu Y, Ikegami M, Whitsett JA. (2009) Deletion of Scap in alveolar type II cells influences lung lipid homeostasis and identifies a compensatory role for pulmonary lipofibroblasts. *J Biol Chem* 284:4018–4030.

Clark JA. (2002) Gaining Ground. Indiana University Press, Indiana.

Lambertz M, Grommes K, Kohlsdorf T, Perry SF. (2015) Lungs of the first amniotes: why simple if they can be complex? *Biol Lett* doi: 10.1098/rsbl.2014.0848.

12

Yet Another Bite of the "Evolutionary" Apple

Introduction

There is no unifying "central theory" of biology, so in its stead we continue systematically collecting information in the name of knowledge, just as Linnaeus did back in
the eighteenth century at the inception of biology. Ernest Rutherford said that "All
science is either physics or stamp collecting." Nowadays, the practice of amassing
information as "knowledge" is validated by informatics, which espouses that even the
most complex problems, including that of evolution, can be solved given enough data.
As a result, contemporary genomic, proteomic, and interactomic analyses are merely
collections of data that are hypothesis-generating, not predictive. In contrast to this,
Dmitri Mendeleev's periodic table organized the chemical elements using their atomic
numbers as the organizing principle, resulting in a predictive algorithm. Evolution
has no equivalent. DNA is just a further reduction of the problem, but it is not a solution. Solving the riddle of evolution as "all of biology," as famously stated by Ted
Dobzhansky, would provide us with such a tool, but it has eluded us now for centuries.
The prevailing theory of evolution is based on metaphors such as natural selection,
survival of the fittest, and descent with modification. These descriptive properties of
evolution are appealing to our common senses, but they lull us into a false sense of
knowledge, rationalizing that we have actually figured out the process, when in reality
the mechanism of evolution lies elsewhere. Gerhart Wiebe referred to this as the
"well-informed futility syndrome" – the more informed you are, the more you think
of knowledge as power. Metaphors are useful as space savers, but as Denis Noble
states, such "ladders" should be removed from under us so that knowledge can
advance. For example, Darwin didn't know about genes, DNA, or soluble growth factors and their cell-surface receptors, the latter not being discovered until 1978, more
than one hundred years after the publication of *On the Origin of Species*. As a result,
instead of enabling the discovery of the basis for physiology, evolution theory con-
fuses and misguides our scientific understanding of its true nature. Witness the popularity of intelligent design, the religiously based diversion that more often than not
supersedes evolution theory. In reality, it should have no credibility in an age of reason and evidence-based science and medicine. But it is viewed by many as the correct
way of thinking about the origins of life. In a recent essay entitled "Moon Man: What
Galileo Saw," Adam Gopnik refers to "smart accommodationists in favor of evolution"
for whom evolution is not an alternative to intelligent design; evolution is intelligent

Evolution, the Logic of Biology, First Edition. John S. Torday and Virender K. Rehan.
© 2017 John Wiley & Sons, Inc. Published 2017 by John Wiley & Sons, Inc.

design seen from the vantage point of a "truly intelligent designer." This is a failure of our commitment to science as our only means of knowing what we don't know.

In the midst of the sea change we are now experiencing in the post-genomic era, it is essential that we step back and evaluate our contemporary perspective on biology. For example, the take-home message of the Human Genome Project was that humans have fewer genes than a carrot (19,000 vs 40,000, respectively), whereas it had been predicted that we would have at least 100,000 genes, based on the number of genes found in worms, flies, and other model organisms – so much for a predictive paradigm. This is yet another glaring example of the absence of a logic for biology. The fact that we humans have fewer than the predicted number of genes obviously doesn't mean we are "simpler." It is more likely that we have used fewer genes more effectively to adapt to our environment as a result of heretofore unidentified evolutionary processes. In light of this, our laboratory has gained great insight to mechanisms of evolution using an alternative cell-molecular level approach.

For evolution to be scientifically testable, we need specific mechanisms, not just a *deus ex machina*, like natural selection or survival of the fittest. Since embryology generates form and function through well-known cell-molecular mechanisms, that is a logical place to look. Evolutionary developmental biology, or evo-devo, is an attempt to do so, but for historic reasons, evolution theory has rejected cell biology, which is the *sine qua non* for developmental mechanisms. We have proposed a cellular-molecular approach to evolution that effectively integrates biology by focusing on homeostasis as the underlying selection pressure – since it was the initial reduction in entropic energy within primitive cells, sustained and perpetuated by homeostatic control, that fostered life. Although Walter B. Cannon coined the term "homeostasis," he did not reduce it to the cellular level. Homeostasis is a fundamental principle of biology, without which there could be no Linnaean hierarchy of species, or the ability to recognize discrete species, despite development under a variety of environmental conditions. To my knowledge, this is the first such proposed mechanism for evolution that utilizes contemporary biologic principles of development that have been known for more than 50 years, begging the question as to why that is the case, since homeostasis is universally accepted as the fundament for physiology.

We would like to speculate that this is because soft tissues in general, and visceral organs in particular, are not studied evolutionarily for two reasons: there is no fossil record, and internal selection has been out of favor for centuries. We began thinking about lung evolution in the context of physiology because one of us (J.T.) had experimentally stumbled onto the process of neutral lipid trafficking as the basis for alveolar homeostasis (Figure 12.1). In brief, lipid substrate for lung surfactant is actively recruited from the alveolar circulation by lipofibroblasts by expressing adipocyte differentiation related protein (ADRP); ADRP is regulated by the stretch-mediated epithelial cell secretion of parathyroid hormone-related protein (PTHrP), which binds to its receptor on the lipofibroblasts, and stimulates the uptake of lipid; lipofibroblasts secrete the lipid in response to prostaglandin E_2 produced and secreted by alveolar epithelial type II (ATII cells); lipofibroblasts also synthesize and secrete leptin, which stimulates surfactant phospholipid synthesis by ATIIs. ADRP mediates the uptake of lipid by the ATIIs, ensuring efficient mobilization of substrate for surfactant production. It would have taken more than 9×10^{16} years for this mechanism to have occurred by chance alone, which is longer than the age of the Earth. Therefore, by deduction this mechanism must have evolved through serial selection pressures (see below).

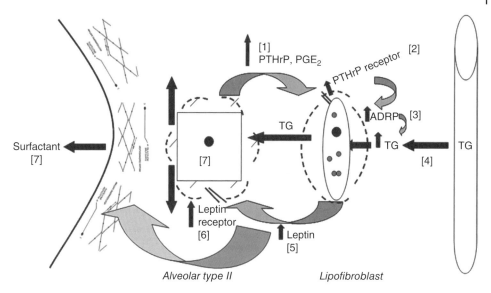

Alveolar type II   *Lipofibroblast*

Figure 12.1 Schematic for paracrine determinants of alveolar homeostasis and disease. We have observed coordinating effects of stretch on alveolar type II cell expression of parathyroid hormone-related protein (PTHrP) [1], the lipofibroblast PTHrP receptor [2], its downstream effect on lipofibroblast adipocyte differentiation related protein (ADRP) expression [3], and triglyceride (TG) uptake [4], and on the interaction between lipofibroblast-produced leptin [5] and the alveolar type II cell leptin receptor [6], stimulating de novo surfactant phospholipid synthesis and secretion by alveolar type II cells [7]. From Torday and Rehan, 2007. Reproduced with permission of Nature Publishing Group.

It is paradoxical to think of "stasis" and change simultaneously giving rise to novel structures and functions. On this subject, R.G.B. Reid said "If homeostasis is characterized as constancy, and evolution as change, how could the homeostatic condition possibly evolve?" Moreover, homeostatic evolution is dependent on internal selection, which is rejected by evolutionists, though several notables have extolled its merits – among them Aristotle (entelechy), Lancelot L. Whyte (coordinative conditions), Rupert Riedl (burden), and the orthogonalists Adolf Remane and Bernard Rensch. In particular, L.L. Whyte was a major proponent of internal selection. He thought that cells were hierarchically ordered systems, and in his *The Unitary Principle in Physics and Biology* Whyte surmised that "all mutations to new stable patterns may necessarily possess favorable or unfavorable properties in relation to the self-stabilizing organization of the system." But without knowing about internal physiologic regulation at the cellular-molecular level, there was no scientific evidence for such speculation.

Similarly, in 1945 Norman Horowitz rationalized the evolution of biochemical pathways based on metabolic change through interactions between the external and internal environments. He was able to envisage an organism that could not synthesize a particular biochemical substance essential for life, forcing the organism to obtain it from the environment, or become extinct. When that substance in the environment was depleted as a result of reproductive success, those organisms that possessed the last enzyme in the biosynthetic pathway made use of its immediate precursor, converting it to the end product, until the supply of the immediate precursor was also exhausted, and so on and so forth. Only then could those organisms that had the next-to-last enzyme survive, iteratively, until the complete biosynthetic pathway was in place. This inventive, descriptive analysis relies on natural selection, but fails to provide a biologic mechanism for the serial homologies. Essentially, it describes how selection through cellular interactions for homeostasis functions, but without actually determining the nature of the molecular intermediaries selected for by the effects of external and internal stresses.

Now, knowing the cell-molecular basis for embryologic development and homeostasis, we have provided empiric evidence for such interactions between internal and external selection pressures to understand the ontogeny and phylogeny of the physiologic basis for lung evolution. That perspective was facilitated by the discovery of soluble growth factors as the mediators of organogenesis during embryologic development, a principle bypassed by the evolutionists, who favored mutation and selection. This is a historical glitch, since the evolutionists parted company with embryologists (as the forerunners of cell biologists) at the end of the nineteenth century, and have never looked back. Without a working knowledge of cell biology, such a cellular premise for evolution is moot.

Biologic Cell–Cell Signaling is Analogous with Chemical Bonding

In the aftermath of the Big Bang, chemical bonds rapidly formed, creating physico-chemical stability in the Universe. Biologic "cellular bonds" (Figure 12.2) were largely formed after the Cambrian Burst. Chemical bonds form by sharing electrons. Similarly, cellular "bonds" are formed by soluble growth factors and their receptors generating structure and function during embryo development, culminating in homeostatic regulation by the same growth factors. A reduction in entropy is common to both, but with one fundamental difference: inorganic chemical bonds are inert, whereas biologic bonds are generative, regenerative, and evolutionary.

Moreover, when you alter atomic structure by changing electrons, protons, and neutrons, elements change their identities from one form to another. This is the domain of quantum mechanics. In contrast, when biologic homeostasis is altered, it responds and adapts based on its ontogenetic and phylogenetic cell–cell signaling history, which is its "quantum mechanics." So there is a fundamental difference in the ways physics and biology respond to change, the former obeying fixed rules, the latter making up its own rules as it goes along, pragmatic and existential.

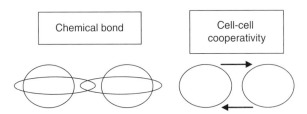

Figure 12.2 Contrasting chemical and cellular bonding. Both bonding processes result in reduction in entropy, but unlike inert chemical compounds, biologic bonds promote development, homeostasis, regeneration, reproduction, and evolution.

Homeostasis as the Universal Underlying and Overarching Mechanism for Evolution

Both inert chemical bonds and biologically active "bonds" result in reduced entropy by moving to a non-equilibrium state. Chemical bonds result in stasis, whereas in the case of biologic bonds the reduction in entropy is generated by chemiosmosis, maintained and perpetuated by homeostasis. Having said that, one major tenet of contemporary evolutionary biology – that there are both proximate and ultimate causes in evolutionary biology – is threatened by viewing evolution as a continuum from a cellular homeostatic perspective. If that precept, which Ernst Mayr used as a canard to maintain the independence of biology from physics and chemistry, could be eliminated it would greatly aid in the advancement of evolution theory, moving it away from dichotomous mutation and selection toward a unified theory of biology.

Homeostasis is not static, it is the dynamic basis for all of life. Therefore, as the site for evolutionary selection pressure, this concept applies universally to all living organisms.

The Ultimate Mechanism of Evolution Transcends Time and Space

We have previously shown that by viewing lung ontogeny and phylogeny from a common denominator of cell–cell interactive paracrine signaling, they are one and the same mechanism of morphogenesis, suggesting that time is superfluous to the understanding of vertebrate evolution – it is an artifact of descriptive biology. Therefore, when the processes of lung ontogeny and phylogeny are merged together, lung evolution is reducible to the facilitation of gas exchange, beginning with the introduction of cholesterol into the cell membranes of unicellular organisms. This property is seemingly exclusive to eukaryotes, since prokaryotes are devoid of sterols; this fundamental structural difference in the cell membranes of eukaryotes facilitated the interactions between the external and internal environments of the cell. You can visualize the unicellular origins of vertebrate evolution by focusing on cholesterol, which Conrad Bloch considered to be a "molecular fossil" – the synthesis of cholesterol and its insertion into the cell membrane rendered the membrane deformable, allowing increased gas exchange due to the thinning of the cell membrane (*respiration*), facilitating both endocytosis and exocytosis (*metabolism*), and enhanced cell movement (*locomotion*) through cytoplasmic streaming. Notably, metabolism, respiration, and locomotion are the three driving forces behind vertebrate evolution. Consequently, the unicellular cell membrane is a functional homolog for the skin, lung, gut, kidney, and brain of metazoans. Functional homologies for all of these complex physiologic structures are linked molecularly through cholesterol utility, either directly (lung – surfactant; brain – myelinization; skin – stratum corneum), or derivatively, through the barrier function for all of these traits. Therefore, the unicellular state is the biologic life form that metazoans are derived from. And the morpho-space that is filled by the biota, like time, is also an epiphenomenon that distracts us from understanding the ur-mechanisms of vertebrate evolution.

Seeking a Universal Language for Biology, Medicine, and Evolution

In the current modality, biologic phenomena are anecdotal, giving rise to descriptive medicine and evolution. As a result, the languages of all three disciplines have constructed a "Tower of Babel." Finding a common language is daunting, yet Thomas Kuhn characterized a paradigm shift as a change in the language. The cellular approach to evolution greatly simplifies this problem, leveling the differences between ontogeny and phylogeny as the common history (short and long, respectively) of the organism, providing a deeper understanding of "how and why" metazoan structures and functions have evolved from their unicellular origins. When physiologic traits are reduced to their cellular and molecular components – independent of species, age, and gender – the differences between them dissipate, allowing for a new perspective on molecular homologies.

The language of cell communication is universal. For example, epidermal growth factor signaling between cell-types as it applies to development, homeostasis, regeneration, and repair – how and why it is used during the history of the organism – can be mapped out, annotated, and integrated with other upstream and downstream signaling pathways, such as Notch, Fox, transforming growth factor beta, Wingless/int, β-catenin, and bone morphogenetic proteins, using the functional phenotypes as templates for ultimately determining the rules of organization.

Genetic Assimilation: A Case in Point

Genetic assimilation is a classic example of how reducing evolution to cellular homeostasis can change our perspective, impacting on the language of evolutionary biology. Genetic assimilation describes how an organism's phenotype can change across environments (phenotypic plasticity), and that selection can operate both on the expression of traits within particular environments, and on the shape of the reaction norm itself. Terms like canalization, genetic landscape (cell–cell signaling), adaptive peaks and valleys (cell–cell signaling for homeostatic set-points), plasticity, adaptive inactivation of the canalizing system under environmental stress, reaction norms, directed preadaptation, non-directed preadaptation, atavism, and the "cost" of plasticity (homeostasis as energetics) all describe environmental effects on phenotypes that assume genes are the underlying determinants of these manifestations. However, the genetics of genetic assimilation are rarely directly examined, either observationally or upon experimental manipulation (gain and loss of function). It is assumed that is because the genetic changes are the result of spontaneous mutations and selection for such mutations. Consequently, all of this terminology is non-mechanistic, derivative, descriptive, *a posteriori* thinking.

In contrast, the cell-molecular homeostatic model for evolution and stability addresses how the external environment generates homeostasis developmentally at the cellular level; determines homeostatic set-points in adaptation to the environment (the reaction norm) through specific environmental effectors (growth factors and their receptors, second messengers, inflammatory mediators, shear stress, biochemistry, mechanotransduction, apoptosis, stem cells, DNA repair, crossover mutation, gene duplication, etc.) that may or may not alter the homeostatic set-point (reaction norm). This is a highly mechanistic, heritable, plastic process that lends itself to understanding evolution at the cellular, tissue, organ, system, and population levels mediated by physiologically linked mechanisms throughout, without having to invoke random chance mechanisms to bridge different scales of evolutionary change, that is, an integrated mechanism that can often be traced all the way back to its unicellular origins.

Utility of the Approach

By adopting the proposed cellular communication approach for "problem solving" in biology and/or medicine, there would be no boundaries, intellectual or otherwise – one would no longer be monolithically restricted to either one discipline, cell-type, tissue, organ, or species; or to just development, homeostasis, pathophysiology, or regeneration. All data related to the question at hand could be brought to bear on the broadest context of biology. This approach would also allow understanding of causality based on *a priori* principles rather than on relativistic phenomena – health as the absence of disease, pathology as signs and symptoms, biologic traits as monogenetic. Biology and medicine would now be based on hard science instead of approximations. On the contrary, how a particular gene forms structure-function relationships in all of these contexts would now be accessible and useful to understanding its ultimate role(s) in homeostasis, physiology, and disease processes. As a result, this *a priori* approach to biology and medicine is predictive, ubiquitous, testable, and refutable. The results of such a comprehensive analysis are universal, durable, and falsifiable. There are so many functional components that could be tested in a wide variety of conditions, both homeostatic and pathologic, to determine if they comply with the predicted parallelisms within and between traits.

This model of biology allows for the consideration of such information as: (i) knockouts that do not produce altered phenotypes resulting from adaptive compensatory mechanisms useful in treating disease; (ii) understanding the phenomenon of cryptic genes that emerge during disease processes as the evolutionary recapitulation of ontogeny and phylogeny in service to retrograde control of homeostasis; and (iii) chronic disease as retrograde evolution, and integrated physiology as the aggregate "history" of the organism. Ultimately this approach would generate a predictive algorithm for functional genomics, proteomics, and interactomics.

Descriptive concepts such as plasticity, evolvability, systems biology, and homology, when looked at from a cellular perspective, are far more comprehendible than when they are seen from the vantage point of superficial description and metaphoric thinking. Plasticity is likely delimited by the range of reaction norms for any given biologic trait; evolvability reflects the nature of the cell-molecular linkage between the external environmental stressor or mutation and the internal deep homology being challenged; systems biology is seen from the cellular perspective on homeostasis, and homologies as cell-molecular properties that sustain and facilitate homeostatic adaptation.

Importantly, this evolutionary model of physiology lends itself to thinking about health and disease as a continuum, rather than as a mutually exclusive dichotomy, revealing the true nature of aging, for example, as an integral part of the life cycle, not merely as a consequence of cumulative pathology. Ultimately, this epistemologic change in our view of biology and medicine would form the basis for bioethics based on logic, rather than on anecdote and subjective speculation. This approach then provides a platform for rational, ethical healthcare policies, and for effective societal resource allocation.

If one starts from the physical Universe, and reduces the problem to biology, you gain one perspective. Conversely, if one starts from the automatous formation of cells, which then provide an internal environment (*milieu*) for the reduction in entropy, you gain a very different perspective. This is not merely a philosophical problem, it determines how we understand our place in the Universe, as either predetermined or having free will. And yet without free will you and I could not contemplate the nature of evolution! That realization carries with it the burden of our ethical responsibility for Nature.

On a grand scale, this formulation lends itself to the creation of a comprehensive algorithm that would functionally link biology, physics, and chemistry together as a robust, interactive, and predictive cipher for all of the natural sciences, not unlike what E.O. Wilson referred to in his book *Consilience* as the consequence of reducing all of the world's knowledge to ones and zeros.

The Cellular "View" of Evolution is Simple(r) and Predictive

The evolution of the lung was a *sine qua non* for the emergence of vertebrates from water to land. Focusing on the biosynthesis of lung surfactant, both developmentally and phylogenetically, has elucidated the cellular-molecular foundations of this process. The development of lung surfactant is the result of mesenchymal–epithelial interactions mediated by soluble growth and differentiation factors and their signaling receptors, such as PTHrP and leptin, and by their mediators, such as adipocyte differentiation related protein (ADRP) and peroxisome proliferator activated receptor gamma (PPARγ). During this process, the walls of the alveoli thin out, generating primary and secondary alveoli, lined by type I and type II alveolar epithelial cells, the latter producing surfactant. At the time of full-term birth, the lung has the capacity to exchange gases across the alveolar wall and remain expanded due to the surface tension-reducing property of the surfactant. Homologous stages of lung structure and function can be seen in the phylogenetic changes from the swim bladder of fish to the lungs of amphibians, reptiles, mammals, and birds. In parallel, the surfactant composition becomes progressively more complex, starting with cholesterol as the simplest surfactant in the swim bladder, followed sequentially by progressively more complex phospholipid mixtures and surfactant apoproteins. Thus, the cellular reduction of the physiology of the alveolus, in combination with the molecular mechanism for surfactant production, reveals the congruence of lung ontogeny and phylogeny in adaptation to atmospheric oxygen for metabolic drive.

The Next-Generation Zygote is the Level of Evolutionary Selection

Contrary to popular belief, recent scientific evidence demonstrates that the epigenetic "marks" in both the gonads and somatic cells accumulate during the life cycle, and are not wiped clean during meiosis. Therefore, evolution is actually a mechanism for gleaning information from the environment during the life cycle to inform future generations, obviously not benefitting the reproductive success of the parent, but for the future success of the zygote-as-adult.

Philosophical questions and quandaries aside, it is interesting to ponder this question, if nothing else, as a way of consciousness-raising about what the evolutionary process actually signifies. The hypothesis that the zygotic or unicellular state is the phenotype being selected for is attractive since it is at that phase of the vertebrate life cycle that the skin, lung, kidney, skeleton, and brain phenotypes all (re)coalesce in the unicellular plasma membrane, both phylogenetically and ontogenetically. The developmental recapitulation of these functions may represent an evolutionary failsafe mechanism for any epigenetic or mutational modifications acquired during the prior life cycle, ensuring the effective, overall vertical integration of all of the homeostatic mechanisms that have evolved from that mutation, by putting it into the ontogenetic/phylogenetic/homeostatic "context" of the developing embryo. This may actually be an atavistic trait, harking back to the original binary fission method of reproduction manifested by our unicellular ancestors for the first 4.5 billion years of life on Earth – nothing succeeds like success. Perhaps this is why "ontogeny recapitulates phylogeny," and Haeckel was right after all! The fact that even the asexually reproducing slime mold, as well as plants, acquire epigenetic marks from their environment encourages us to think that this Lamarckian evolutionary strategy may be universal among biota.

A novel perspective on evolution in which the primary selection pressure is for the zygote rather than the adult stage of the life cycle is analogous to the conceptual shift of the center of the Solar System from the Earth to the Sun. The recognition and recalibration of this and other anthropocentrisms such as the "rights of man" and the "great chain of being" have proven important for the advancement of the human species, though of late our intellectual growth in the wake of the Age of Enlightenment has been challenged by social, nutritional, cultural, technical, and, most recently, climatic forces – all of which are extensions of human evolution as sociobiology. The realization that we humans have merely evolved our big brains as our answer to survival of the fittest is no different from other species evolving eyesight, smell, running ability, flying, swimming, and so forth. All species are the products of their respective environments, which are ever-changing, and as such we are all equals in the eyes of Nature.

The Darwinian Biologic Space-Time Continuum

Darwin saw a continuum of speciation based on principles of natural selection, not the anthropocentric "great chain of being." Darwin's explanation for the biologic patterns he observed was survival of the fittest, which is a metaphor for the evolutionary process, but does not provide a paradigm for drilling down to the cell/molecular origins of life. A causal mechanistic model is essential if we are to take full advantage of the human genome and the genomes of other model organisms. For example, the cell-molecular mechanism of lung evolution based on the ontogeny and phylogeny of the evolution of pulmonary surfactant implies that there is a cellular "continuum" from development to homeostasis and regeneration/repair. This concept of the process of lung evolution, like a cladogram, also implies a vectorial direction and magnitude of change. That perspective is not unlike Einstein's vision as a 16-year-old of traveling in tandem with a light beam through space, which gave him the insight to the physical continuum from Brownian movement to the photoelectric effect and relativity theory in his "miracle year" of 1905. Einstein was severely criticized for his Baconian intuition, yet his thought process has been borne out by Popperian experimentation. The space-time continuum that emerged from that epiphany has similarities to the space-time continuum of lung biology (Figure 12.3); that is, seen from a cell–cell signaling perspective, lung ontogeny, phylogeny, homeostasis, and regeneration (as evolution in reverse) are a series of simultaneous equations that form the continuum of lung evolution as the "solution" for all of the equations. And since this model is largely based on universal developmental principles, the evolution of all other tissues and organs is also amenable to the same analytic approach. This analysis assumes that there are enough "molecular fossil" data to solve the "equation," since mathematically there must be as many variables as equations, hence the utilization of all of the available, relevant phenotypic functional genomic data sets.

Reverse-Engineering Physiologic Traits as a Portal for Viewing Evolution

The rationale for the approach we have taken to evolution is that by tracing the ligand–receptor cell communications that determine the pulmonary surfactant "phenotype" backwards in time and space developmentally, both within and between species, ontogenetically and phylogenetically, we would be able to understand the

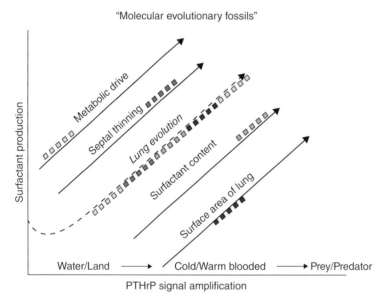

Figure 12.3 Lung evolution expressed as a series of simultaneous equations. Lung evolution (middle line), as the aggregate of cell–cell signaling mechanisms for ontogeny, phylogeny, homeostasis, and repair (evolution in reverse), mediated by parathyroid hormone-related protein (PTHrP) signaling. Phenotypic data sets are expressed as simultaneous equations, forming the continuum of lung evolution as the mathematical "solution." Colored boxes represent strings of incomplete "molecular fossil" data.

cell-molecular mechanisms that have fashioned the lung through external and internal environmental selection pressures. We refer to this as a "middle-out" approach, in contrast to the traditional top-down or bottom-up strategies. As mentioned above, Horowitz had formulated a similar approach to the evolution of biochemical pathways by assuming a retrograde mode of evolution. That approach *describes* the functional phenotype for the evolution of a biosynthetic pathway, whereas the cell-molecular, middle-out approach would provide the *mechanistic* series of ligand-receptor homeostatic mechanisms that determined those biosynthetic pathways from phenotypes to genes, providing a way of understanding physiology from its origins. Selection pressure on such ligand-receptor gene regulatory networks (GRNs) generated both evolutionary stability and novelty through gene duplication, gene mutation, redundancy, alternative pathways, compensatory mechanisms, positive and balancing selection pressures, and so forth. Such genetic modifications were manifested by the structural and functional changes in the blood–gas barrier, primarily its thinning out as a direct result of adaptive phylogenetic changes in the composition and physiologic regulation of lung surfactant, as described eloquently in a series of studies by Sandra Orgeig and Chris Daniels. The reverse-engineering of these phenotypic changes in the blood–gas barrier form the basis for our functional genomic approach to lung evolution.

Communication Between Cells as the Basis for the Evolution of Metazoans

Based on the middle-out cell communication model, how might biologic evolution have begun? One school of thought is that cellular organisms emerged as a result of the wetting and drying of lipids due to the diurnal rhythms of the sun, abetted by the waxing and waning of the moon, which gave rise to micelles, semi-permeable lipid membranes that would have provided a protected environment for the reduction of entropy through enzymatic catalysis. Over the ensuing 4.5 billion years, unicellular organisms have evolved in adaptation to their physical surrounds, eukaryotes evolving from prokaryotes, the former being distinguished from the latter by the presence of a nuclear envelope. That perinuclear membrane evolved to protect the eukaryotic nucleus against invasion by prokaryotes. And the eukaryotic acquisition of mitochondria from prokaryotes, known as the endosymbiosis theory, similarly occurred as a result of the ongoing competition between prokaryotes and eukaryotes. Subsequently, about 500 million years ago, unicellular organisms began cooperating with one another metabolically. For prokaryotes, this took the form of such phenomena as biofilm formation and quorum sensing. Those adaptations, in turn, may have been the positive selection pressure for eukaryotes also to cooperate (or become extinct) by evolving the cell–cell signaling mechanisms that we recognize today as the soluble growth factors and cell-surface cognate receptors that universally mediate morphogenesis, homeostasis, regeneration, and reproduction. However, it should be remembered that this cascade began with a decrease in entropy, defying the second law of thermodynamics, fending off physical forces in the environment (gravity, oxygen, etc.), and eukaryotes fending off prokaryotes, evolving the chemical balance of physiology through homeostasis. But you cannot defy the laws of physics forever. There has got to be some trade-off since matter and energy cannot be created or destroyed. Unicellular organisms are "immortalized" by reproducing through binary fission, whereas eukaryotes evolved sexual reproduction as a means of communicating their genetic information from one generation to the next. Since the overall evolutionary selection pressure for vertebrate evolution is for reproductive success, the distribution of bioenergetics is asymmetrically distributed throughout the life cycle, being biased in favor of the reproductive phase. The trade-off is that the cellular machinery must ultimately fail due to the omnipresence of bacteria, oxidative stress, and all the other environmental forces that initiated the evolutionary strategy. The result is a decrease in "bioenergetics" after the reproductive phase, which we recognize descriptively as aging, resulting in such phenomena as increased oxidative stress, lipid peroxidation, protein misfolding, endoplasmic reticulum stress, and failure of other such metabolic mechanisms, which are thought to be the causes of aging based on the descriptive, top-down approach. In contrast, the cell–cell communication model predicts that the decline in bioenergetics causes decreased cell communication as an energy-requiring process, ultimately culminating in the catastrophic failure of signaling, or death. But the gene pool is "immortalized" by the communication of DNA from one generation to the next. So, in the final analysis, each phase of this perspective on the "how and why" of evolution is one of "cell communication," initially between unicellular organisms and their physical environment, followed by cell communication as the basis for metazoan structure and function, and ultimately reproduction as communication of the environmental knowledge gleaned from generation to generation for adaptation and survival.

Cell Communication as the Essence of Evolution

The metabolic cooperativity that underlies endosymbiosis in the emergence of eukaryotes has evolved from processes to cellular forms that have been recapitulated throughout the evolution of multicellular organisms as phylogeny and ontogeny. Take, for example, the epithelial–mesenchymal interactions that form tissues and organs during embryogenesis. Such interactions are necessary for both the formation of the liver, as well as its homeostatic control of lipids, which shuttle back and forth between stellate cells and hepatocytes. The epithelial-mesenchymal cell–cell interactions that control development and regulation of endocrine tissues such as the adrenals, gonads, prostate, and mammary gland can be viewed similarly.

In the cell–cell communication model of lung development and homeostasis that we have formulated, lipids form the basis for the structural integrity of the alveoli. Lung surfactant, a lipid-protein complex, is produced by epithelial type II cells in the corners of the alveoli. As lung volume increases and decreases during breathing excursions, physical force (or stretch) on the alveoli regulates surfactant production and secretion. The specialized connective tissue cells of the alveolar wall, or lipofibroblasts, actively recruit lipids from the circulation and transfer them to the epithelial type II cells for surfactant phospholipid synthesis.

Lipofibroblast lipid uptake and storage, or neutral lipid trafficking, is mediated by adipocyte differentiation related protein (ADRP), which is under the control of the parathyroid hormone-related protein (PTHrP) signaling pathway. This series of functionally interrelated proteins is expressed compartmentally; PTHrP, surfactant, and the leptin receptor in the epithelium; PTHrP receptor, ADRP, and leptin by the adjacent lipofibroblasts in the alveolar wall. Interrupting this cellular homeostatic cross-talk causes epithelial and mesodermal cells to readapt in a process we recognize as disease.

It is hard to imagine that such a highly integrated and complex cell-molecular communication mechanism could have occurred purely by chance, as proposed according to the neutral theory of evolution: this entails the appearance of an adipocyte-like cell-type in the alveolar wall, flanked by the vasculature on one side, the epithelium on the other, trafficking lipid from the circulation to the alveolar space under stretch-regulation by PTHrP and its receptor, leptin and its receptor, and prostaglandin E_2 and its receptor. We have speculated that for such a sequence of events to have occurred by chance would have taken longer than the 5 billion years that the Earth has existed, if ever. Moreover, it is hard to ignore the fact that the direction and vectorial trajectory of lung ontogeny, phylogeny, and pathophysiology (as reverse evolution) are all consistent with the evolution of this process. And importantly, because these fundamental relationships are linked by specific cellular-molecular mechanisms, the model is experimentally testable and refutable.

Similar chains of events occur in all structures that exhibit such developmental cell–cell interactions. The recognition that ontogeny, phylogeny, physiology, and pathophysiology are a continuum of cell–cell communications infers that such motifs represent the "rules" for the "first principles of physiology," serving as the basis for constructing a biologic "periodic table."

An Integrated, Hierarchical Mechanism for Evolution and Physiology

In the post-genomic era, the biggest challenge we face is to effectively integrate functionally relevant genomic data to determine *physiologic first principles*, and how they can be used to decode complex biologic traits. The consensus is a stochastic systems biology approach based on the bioinformatic premise that if you have not solved the problem, you need more data, analyzing large genomic data sets to identify genes associated with structural and functional phenotypes; whether they are causal or not is seemingly ignored. This approach is an extrapolation from descriptive systematic biology, beginning with Linnaeus's invention of binomial nomenclature.

Systems biology can be viewed at several different levels: the gene, the transcript, the protein, the cell, organ, organ system, or population; evolution can impact the underlying processes at any one of these levels. There are many such descriptive analyses in the literature, but they don't provide hierarchically integrated, functional genomic, evolutionary mechanisms that lead to novel insights to the process, let alone further hypothesis testing, and ultimately to prediction. Those who have attempted such integrations have used either a "top-down" or "bottom-up" approach, but selection pressure – intrinsic, extrinsic, or both – must be applied at a level where it can have the necessary homeostatic effect for adaptation, that is, at the level where the genetic expression is functionally integrated within the phenotype. Based on this precept, we have elected to take a unique "middle-out" approach (in contrast to Sydney Brenner's nominal cellular approach to the problem, which has led him to a project describing the complete inner workings of the cell), focusing on functional nodes defined by ligand–receptor interactions that establish phenotypes during development, sustain them physiologically, and recapitulate them in injury-repair processes and regeneration. Unlike the aforementioned top-down or bottom-up approaches, starting in the middle offers the advantage of minimizing *a posteriori* assumptions by focusing on the GRNs that generate form and function, particularly those that have "evolved" using the same ontogenetic/phylogenetic, homeostatic, and regenerative cell-molecular motifs.

Those vertically integrated, cell-to-functional phenotypic mechanisms that best represent physiology across species and development, particularly the lung, are *archetypes* for the analytic approach being advocated. Incorporating the role of the external forces of natural selection shaping physiology provides a way of understanding the apposition of the lipofibroblast (which recruits and stores neutral lipid) and the epithelial type II cell in the alveolar wall to produce surfactant phospholipid from those neutral lipids. Cell-types derived from different embryonic germlines (mesoderm, endoderm) have evolved to coordinately regulate both surfactant production and alveolar capillary perfusion through a "stretch-regulated" mechanism, from fish to human (Figure 12.4). The rise in oxygen in the Phanerozoic eon gave rise to these alveolar lipofibroblasts (see Figure 1.4), beginning with the observation by Barbara Wold's laboratory that muscle stem cells spontaneously differentiate into adipocytes in 21% oxygen, but not in 6% oxygen, likely due to the effects of oxygen on fibroblast differentiation. And perhaps this phenomenon explains the positive selection for the cytoprotective effect of the neutral lipids stored in lipofibroblasts as an adaptation to the rising atmospheric oxygen tension during the Phanerozoic. This adaptation was followed by the stretch-regulated production of leptin by these cells, perhaps in response to positive selection for endothermy by somatic fat cells – leptin is a molecular homolog of interleukin-6, an inflammatory cytokine thought to have fostered endothermy. The increase in body temperature from 25 °C (ambient temperature) to 37 °C (body temperature) would have rendered lung surfactant 300% more surface-active, leading to selection pressure for a PTHrP stretch-regulated mechanism for the integration of surfactant production and alveolar capillary perfusion, since PTHrP is a potent vasodilator. This cell-molecular series of evolved homologies, coupled together by alternating external and internal selection pressures, is well recognized in conventional descriptive physiology as ventilation-perfusion (V/Q) matching. Furthermore, it is known that the stretch-regulated mechanism for PTHrP expression is intrinsic to alveolar epithelial type II cells, because in a microgravity environment these cells will contract, resulting in decreased PTHrP expression. This trait may have originated as selection pressure for the expression of PTHrP in the fish swim bladder in adaptation to gravity (buoyancy) for efficient feeding; the functional homology between gas exchange for buoyancy and respiration uses the same genes expressed within the epithelium and mesenchyme derived from the esophagus for both structures.

Tiktaalik, the Fossil Evidence for the Vertebrate Water-to-Land Transition: An Object Lesson in Cellular-Molecular Evolution

PTHrP signaling provides the mechanistic basis for the evolution of fish into tetrapods, like Neil Shubin's *Tiktaalik*, discovered in 2004. All of the essential water-to-land adaptations – lung, skin, kidney, gut, and brain – would have been facilitated by a timely gene duplication of the PTHrP receptor that seemingly occurred just as fish evolved into

Oxygen/stretch and positive selection pressure

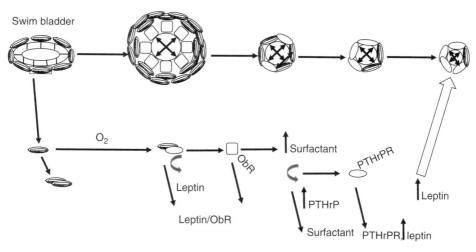

Figure 12.4 External selection for lung evolution. The swim bladder (*far left*) evolved into the vertebrate lung, beginning with atmospheric oxygen stimulating myofibroblast/lipofibroblast (LIF) differentiation; LIF production of leptin; leptin stimulation of surfactant by type II cells having the leptin receptor (ObR) on their surface; stretch upregulation of parathyroid hormone-related protein (PTHrP) by type II cells and leptin by LIFs; thinning of the alveolar wall, and reduction of alveolar diameter, increasing surface area-to-blood volume ratio for increased gas exchange.

amphibians. This event seems fortuitous for vertebrate evolution from water to land, but seen in the context of cellular mechanisms, may have been a direct consequence of the generation of excess oxygen radicals and lipid peroxides, since such substances can be generated locally by vascular wall shear stress in microcirculations like those of the alveolus and glomerulus, causing context-specific adaptive gene mutations. Such extrinsic factors as oxygen and gravisensing implicate population genetic mechanisms for evolutionary selection, underpinned by the cell–cell signaling model for lung evolution.

If adaptation is thought of in the context of internal selection caused by vascular shear stress, the concept of plasticity becomes much more relevant, not to mention being testable, constitutive genes being the ones that were most vulnerable to mutation, since they were the genes being targeted by such selection mechanisms. And perhaps such unconventional internal selection was followed by classic Darwinian population selection for those members of the species that were fittest to regulate those constitutive genes to survive, rendering the newly evolved homeostatic mechanism regulatable. That is precisely what we have observed to have occurred during the process of lung evolution – the transition from PTHrP and leptin as constitutive genes to stretch-regulated genes, both developmentally and in the transition from fish to mammals. Theoretically, this may have been due to the fact that regulated mechanisms would be more resilient, and therefore less likely to generate mutagens than non-regulated constitutive genes. And this may also explain why humans have fewer than the originally predicted number of genes.

There have been numerous attempts to reconstruct biology from its component parts, beginning with Darwin, who was a master at seeing the "forest-and-trees" connections within and between species, and his pronouncement of a mechanism for descent with modification, namely natural selection, which is clever and facile, but not mechanistic in the context of molecular genetics.

Darwinian thought then fostered the works of Ernst Haeckel, Conrad Waddington, Rupert Riedl, Adolf Seilacher, and Stephen Jay Gould, to name only a few of those who have attempted to further our insights to evolution. And more recently, Harold Morowitz and Geoffrey West have gained much notoriety by formulating comprehensive analyses of physiology, but their reductionist/synthetic approaches are similarly *a posteriori* descriptive metabolic pathways and flow patterns in physiology. The problem with these approaches is that they reason from existing structures and functions backwards, consistent with contemporary biology and medicine, *but they do not predict the changes that have occurred over the course of evolution*, even given all the "moving parts," leaving biology as loosely linked anecdotes, and medicine virtually non-predictive and ultimately incomplete in its philosophic and functional scope.

Systems biology is an effort to implement the fruits of reductionism to provide a mechanistic, causal, and explanatory linkage between genotypes and their associated phenotypes by studying the structure and dynamics of convoluted molecular interactive networks that regulate cells and tissues in development, homeostasis, and aging. Systems biology aims at developing and integrating experimental and mathematical techniques in pursuit of principles that would make the nature of cellular phenotypes more intelligible, and their control more deliberate. This effort is motivated by the practical need and desire to cure diseases, or at least make the symptoms go away. However, it also reflects a desire for a theoretical framework by which to deconvolute the complexities of the cell and the organism.

It Takes a Process to Decipher a Process

Evolution is a biologic puzzle. For example, the "solution" for the reassembly of the Dead Sea scrolls was also a puzzle: You are given a box containing the remaining 10,000 fragments of the parchment scrolls. How do you reassemble them based on some mechanism or guiding principle? It takes a process to understand a process, because you need as many equations as variables to solve such complex algebraic problems. The inspiration for the solution to the reassembly of the Dead Sea scroll shards puzzle came from the insight by scientists at the Hebrew University's Koret School for Veterinary Science near Rishon Le Zion that the fragments of each scroll found decades earlier in Qumran were parts of one parchment made from one goat skin. Reasoning in the forward direction, from means to ends, these investigators used molecular biology to identify the fragments that were genetically related to the goat skin from which the parchment was made. So a scriptural puzzle had a biologic solution. The solution to the evolutionary biologic puzzle is even more counterintuitive but must likewise be reasoned from means to ends. The Dead Sea scrolls were reassembled using the DNA signature that was the molecular basis for creating the original parchment. Molecular biology can also be used to decipher physiology, but it must be applied in a way that is consistent with the process being evaluated. The evolution of complex physiologic traits was not an acellular, random, statistical event. It was the result of selection pressure for adaptation to the environment, communicated from generation to generation, over eons.

An Epistemologic "Forest-and-Trees" Problem

Perhaps this unique solution to the reassembly of the Dead Sea scrolls is a cipher that may help us overcome the current stagnation in research in biology and medicine, particularly considering all of the powerful technologies to which we now have access. A recent Blue Ribbon Panel of the American Academy of Arts and Sciences charged with determining how to ameliorate the crisis in US funding for biomedical research recommended investing in young scientists and in high-risk, high-reward research. But our problem is far more fundamental. It is due to the lack of an effective and accessible algorithm for readily translating genes into phenotypes, or biomolecules into a parchment scroll. The problem is readily apparent compared with the advances in physics over the past 150 years, starting with the Mendeleev version of the periodic table, followed by quantum physics and Einstein's formulation of $E = mc^2$.

Ironically, Darwin set us off in search of our evolutionary origins at about the same time that Mendeleev formulated his periodic table of elements. That contrast is now underscored by the publication of the draft human genome in 2000, which sorely lacks an algorithm like the periodic table to convert genes into phenotypes.

To some extent, the failure to advance biomedicine is due to the false hope raised in the wake of the Human Genome Project (HGP) by the promise of systems biology as a ready means of reconstructing physiology from genes. Like the atom in physics, the cell, not the gene, is the smallest completely functional unit of biology. Trying to reassemble GRNs without accounting for this fundamental feature of evolution will result in a genomic atlas, but not an algorithm for functional genomics. Indeed, the reductionist premise of systems biology is reflective of a recurrent pattern in evolutionary biology, vacillating between genes and phenotypes over its stormy history, failing until recently to show that morphogenetic fields exist experimentally, and how they do in fact generate structure and function. The scientific validity of morphogenetic fields has been borne out by contemporary molecular embryology, beginning with the breakthrough discovery of homeobox genes, demonstrating the homologies across phyla first predicted by Étienne Geoffroy Saint-Hilaire in the nineteenth century.

A Path Through the Forest and Trees

Our intention is largely to convey a mechanistic approach for understanding the "first principles of physiology" based on evolutionary precepts, one that challenges the prevailing descriptive paradigm. We are motivated by recently published novel insights to the cell-molecular mechanisms of lung evolution, which are a common denominator for the cell–cell signaling mechanisms of embryogenesis, homeostasis, and regeneration.

The evolutionary literature is replete with metaphors that have sustained interest in this esoteric, hermeneutic topic for decades. But such metaphoric thinking has bogged evolutionary biology down in description ever since Darwin first coined the term natural selection to provide a proximate mechanism for evolution.

There have been many attempts to systematize the formation of complex physiologic systems. For example, Walter Cannon formulated the concept that biologic systems were designed to "trigger physiological responses to maintain the constancy of the internal environment in face of disturbances of external surroundings," which he termed homeostasis. He emphasized the need and goal for reassembling the data being amassed for the components of biologic systems into the context of whole organism function. Hence, Ewald Weibel, C. Richard Taylor, and Hans Hoppeler tested their theory of symmorphosis, the idea that physiology has evolved to optimize biologic function. Harold Morowitz is a proponent of the concept that the energy that flows through a system acts to organize that system. Geoffrey West, James Brown, and Brian Enquist have derived a general model for allometry, including a mathematical model demonstrating that metabolism complies with the $M^{3/4}$ rule. Horowitz has suggested that all of biochemistry can be reduced to hierarchical networks, or "shells." The significance of all of these observations is that the investigators acknowledge that there are fundamental rules of physiology, but they do not address how and why they have actually evolved. Here we apply the mechanism of cell–cell signaling in sustaining and perpetuating homeostasis, starting with the reduction in intracellular entropy as the organizing principle for metazoan evolution.

Even to the untrained observer it is intuitively obvious that there are patterns of size and shape in biology. Darwin was a master at tracing these patterns, and formulating a process by which they may have evolved through descent with modification, as well as a descriptive mechanism, natural selection. But unfortunately such metaphors are grossly inadequate in the age of genomics since they do not generate testable, refutable hypotheses. Without an understanding of how and why evolution has occurred, we cannot take advantage of the underlying principles, particularly as they might apply to human physiology and medicine. This problem arises over and over again in various ways that are referred to euphemistically as "counterintuitive," which is an expedient way of dismissing observations that cannot be explained using the prevailing descriptive paradigm, particularly when the illogic of biology becomes the rule, not the exception. For example, why is it that organ systems have coevolved by linking lipid metabolism and respiration (alveolar surfactant and gas exchange), photoreception and circadian rhythms (the pineal as the "third eye"), blood volume control and erythropoiesis, or why ear ossicles evolved from fish jaws? This may be due to the lack of a functionally relevant perspective on the process of evolution.

Alternatively, with the aid of genomics as the basis for biologic analyses, we have reconsidered the process of evolution from a cellular-molecular signaling perspective, because that is where this process emanated from and evolved to. Such a Kuhnian paradigm shift would allow us to distinguish "forest and trees," and how an understanding of the evolution of structure and function lends itself to the application of genomics to medicine. It seems intuitively obvious that there are fundamental commonalities between ontogeny and phylogeny, given that both start from single cells and form progressively more complex structures through cell–cell interactions mediated by growth factors and their cognate receptors. By systematically focusing on such cell-molecular developmental mechanisms as serial homologies across vertebrate classes, as implied by cladograms, it ultimately may be possible to determine the mechanisms of evolution.

The networks of genes that derive from the proposed algorithm can be used to generate a self-organizing map, offering dynamic new ways of thinking about how the genomic "elements" of physiologic systems are recombined and permutated through evolution to generate novelty based on cellular principles of phylogeny and development, rather than on static descriptions of structure and function. This is analogous to the periodic table being constructed based on atomic number as an independent "self-organizing principle" for the physical elements. And like the periodic table of elements, which predicts new elements, the biologic algorithm will predict novel GRNs. Ultimately, this biologic space-time hologram will reveal the underlying rules for the "first principles of physiology." Our laboratory has devised several models with which to test this evolutionary cell-molecular concept: the developing rat and mouse, the embryonic chick, and the *Xenopus* tadpole. These models offer a concerted developmental and phylogenetic approach for determining specific functional GRNs across phyla.

Ontology

Life forms have inhabited the Earth for billions of years, starting with primitive cells, which gave rise to unicellular organisms over the course of the first 4.5 billion years of the Earth's existence. Evolution of progressively more complex biologic organisms began simply by entraining entropy, the "vital" force, followed by endosymbiosis, and subsequent internal selection for the organelles that furthered homeostasis in defiance of the second law of thermodynamics. The diversification of unicellular organisms fostered metabolic cooperativity, mediated by ligand–receptor interactions for cell–cell communication. The ensuing selection mechanism generated increasing complexity, assuming that there is no difference between the proximate and ultimate causes of evolution, since they are both founded on the same physical laws and principles. Conversely, failed homeostatic signaling causes structural simplification, consistent with "reverse" phylogeny and ontogeny, suggesting that pathology is the inverse of phylogeny and ontogeny.

How life on Earth actually began can only be speculated, unless we can witness it unfolding on "other Earths." And even if we could do so, the process might differ qualitatively and quantitatively from what transpired on Earth since it was contingent on the initial conditions, catalyzed by such external forces as the angle of the Earth's axis, and the formation of the Moon – like the ouroboros, a snake catching its own tail as a symbol of cyclicity in the sense of something constantly recreating itself, and other events perceived as cycles that begin anew as soon as they end.

Aleksandr Oparin was the first to formally conceptualize the beginnings of life on Earth, followed shortly thereafter by John Haldane. They speculated that the early Earth's environment lacked atmospheric oxygen, so a variety of organic compounds could have been synthesized in reaction to energy from the Sun, and by electrical discharges generated by lightning. Haldane thought that in the absence of living organisms feeding on these putative organic compounds, the oceans would have attained a hot, soupy consistency.

Metabolic theories for the origins of life, such as those of Oparin and Sidney Fox, assume the existence of a primitive cell-like form, or protocell, in which metabolism may have emerged. Metabolism caused the growth of the cell and its division into daughter cells when its physical limits of gas and nutrient exchange had been reached or surpassed. One way in which cellular life has been postulated to have originated was through the well-known process by which the repeated wetting and drying of lipids naturally generates liposomes, which are lipid spheres composed of semi-permeable membranes. Perhaps this occurred on the shores of the primordial oceans, with waves depositing lipids derived from plant life at the water's edge (algae have been around for 3.5 billion years and are composed of as much as 73% lipid); this material was wetted and dried again and again, repeatedly over eons. Within these primitive cells, catalytic reactions that would have reduced entropy within them could have resulted from random interactions between molecules generated by the electrical discharges during thunderstorms passing through the primordial atmosphere.

Stanley Miller and Harold Urey were the first to test this hypothesis by passing an electrical charge through a reaction vessel containing water, methane, ammonia, hydrogen, and carbon monoxide, modeling the composition of the prebiotic Earth atmosphere. After running the reaction for several days, the vessel was opened and the reaction products were analyzed. They identified a wide variety of organic compounds, including amino acids (the building blocks of proteins), sugars, purines and pyrimidines (the building blocks for DNA), fatty acids, and a variety of other organic compounds, inferring that the conditions in the primitive Earth atmosphere gave rise to the origins of life.

Günter Wächtershäuser refined this concept by suggesting that chemical reactions may have taken place between ions bonded to a charged surface. Advocates for this school of thought maintain that the emergence of such a structure that walled itself off from its environment by a membrane gave rise to the partitioning between life and non-life. Membrane proponents focus on the primordial role of lipids in this process, and the fundamental role of membranes in the conversion of light energy into chemical, electrical, or osmotic energy, fostering the growth of protocells through metabolic processes within them. Morowitz suggested that the prebiotic environment contained hydrocarbons, some of which were composed of long chains of carbon and hydrogen. These compounds accumulated on the surface of the ocean, where they interacted with minerals to generate amphiphiles, such as phospholipids, which are molecular dipoles: one end is hydrophilic and the other end is hydrophobic. These molecules condensed into various structures, including mono- and bilayers, or lipid sheets. Amphiphilic bilayers spontaneously form spheres in an aqueous solution, with the polar heads of the two layers pointing outward into the adjoining aqueous phase. The non-polar ends of the bilayer point inward toward the center. This is the basic structure of biologic membranes that form the outer surfaces of all cells, allowing active transport of chemicals across the membrane in conjunction with proteins interspersed in the lipid bilayer.

The synthesis of closed vesicles is the origin of triphasic systems consisting of a polar interior, a non-polar membrane core, and a polar exterior. Morowitz empirically demonstrated that the advent of life processes depended on the properties of amphiphilic vesicles, such as non-polar chromophores absorbing light energy, causing them to dissolve in the lipid core of the membrane, where light energy is converted into electrical energy, driving a variety of chemical reactions, including the generation of even more amphiphiles. In modern-day cells, such reactions are mediated by phosphate bond energy, whereas in their primitive condition these reactions were facilitated by pyrophosphates.

Generation of new amphiphiles through this mechanism increased the vesicle size. Once the vesicle reached a critical size, it fragmented into smaller, more stable vesicles in the same way that soap bubbles do. This process is thought to be the origin of cell division.

Pleiotropy as a Rubik's Cube

Erno Rubik invented his eponymous cube (Figure 12.5) in 1974 as a way of teaching his students about spatial relationships and group theory. By twisting the multicolored cube, you can generate 4×10^{19} permutations and combinations of green, yellow, white, orange, red, and blue squares in space and time. Similarly, as a zygote "twists and turns" in biologic space and time it ultimately generates hundreds of different cell-types to form the human body; moreover, those various cell-types generate tissue-specific homeo-static interactions to accommodate structure and function. The fact that the genes of all the cells are all the same, yet they generate different phenotypes both within and between tissues, is also a "puzzle." The key is that the genes are expressed within a cellular context that confers spatial and temporal knowledge.

Pleiotropy is the expression of a single gene generating two or more distinct traits – much like twisting the Rubik's cube and getting various combinations of colors. In the case of the biologic process, it generates the various cellular phenotypes that compose the body, with equally varied homeostatic interactions. If this mechanism is tracked phylogenetically and ontogenetically, it gives us insight to the mechanisms of evolution, as unicellular organisms gave rise to multicellular organisms under the influence of both internal and external selection pressure. I use the Rubik's cube (see Figure 12.5) as a metaphor for the mechanism of pleiotropy to explain how one gene can affect multiple phenotypes. You will notice that there are histologic images on the faces of the cube associated with different colors. When the cube is twisted to reconfigure the colors, those histologic forms are repermutated and recombined. The implication is that the underlying cellular traits are redistributed as in the process of evolutionary adaptation. The reallocation of genes and phenotypic traits is not due to "random selection," but is dictated by homeostatic constraints within each newly established cellular niche. Furthermore, because the constraints have evolved from the unicellular blueprint, they must be internally consistent both phylogenetically and developmentally; if they are not, they can be compensated for by other genetic motifs, or "silenced." It is such a process that explains why traits are pleiotropically distributed throughout biologic systems.

In the book *Evolutionary Biology, Cell-Cell Communication, and Complex Disease*, we used this pleiotropic property of biology to explain both physiology and pathophysi-ology. In the case of the former, we demonstrated how the alveolus of the lung and the glomerulus of the kidney are homologous structures, even though they seem so func-tionally divergent in principle – one mediating gas exchange between the environment and the circulation, the other mediating fluid and electrolyte balance in the circulation. Yet both of these structures sense pressure, and thereby regulate the homeostasis of the lung and kidney, maintaining overall systemic homeostasis through cellular cross-talk between the epithelial cell and neighboring fibroblast. In the case of the lung, the stretch-regulated gene expression of PTHrP produced by the epithelial type II cell feeds back to its receptors on neighboring fibroblasts to regulate lung surfactant, lowering surface tension to maintain the alveolus in a functionally "open" position; in the case of the kidney, the same PTHrP molecule is produced in the podocytes that line the glo-merular fluid space, regulating the fibroblastic mesangium, the pressure sensor that monitors and determines fluid and electrolyte balance in the circulation.

Similarly, innate host defense genes expressed in the skin and lung account for both protective coloration and asthma. Asthmatic patients have a skin disease known as atopic dermatitis, caused by a mutation in one of the beta-defensin genes of the innate host defense system of the skin; that gene, in turn, determines coat color in dogs, which also develop asthma. The mutation of the beta-defensin in the lung airway causes

Figure 12.5 Rubik's cube. By twisting the multicolored cube, you can generate 4×10^{19} permutations and combinations of green, yellow, white, orange, red, and blue squares in space and time. Similarly, as a zygote "twists and turns" in biologic space and time it ultimately generates hundreds of different cell-types to form the human body; moreover, those various cell-types generate tissue-specific homeostatic interactions to accommodate structure and function. (*See insert for color representation of the figure.*)

asthma; in the skin, the same beta-defensin mutation supersedes the asthma phenotype by providing for both protective coloration and reproductive mate selection.

The mechanistic selection advantage for pleiotropy – when one gene influences multiple traits – has never been determined, though it has been invoked to explain the aging process. George C. Williams first proposed antagonistic pleiotropy as the mechanism of aging in 1957, reasoning that genes that were advantageous prior to reproduction became disadvantageous in later life after the organism had reproduced, causing aging – an interesting concept, but without a specific cellular mechanism to either explain the process, or to test the hypothesis. In contrast, we have postulated that there was positive selection for cell–cell communication to sustain the reduction in entropy generated by unicellular organisms, but that the cost-shift allocating bioenergetic resources to favor reproduction in the earlier phase of life resulted in a breakdown in bioenergetics in later life, causing failure of cell–cell communication as the ultimate process of aging, not the cumulative disease processes associated with aging. There are far more data for the latter than the former, yet the misfolding of proteins and build-up of oxidized waste products may be epiphenomena of aging as the breakdown in homeostasis, masking the true mechanism. The manifold mechanisms of cell–cell communication for homeostasis also entail pleiotropy, providing a mechanistic explanation for the breakdown in aging based on loss of cellular metabolic control, shifting from a "Just So Story" to a scientifically testable mechanism for both integrated physiology and pathophysiology as one continuous, iterative process.

Food Deprivation-Induced Metabolic Syndrome as Proof of Principle

Metabolic syndrome – diabetes, hypertension, obesity – can be induced in rat offspring simply by depriving the mother rat of food during pregnancy. This experiment gives insight to the relationship between the life cycle, reproduction, chronic disease, and aging. Under the banner of descriptive biology, all of these features of biology are seen as independent of one another. Yet when thought of in evolutionary terms, what is the selection advantage of fetal growth retardation and the onset of diseases associated with the process of aging? The key to these interrelationships may lie in the fact that fetal food deprivation also causes precocious puberty in females, likely due to increased body fat. Seen in the light of advanced reproductive capacity, the whole life cycle is compressed, perhaps to speed up the intervening life cycle experienced in a low-abundance food environment. So here we see that the life cycle is subordinated to the rate of transit from zygote to zygote, suggesting that it is the fertilized egg that is the primary phenotype of the life cycle, not the adult, which is being selected for.

In an article in the *New York Times* Sunday Magazine section (December 2, 2012), entitled "Forever and Ever," Nathaniel Rich reported on the so-called immortality of the jellyfish *Turritopsis dohrnii* based on its ability to return to its developmental origins when under stress or injury. In our opinion, this is yet another example of the anthropocentric attitude of humanity toward our place in the Universe, following in the great self-delusional tradition of geocentrism, the "great chain of being" and the "anthropic principle." The supposition that the jellyfish medusa is immortal because it can reverse its life cycle is a conceit. In our recently published book (Torday and Rehan, *Evolutionary Biology, Cell-Cell Communication and Complex Disease*; Wiley-Blackwell, 2012), we make the case for multicellular organisms evolving from unicellular organisms, and that it is the unicellular stage of the life cycle that is the ultimate mediator of evolutionary selection pressure, not the adult. Evolution is the mechanism that mediates the monitoring of the environment and allows us to genetically adapt through epigenetics – the biochemical modification of DNA. We tend to see the adult organism as the focus of evolution because like them, we are the "sentient state" of the life cycle, but that is a projection of human vanity. Unicellular organisms are "immortal" because they divide into daughter cells by binary fission. For example, a slime mold exists as a unicellular organism, but can form colonies when it senses oncoming famine.

Perhaps this perspective explains why there is such huge variation in the length of the adult phase of the life cycle, from 1 day in the case of the mayfly, to thousands of years in the case of the giant sequoia, and why organisms make such a huge investment in their reproductive strategy. It is their means of perpetuating the process of evolution, not immortalizing their DNA. So it is not necessarily a fixed sequence of reproduction, birth, life, and death. There are other ways of sustaining and perpetuating the gene pool.

Fossils, Molecular Clocks, Evolution, and Intelligent Design

In a recently published paper, Matthew J. Phillips traced placental mammals back to a common ancestor using fossil data in combination with molecular sequences. Interestingly, the molecular data indicated an earlier time for the evolution of placental mammals than did the fossil record, which is a common finding in such analyses. Why this consistent disparity between the fossil and molecular histories?

In contrast to conventional thinking, if internal selection is essential to the mechanisms of evolution, such discordance would not only be unsurprising, it should be expected. Yet, because of the disdain evolutionists hold for internal selection, such a scenario is not even mentioned in that study or the accompanying commentary, indicating that it is nowhere on the evolutionists' radar screen. On the other hand, if one thinks of the practical and conceptual implications of internal selection, such disparities between fossil and molecular data would explain punctuated equilibrium, since the "missing" intermediate phenotypes would be "invisible" to the fossil record; recognizing that there are such molecular intermediates would encourage looking at evolution from a functional cellular-molecular perspective, seeking molecular homologies in coevolved species, and/or during embryogenesis, or homeostasis, or regeneration and repair. In order to understand "how and why" form and function have evolved, it is necessary to resolve the branches of the evolutionary tree at the molecular homologic level in order to determine the initiating event for placentation. To dismiss the lack of fossil data as "ghost lineages" circumvents the methodologic problem, evading the opportunity to understand the underlying evolutionary mechanisms involved. In contrast, we have deconvoluted the evolution of the lung by focusing on the interrelationship between lung surfactant synthesis and the cellular events that facilitated the thinning of the gas-exchanger from fish to human, for which there were only "molecular fossils." Unless and until we begin to address such molecular questions by producing "hard predictive evidence," evolution will remain descriptive and vulnerable to unwarranted dismissal. Worse yet, it will prevent us from unearthing the origins of physiology, which would allow us to decipher problems as diverse as disease, aging, and bioethics.

Understanding Lung Evolution Using the Middle-Out Approach

The greatest challenge in the post-genomic era is to effectively integrate functionally relevant genomic data in order to derive physiologic first principles, and determine how to use them to decode complex physiologic traits. Currently, this problem is being addressed stochastically by analyzing large data sets to identify genes that are associated with structural and functional phenotypes –whether they are causal is largely ignored. This approach is merely an extrapolation of the *Systema Naturae* published by Linnaeus in 1735. The reductionist genetic approach cannot simply be computed to generate phenotypes. Evolution is not a result of chance; it is an "emergent and contingent" process, just like the formation of the Universe. As Einstein famously stated, "God does not play dice with the Universe." And ironically, the cosmologist Lee Smolin has applied Darwinian selection to stellar evolution, hypothesizing that there is a mechanistic continuum from elementary particles to the formation of black holes.

In our current and future research environment, we must expand our computational models to encompass a broad, evolutionary approach – as Dobzhansky has famously said, "Nothing in biology makes sense except in the light of evolution." Elsewhere, we have formally proposed using a comparative, functional genomic, middle-out approach to solve for the evolution of physiologic traits. The approach engenders development, homeostasis, and regeneration as a cluster of parallel lines that can be mathematically analyzed as a family of simultaneous equations. This perspective provides a feasible and refutable way of systematically integrating such information in its most robust functional genomic form to retrace its evolutionary origins. Among mammals, embryonic lung development is subdivided into two major phases: branching morphogenesis and alveolization. Fortuitously, Rubin *et al.* have observed that deleting the PTHrP gene results in failed alveolization. The generation of progressively smaller, more plentiful alveoli with thinning walls for gas exchange was necessitated by positive selection pressure for the water-to-land transition. It was during this transition that the PTHrP receptor underwent a gene duplication event, supporting the evolution of the lung, kidney, skin, gut, and brain, all of which exhibit ontogenetic, phylogenetic, and homeostatic co-option of PTHrP signaling for development, homeostasis. and regeneration. Both PTHrP and its receptor are highly conserved, stretch-regulated (adapted for gravity), and form a paracrine signaling pathway mechanistically linking the endodermal and mesodermal germ layers of the embryo with the microvascular capillaries. This observation compels us to exploit this key transitional GRN to further our understanding of physiology based on first principles.

This model transcends lung ontogenetic and phylogenetic principles. PTHrP produced by the lung epithelium regulates mesodermal leptin through a receptor-mediated mechanism. We have implicated leptin in the normal paracrine development of the lung, demonstrating its effect on lung development in the *Xenopus* (frog) tadpole, for the first time providing a functional, cell-molecular mechanism for the oft-described and widely accepted coevolution of metabolism, locomotion, and respiration that form the basis for vertebrate evolution. These experiments have led to the question as to why the lipofibroblast appears in vertebrate lung alveoli, beginning with reptiles: the lipofibroblast, an adipocyte-like mesenchymal derivative of the splanchnic mesoderm, could have evolved as an organizing principle for PTHrP/PTHrP receptor-mediated alveolar homeostasis as follows: muscle stem cells will spontaneously differentiate into adipocytes in 21% oxygen (room air) but not in 6% oxygen, suggesting that as the atmospheric oxygen increased over evolutionary time, lipofibroblasts may have formed spontaneously. Consistent with this hypothesis, Torday *et al.* have previously shown that lipofibroblasts protect the lung against oxidant injury. Leptin is a ubiquitous product of adipocytes, which binds to its cognate receptor on the surface of the alveolar epithelium of the lung, stimulating surfactant synthesis, thereby reducing surface tension and generating a progressively more compliant gas-exchange surface area on which selection pressure could ultimately select for the stretch-regulated PTHrP co-regulation of surfactant and microvascular perfusion. This mechanism could have given rise to the mammalian lung alveolus, with maximal surface area resulting from stretch-regulated surfactant production and alveolar capillary perfusion, thinner alveolar walls due to the apoptotic effect of PTHrP on fibroblasts, and a reinforced blood–gas barrier due to the evolution of type IV collagen. This last feature may have contributed generally to the molecular bauplan for the peripheral microvasculature of evolving vertebrates. And in some individuals this atavistic trait causes Goodpasture's syndrome, characterized by the simultaneous collapse of the alveolus and glomerulus due to failure of basement membrane collagen in both beds. This collagen evolved sometime between the emergence of fish and amphibians through selection pressure for specific amino acid substitutions that rendered it more hydrophobic and negatively charged, physically preventing the exudation of water and proteins from the microcirculation into the alveolar space. Goodpasture's syndrome is an autoimmune disease caused by pathogenic circulating autoantibodies targeted to a set of discontinuous epitope sequences within the non-collagenous domain 1 (NC1) of the α3 chain of type IV collagen [α3 (IV) NC1], referred to as the *Goodpasture autoantigen*. Basement membrane extracted NC1 domain preparations from *Caenorhabditis elegans*, *Drosophila melanogaster*, and *Danio rerio* do not bind Goodpasture autoantibodies, while frog, chicken, mouse, and human α3 (IV) NC1 domains bind autoantibodies. The α3 (IV) chain is not present in worms (*C. elegans*) or flies (*D. melanogaster*), but is first detected in fish (*D. rerio*). Interestingly, native *D. rerio* α3 (IV) NC1 does not bind Goodpasture autoantibodies. In contrast to the recombinant human α3 (IV) NC1 domain, there is complete absence of autoantibody binding to recombinant *D. rerio* α3 (IV) NC1. Three-dimensional molecular modeling of the human NC1 domain suggests that evolutionary alteration of electrostatic charge and polarity due to the emergence of critical serine, aspartic acid, and lysine amino acid residues, accompanied by the loss of asparagine and glutamine, contributes to the emergence of the two major Goodpasture epitopes on the human α3 (IV) NC1 domain, as it evolved from *D. rerio* over 450 million years. The evolved α3 (IV) NC1 domain forms a natural physico-chemical barrier against the exudation of serum and proteins from the circulation into the alveoli or glomeruli, due to its hydrophobic and electrostatic properties, respectively, which were more than likely the molecular selection pressure for the evolution of this protein, given the oncotic and physical pressures on the evolving barriers of both the lung and kidney.

The Cell Communication Model of Evolution Guides Us Backwards from Current to Ancestral Phenotypes

We have more recently overarched the developmental and comparative aspects of the leptin mechanism by applying it to frog lung development. Erica Crespi and Robert Denver had shown that leptin stimulates tadpole limb development, a provocative observation because it provided a pleiotropic mechanism for the evolution of land vertebrates, since metabolism, locomotion, and respiration are the driving forces behind this process. To test the hypothesis that leptin biology might be a working model for vertebrate evolution, we treated *Xenopus* tadpole lungs in culture with frog leptin, and surprisingly, found that it had much the same effect that it does in mammalian lungs – it stimulated thinning of the blood–gas barrier in combination with increased expression of surface-active surfactant phospholipid and proteins. The effects of leptin on surfactant were counterintuitive because the frog lung alveolus, termed a faveolus, is so large and muscularized that it does not require surface tension-reducing activity physiologically. However, leptin stimulates expression of surfactant protein A, which is an antimicrobial protein. This, and the fact that antimicrobial peptides are expressed in the gut and skin, suggests that the original selection pressure was for host defense, which was co-opted for barrier expansion of the foregut respiratory pharynx, lung, and skin. That scenario is of interest in light of our studies of the effects of bacterial infection on lung development. We observed that the bacterial wall constituent lipopolysaccharide (LPS) had the same stimulatory effect on developing lung epithelial cells that leptin has, suggesting that the original stimulus for the epithelial–mesenchymal interaction may have been extrinsic bacterial stimulation, followed evolutionarily by the intrinsic leptin mechanism. These interrelationships may relate back to the swim bladder origins of the lung, since the lung is homologous to the swim bladder of physostomous fish, which have a tube connecting the swim bladder to the gut, like a trachea, creating access to the swim bladder for bacterial infiltration. This is in contrast to physoclistous fish, in which the swim bladder has no physical connection to the esophagus.

Predictive Value of the Lung Cell Communication Model for Understanding the Evolution of Physiologic Systems

Unlike the classic pathophysiologic approach to disease, which reasons backwards from disease to health, the evolutionary-developmental approach reasons from the cellular origins of physiology, resulting in prediction of the cause of chronic disease, as we have shown for the lung, and as Robert Bacallao and Leon Fine have shown for the kidney, specifically stating that "Regeneration seems to follow the same pattern of sequential differentiation steps as nephrogenesis. The integrity of the epithelium is restored by reestablishing only those stages of differentiation that have been lost. Where cell death occurs, mitogenesis in adjacent cells restores the continuity of the epithelium and the entire sequence of differentiation events is initiated in the newly generated cells."

As a proof of principle, we will cite other examples of the fundamental difference between a pathophysiologic and an evolutionary approach to disease. As indicated above, we have found that PTHrP is a stretch-regulated gene that integrates the inflation and deflation of the alveolar wall with surfactant production and alveolar capillary perfusion. PTHrP is classically thought of as a bone-related gene that regulates calcium flux. With this and the stretch effect in mind, we recalled that astronauts develop osteoporosis due to weightlessness. Others have pursued a more conventional pathophysiologic tack to this phenomenon, reasoning that post-menopausal women develop osteoporosis and are estrogen deficient, therefore estrogen replacement would be the appropriate treatment for osteoporosis. We, on the other hand, have tested the hypothesis that microgravity would inhibit PTHrP expression in bone and lung cells. One of us (J.T.) observed a decrease in PTHrP expression by osteoblasts and lung epithelial cells in free fall attached to dextran beads, mimicking $0 \times g$. When these cells were returned to unit gravity, the expression of PTHrP returned to normal levels in both cell-types. We subsequently examined the PTHrP expression by the weight-bearing bones of rats flown in deep space for 2 weeks. Here too, we observed a significant decrease in PTHrP expression compared to ground-based littermate controls. The weightlessness effect was not seen in the parietal bone, which is non-weight bearing, consistent with the effect of unloading of the weight-bearing bones due to $0 \times g$.

We have also taken an unconventional evo-devo biologic approach to chronic lung disease. We have pursued the concept that there is an evolutionary continuum from development to homeostasis and regeneration mediated by soluble growth factors. Based on that approach, we have discovered that the cell communication between the epithelium and mesoderm is critically important for the development and maintenance of the alveolar lipofibroblast, andwhen that signaling mechanism fails, the lipofibroblast "defaults" to its cellular origin as a muscle cell, or myofibroblast, the signature cell-type for fibrosis. This approach has given us insight not only to the multifactorial causes of bronchopulmonary dysplasia (BPD) – pressure, oxygen, infection, maternal smoking (nicotine) – but also to a novel treatment for BPD based on the use of PPARγ as the nuclear transcription factor that determines the lipofibroblast phenotype. Thiazolidinediones are potent PPARγ agonists, and we have found that they can prevent or reverse the effects of all of the BPD-inducing agents we have studied, ranging from pressure to oxygen, infection, and nicotine. This evolutionary-developmental approach may be far more successful in the treatment of BPD than more traditional, generic anti-inflammatory agents such as antenatal or post-natal steroids, or prophylactic surfactant therapy.

As indicated earlier, evolutionary selection pressure generated metazoa through cell communication, leading to reproduction, aging, and death. There is accumulating evidence for the loss of cell communication in aging rats to support this perspective on the life cycle. By inference, the selection pressure for reproductive success may optimize for cell communication, and there is accumulating evidence in this regard as well. Therefore, one could devise strategies for "healthy aging" based on this premise, rather than accepting the inevitability of aging as a slow, degradative pathologic process. It has recently been shown, for example, that there is a subset of aging humans who experience a precipitous death rather than experiencing the slow loss of biologic function over years. These data suggest that what we conventionally think of as aging and death is pathology, not evolved biology.

Conclusion

With the aid of the human genome, we must address the evolutionary origins of human physiology based on phylogenetic and developmental mechanisms. The approach we have proposed may not directly identify such first principles because we are missing intermediates from the "molecular fossil record" that failed to optimize survival, resulting in extinction. But some aspects of those "failures" were likely incorporated into other existing functional phenotypes, or into other molecularly related functional homologies, like those of the lung and kidney, photoreceptors and circadian rhythms, and the crystallins of the lens of the eye and liver enzymes. What this approach does provide is a robust means of formulating refutable hypotheses to determine the ultimate origins and "first principles" of physiology by providing candidate genes for phenotypes hypothesized to have mediated evolutionary changes in structure and/or function. It also forms the basis for predictive medicine and even for bioethics – this is an epistemologic cause-effect problem. For if one starts from the chemistry of the Universe, and reduces the problem to biology you get one perspective. If one starts from the spontaneous formation of cells, which then provide an environment for the reduction in entropy, you gain a very different perspective. This is not merely a philosophical problem, it determines how we see ourselves in the Universe, either as having free will or as being determined by physics and chemistry.

This Chapter proposes a paradigm shift in evolution theory toward a cellular perspective founded on homeostasis as the mechanism for evolution. We have suggested that such a shift would revolutionize biology and medicine, extricating us from descriptive biology based on DNA as its mechanism, toward cellular homeostasis mediated by cell–cell interactions. Such an approach would be in keeping with contemporary mechanisms of biology and medicine, leading to a predictive model for each rather than the current post-dictive model, replete with paradoxes and internal inconsistencies. The scale-free simplifying characteristics of a cellular approach to evolution alone would justify its consideration, not to mention its Popperian refutability.

When Copernicus recentered the Solar System on the Sun, humanity began seeing itself differently. Recentering evolution theory on the cell would be a similar game-changer. An effective evolution theory based on its founding mechanistic principles would effectively counter intelligent design, foster novel ideas in biology, and form the basis for predictive and preventive medicine. Moreover, by reducing biology to its principal elements, it can functionally interface with chemistry and physics, providing the opportunity to generate a common database for all of the natural sciences.

Evolution could now interface with all the other biologic disciplines through common ground in cell biology. Inevitably, biology and medicine would also find common ground, extricating us from the current malaise generated by the prevailing view that DNA is the alpha and omega of biology.

This reorientation of evolution theory is revisionist in rehabilitating both Haeckel and Jean-Baptiste Lamarck. Lamarck proclaimed the direct effects of the environment on evolution a few hundred years before scientific evidence for epigenetic inheritance was extant and growing. And from a cellular perspective, Haeckel's biogenetic law – ontogeny recapitulates phylogeny – now makes sense. If a novel epigenetically inherited change should occur during the course of the life cycle, initially expressing it in the developing embryo at the phase where it "fits" into the phylogenetic scheme would allow testing of its effect on homeostasis where and when it was first expressed phylogenetically, and throughout the rest of its historic physiologic roles in the organism now makes sense. This is why, for example, genetic manipulation can result in embryonic lethality in a developmentally relevant, spatio-temporal context, rather than creating "hopeful monsters."

The cellular approach allows for effective utilization of genomic information beyond merely acknowledging that some gene is highly conserved – what does that signify biologically? By putting such genes into their ontogenetic and phylogenetic contexts, one can trace their evolutionary homologs back to progressively earlier stages as homeostatic mediators. Seeing genes from this vantage point offers the opportunity to integrate them with other genes algorithmically. Once a critical mass of such data is accumulated, other such physiologic relationships will be predicted so that we don't have to laboriously and painstakingly identify each and every component, much the same way that the periodic table of elements is predictive.

In his book *The Structure of Scientific Revolutions*, Thomas Kuhn stated that a scientific paradigm shift is marked by a change in the language. In the course of this chapter we have shown examples of how the current language of evolutionary biology might change as a result of recentering it on cellular homeostasis. Noam Chomsky referred to such simplification of language theory as "explanatory adequacy." John Maynard Smith and Eors Szathmáry consider language to be the epitome of evolutionary biology, so we have come full circle.

Unlike Kipling's *Just So Stories* about how and why the leopard got its spots, the rhinoceros got its tough skin, or the camel got its hump, the cell–cell signaling model of physiologic evolution is not a tautological "just so story." It is based on mechanisms of cell-molecular embryogenesis, linked to phylogenesis through homeobox genes, for example. It is predictive for chronic disease, as we have shown for the lung. And as proof of principle, we have been able to effectively prevent the chronic lung disease of the newborn, bronchopulmonary dysplasia, experimentally, based on principles of lung cell-molecular evolution. Furthermore, this model of physiologic evolution transcends lung biology, for example, by providing a mechanistic, evolutionary link between the lung and kidney in Goodpasture's syndrome, which may be an atavistic trait that was beneficial for a rapid biologic "answer" to the stress of the water-to-land transition by selecting for an isoform of type IV collagen that was water-resistant, but was usurped by subsequent physiologic adaptations of the glomerulus. Further investigation of Goodpasture's syndrome patients might discover some balancing selection that could be exploited, for example. So, like the solution to the Dead Sea scrolls, by using common "threads" in the evolutionary fabric of biology, we can solve such complex problems.

Three thousand years of descriptive biology and medicine has brought us to the threshold of predictive molecular medicine. Now, aided by our knowledge of the human genome, we must address the evolutionary origins of human physiology based on the fundamental commonalities between phylogenetic and developmental mechanisms. The approach we have proposed may fail to readily identify such first principles because we are missing critical intermediates from the "molecular fossil record" that failed to optimize survival. But some vestiges of those "failures" were likely incorporated into other existing functional phenotypes or into other molecularly related functional homologies, like those of the lung and kidney, photoreceptors and circadian rhythms, and the lens and liver enzymes. What this approach does provide is a robust means of formulating refutable hypotheses to determine the ultimate origins and "first princi-ples" of physiology, forming the basis for predictive medicine, rather than merely show-ing associations between genes and pathology, which is unequivocally a "just so story." In this new age of genomics, our reach must exceed our grasp. My hope is to engage you in this new approach to understanding physiology, our "Dead Sea scroll," by tracing the regulatory pathways affecting the basic operating unit for all of biology, the cell.

In the current Chapter we have expanded on the cellular approach to evolution using additional evidence to exemplify and reinforce the value added in this approach. Chapter 13, entitled "On eliminating the subjectivity from biology: predictions," addresses the value added in seeing physiology and evolution from their origins.

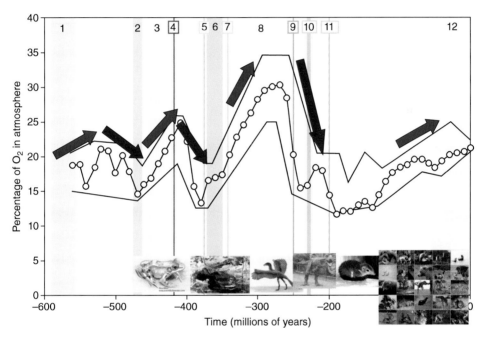

Figure 1.4 Phanerozoic oxygen, hypoxia and physiologic stress. Over the course of the last 500 million years, called the Phanerozoic eon, atmospheric oxygen levels have varied between 15% and 35%, fluctuating fairly drastically. The increases have been shown to foster metabolic drive, but the physiologic effects of the decreases have been overlooked. Hypoxia is the most potent physiologic agonist known, so the episodic decreases in atmospheric oxygen would have fostered remodeling of the internal organs.

Evolution, the Logic of Biology, First Edition. John S. Torday and Virender K. Rehan.
© 2017 John Wiley & Sons, Inc. Published 2017 by John Wiley & Sons, Inc.

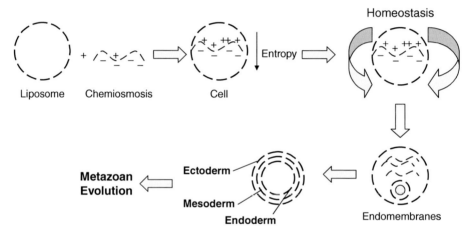

Figure 2.1 On the evolution of form and function. Starting with the formation of micelles and chemiosmosis, life began as the reduction in entropy, maintained by homeostasis. The generation of endomembranes (nuclear envelope, endoplasmic reticulum, Golgi) created intracellular compartments; compartmentation gave rise to the germlines (ectoderm, mesoderm, endoderm), which monitor environmental changes and facilitate cellular evolution accordingly. This is the basis for "fractal physiology."

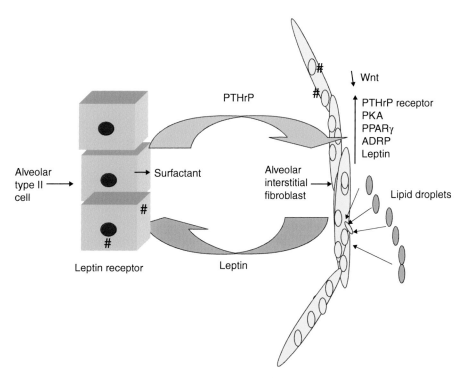

Figure 3.1 Stretch-activated increase in lung surfactant production. Parathyroid hormone-related protein (PTHrP), secreted by the alveolar type II (ATII) cell, binds to its receptor on the adjoining alveolar interstitial fibroblast, activating the protein kinase A (PKA) pathway, which actively downregulates the default Wnt pathway and upregulates the adipogenic pathway through the key nuclear transcription factor, PPARγ, and its downstream regulatory genes ADRP (adipocyte differentiation-related protein) and leptin. Lipofibroblasts in turn secrete leptin, which acts on its receptors on ATII cells, stimulating surfactant synthesis.

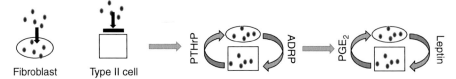

Figure 3.2 Experimental evidence for neutral lipid trafficking. Monolayer cultures of lung fibroblasts actively take up neutral lipids, but do not release them unless they are co-cultured with type II cells (right two images). Type II cells cannot take up neutral lipid (second image from left) unless they are co-cultured with lung fibroblasts. Fibroblast uptake of neutral lipid is determined by parathyroid hormone-related protein (PTHrP) from the type II cell, which stimulates adipocyte differentiation-related protein (ADRP) expression by the lung fibroblast; stretching co-cultured lung lipofibroblasts and type II cells increases surfactant synthesis by coordinately stimulating prostaglandin E_2 (PGE$_2$) production by type II cells, causing release of neutral lipid by the lipofibroblasts, and leptin secretion by the lipofibroblasts, which stimulates surfactant phospholipid synthesis by the type II cells.

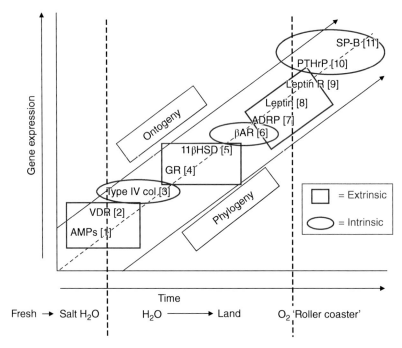

Figure 9.3 Alternating extrinsic and intrinsic selection pressures for the genes of lung phylogeny and ontogeny. The effects of the extrinsic factors (salinity, land nutrients, oxygen) on genes that determine the phylogeny and ontogeny of the mammalian lung alternate sequentially with the intrinsic genetic factors, highlighted by the rectangles and ovals, respectively. **[1]** AMPs = antimicrobial peptides; **[2]** VDR = vitamin D receptor; **[3]** type IV col = type IV collagen; **[4]** GR = glucocorticoid receptor; **[5]** 11βHSD = 11β- hydroxysteroid dehydrogenase; **[6]** βAR = β-adrenergic receptor; **[7]** ADRP = adipocyte differentiation related protein; **[8]** leptin; **[9]** leptin R = leptin receptor; **[10]** PTHrP = parathyroid hormone-related protein; **[11]** SP-B = surfactant protein-B. Along the horizontal axis major geochemical events that caused the cell-molecular changes in lung evolution are depicted: Fresh → salt H_2O = transition from fresh to salt water; H_2O → land = water-to-land transition; O_2 "roller coaster" = fluctuations in atmospheric oxygen tension over the last 500 million years. From Torday, 2013. Reproduced with permission of the American Physiological Society.

Figure 11.2 Positive selection for parathyroid hormone-related protein (PTHrP) signaling during vertebrate evolution. Starting with adaptation to the physical environment by the unicellular cytoskeleton, skeletal adaptation mediated by PTHrP was key to vertebrate water-to-land transition. Positive selection for PTHrP signaling was subsequently adaptive for hypoxia, stimulating the pituitary-adrenal axis for adrenaline production, relieving the constraint on the alveoli by stimulating surfactant secretion. ACTH, adrenocorticotropic hormone.

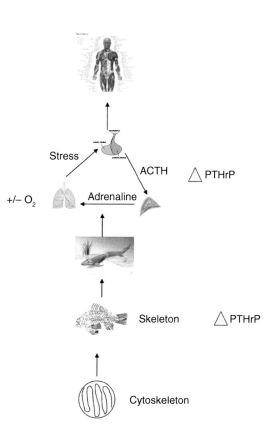

Figure 12.5 Rubik's cube. By twisting the multicolored cube, you can generate 4×10^{19} permutations and combinations of green, yellow, white, orange, red, and blue squares in space and time. Similarly, as a zygote "twists and turns" in biologic space and time it ultimately generates hundreds of different cell-types to form the human body; moreover, those various cell-types generate tissue-specific homeostatic interactions to accommodate structure and function.

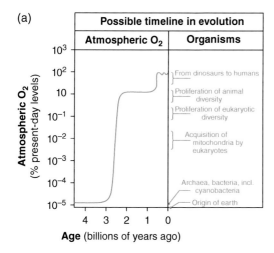

(a)

Possible timeline in evolution	
Atmospheric O₂	Organisms

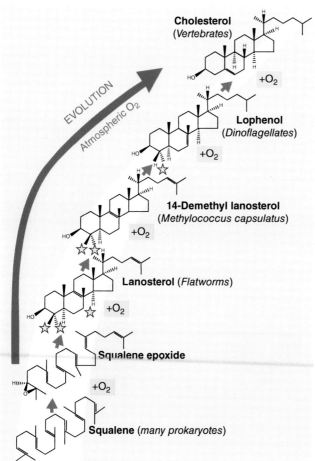

(b)

Figure 18.3 Cholesterol arose in response to a rising oxygen level in the atmosphere. Large amounts of oxygen are necessary for the biosynthesis of cholesterol. **a)** Atmospheric oxygen over the last 5 billion years regressed against corresponding life forms. **b)** The relationship between atmospheric oxygen, and the step-wise evolution of cholesterol biosynthesis.

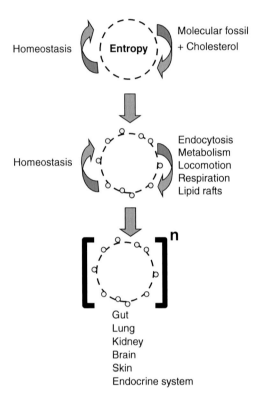

Figure 18.5 Cholesterol in the unicellular cell membrane and evolution of metazoans. The advent of cholesterol in the cell membrane fostered eukaryotic unicellular evolution, followed by multicellular evolution by giving rise to lipid rafts and the endocrine system.

Homeostasis — Entropy — Molecular fossil + Cholesterol

Homeostasis — Endocytosis / Metabolism / Locomotion / Respiration / Lipid rafts

$\left[\right]^n$

Gut
Lung
Kidney
Brain
Skin
Endocrine system

Pleiotropy as a biologic Rubic's cube

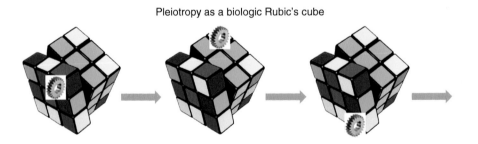

Figure 19.1 Pleiotropy as a Rubik Cube.

13

On Eliminating the Subjectivity from Biology: Predictions

The realization that there are "first principles in physiology," as predicted by the cellular-molecular approach to evolution, is important because of its impact on how we think of ourselves as individuals and as a species, and of our relationship to other species, and to the Earth itself as Gaia. Once we recognize and understand that we have evolved from unicellular organisms, and that this is the case for all of the other organisms on Earth, including plant life, we must accept the fact that we are all interrelated. This kind of thinking has been available in the form of genes that are common to plants and animals alike, but not as part of a larger process of evolution from the physical firmament. This perspective is on par with the reorientation of humans to their surroundings once it was acknowledged that the Sun, not the Earth, was the center of the Solar System, for example. That gave rise to the Age of Enlightenment. In the present day and age, such a frame-shift would provide insight to black matter, string theory, and multiverses.

In retrospect, it should have come as no surprise that we have misapprehended the true nature of our own physiology, given that most discoveries in biomedicine are ser-endipitous and counterintuitive. Medicine is post-dictive, and the Human Genome Project has not yielded any of the breakthroughs that were promised – cures for the diseases that plague humanity. Now, moving forward, we will be able to countenance our own existence as part of the environment, unlike the widely held perspective that validated the anthropic principle (Figure 13.1), that we are in this world rather than being of this world. If what we are about to experience is even more fundamental than heliocentrism, our future holds many prospects thought to be impossible in the old paradigm – telepathic communication, de- and re-molecularization, interstellar space travel, even immortality.

Logic is Subjective: Quantum Logic

"Is logic empirical?" is the title of two articles (by Hilary Putnam and Michael Dummett) that discuss the idea that the algebraic properties of logic may, or should, be empirically determined; in particular, they deal with the question of whether empirical facts about quantum phenomena may provide grounds for revising classical logic as a consistent logical rendering of reality. The replacement derives from the work of Garrett Birkhoff and John von Neumann on quantum logic. In their work, they showed that the outcomes of quantum measurements can be represented as binary propositions, and that these

Evolution, the Logic of Biology, First Edition. John S. Torday and Virender K. Rehan.
© 2017 John Wiley & Sons, Inc. Published 2017 by John Wiley & Sons, Inc.

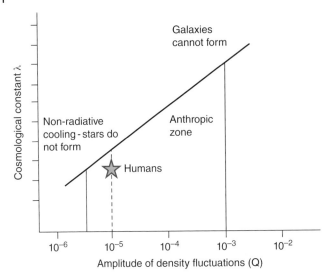

Figure 13.1 The anthropic principle. Graphic depiction of "humans" occupying the space between uninhabitable limits.

quantum mechanical propositions can be combined in much the same way as propositions in classical logic. However, the algebraic properties of this structure are somewhat different in that the principle of distributivity fails. The idea that the principles of logic might be susceptible to revision on empirical grounds has many roots, including the work of W.V. Quine and the foundational studies of Hans Reichenbach.

With the realization that "man is not the measure," what is? Perhaps the answer to this philosophical question will come out of the paradigm shift resulting from the first principles of physiology.

Ad Astra Per Aspera

Vitruvian Man (Figure 13.2) is a well-known drawing by Leonardo da Vinci, completed in 1490. It is accompanied by notes based on the work of the architect Vitruvius. The drawing, which is in pen and ink on paper, depicts a male figure in two superimposed positions with his arms and legs apart and together, simultaneously inscribed within a circle and square. It is an expression of the Renaissance view of Man's place in the (center) of the Universe, a perspective that has persisted for more than 500 years.

In contrast to this, the cellular approach to evolution has formed a continuous physiologic arc from unicellular to multicellular organisms, suggesting our more humble roots deep within the cellular fundament. Understanding the evolution of consciousness, the touchstone for evolutionary physiology, would be proof of principle for this novel approach. Yet it seems intuitively obvious (to us at least) that all organisms, unicellular and multicellular alike, are conscious of their physical environments. Paramecia can be "seduced" by sweets in their proximity, for example. Therefore there must be an ancient biologic "signature" that affects our perception of our surroundings and influences how we perceive them. Once we come to this realization, and can quantify it, we can

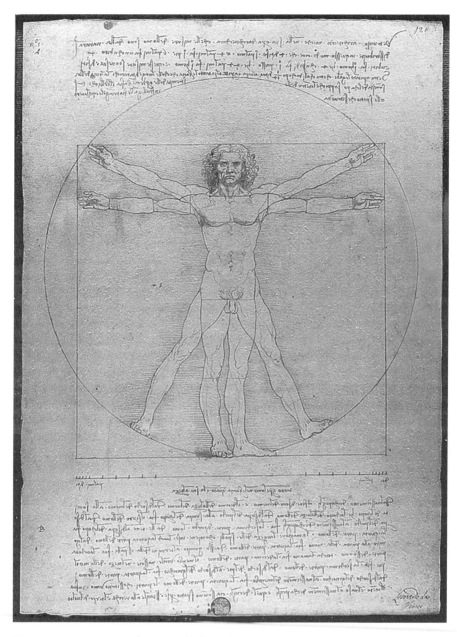

Figure 13.2 Vitruvian Man by Leonardo da Vinci, c. 1490.

free ourselves of that human perceptual subjectivity and more fully appreciate ourselves, our biologic "relatives," and our environment.

Mystics and gurus have been making such claims forever, but without a scientific basis for them. Perhaps by freeing ourselves from our physiologic biases we will be able to understand such phenomena as black matter, string theory, and multiverses. Even better, perhaps, by understanding how we have evolved in reaction to our physical environment we can use it as a manifestation of the physical Universe to iteratively sound the depths of the Cosmos. Carl Sagan told us we were "star stuff," but what does that really mean in biologic terms? This book is dedicated to elucidating what Sagan intuited.

David Bohm and the Implicate Order

David Bohm was a physicist and protégé of Albert Einstein. In his book *Wholeness and the Implicate Order*, Bohm expresses the idea that what we think of as reality is a figment of our evolved senses, which he calls the "explicate order." In contrast to that, he describes a physical field like that which is described for gravitational forces or electromagnetic fields, as the real "implicate order" of things. By analogy, the biologic realm that we think of as reality is derivative of the implicate order.

Predictive Power of the Cellular-Molecular Approach to Evolution

Starting with the unicellular perspective on the life cycle as the primary level of selection pressure, and the necessity of returning to it as the adaptive strategy for epigenetic inheritance, the cellular-molecular approach is highly predictive in comparison to the conventionally dogmatic descriptive view of biology that we have held for thousands of years. The recognition that the unicellular cell membrane is the homolog for all complex physiologic traits forms the basis for understanding the first principles of physiology. And by focusing on the mechanistic transition from the unicellular state to the multicellular organism during both ontogeny and phylogeny, such seemingly insoluble properties of life as pleiotropy, the stages of life, and the aging process can all be understood as one continuous process in service to emergence and contingence.

In the aggregate, this means that the biologic imperative is not for food, water, shelter, and reproduction, which are epiphemonena; it is for emergence and contingence.

The hypothesized evolutionary physiologic interrelationship between stress, metabolism, and endothermy may underlie the effect of meditation on hypometabolism. It has long been known that yogis have the capacity to regulate their metabolism at will, and formal study of this phenomenon has validated it scientifically. Functionally linking to ever-deeper principles of physiologic evolution through meditation and biofeedback may prove to be of wider benefit in healing, both conventional and self-healing alike.

Perhaps more to the point, it has recently been hypothesized that among amniotes, the alveolar lung of mammals may have been the earliest adaptation for land life, followed by its simplification in snakes and lizards. There is no mechanistic basis for such speculation, interesting as this idea is. We have previously pointed out the systematic error being made in showing associations in evolution without offering a causal relationship with environmental factor(s), particularly at the cellular-molecular level, in order to determine relationships to other related evolutionary mechanisms. In that spirit, we have applied the hypothetical role of physiologic stress in mammalian lung evolution to other amniotes with "simple" lungs. As can be seen in Table 13.1, the simple sac-like lungs of other amniotes are associated with a lack of an adrenaline (epinephrine) response to corticoid-mediated stress due to the fundamental difference in the configuration of the adrenal

Table 13.1 Relation between lung phenotype, configuration of the adrenal glands, and catecholamine production. The alveolar lung phenotype, as found in mammals, is associated with an adrenaline (epinephrine)-induced response to corticoid-mediated stress, whereas in other amniotes, with simple sac-like lungs, there is no such response. This is due to the fundamental difference in the configuration of the adrenal glands in mammals vs other amniotes.

Lung phenotype	Adrenal phenotype	Catechol effect
Alveolar	Cortex/medulla	+
Simple	Interspersed	−

glands in mammals versus other amniotes. In mammals, the adrenal cortex is apical to the medulla, and the corticoids secreted by the cortex pass down through the medulla, amplifying adrenaline production by stimulating catechol-*O*-methyltransferase, the rate-limiting step in adrenaline synthesis. In the other amniotes, the chromaffin cells that synthesize catecholamines are interspersed within the cortical tissue, and they are not effectively stimulated by the production of corticoids. Clearly, these organisms found another mechanism to cope with stress, and seemingly as a consequence, their adaptation for breathing as well. The comparators are birds, which have a "stiff" lung composed of air sacs. The lungs are attached to the dorsal wall of the thorax, and this is necessary for air breathing since surgically separating the lung from the chest wall results in death. Furthermore, air entering the lung flows in only one direction, unlike the reciprocating nature of the mammalian lung, suggesting a fundamentally different way of adapting to air breathing in birds. This is confirmed and extended by the fact that birds have blood glucose levels 10–15 times higher than in mammals, suggesting that instead of secreting fatty acids from fat stores in response to adrenaline for metabolic "fuel" on an "as-needed" basis via the fight-or-flight mechanism used by mammals, birds are constantly in a "metabolically always-on" mode.

Moreover, it is noteworthy in the context of metabolism that both birds and humans are bipedal, which may have been a consequence of their both being endotherms. Standing upright is metabolically costly, but by increasing their body temperature in adaptation to a terrestrial lifestyle, both birds and humans have become much more metabolically efficient – cold-blooded organisms require multiple isoforms of the same metabolic enzyme to survive at ambient temperatures, whereas endotherms require only one form. The trade-off may have been bipedalism, freeing the forelegs to evolve into wings and into hands with prehensile thumbs through the exaptation of common genetic traits.

Conclusion

By focusing on the necessity and utility of lipids in initiating and facilitating the evolution of eukaryotes, a cohesive evolutionary strategy becomes evident. In fostering metabolism, gas exchange, locomotion, and endocytosis/exocytosis, cholesterol in the cell membrane of unicellular eukaryotes formed the basis for what was to come.

The basic difference between prokaryotes and eukaryotes is the soft, compliant cell membrane of the latter, interacting with the external environment, adapting to it by internalizing it using the endomembrane system as an extension of the cell membrane. This iterative process was set in motion in competition with prokaryotes, which can emulate multicellular behaviors in the form of biofilm formation and quorum sensing. All of the examples cited in this book – peroxisomes, water–land transition, lipofibroblasts, endothermy/homeothermy – are functional homologs of the originating principle of lipids counterbalancing calcium in service to the evolution of eukaryotes.

Following the course of vertebrate physiology from its unicellular origins instead of its overt phenotypic appearances and functional associations provides a robust, predictive picture of how and why complex physiology evolved from unicellular organisms. This approach lends itself to a deeper understanding of such fundamentals as the first principles physiology. From these emerge the reasons for life cycles and why all organisms always return to the unicellular state, pleiotropy, homeostasis. A coherent rationale is provided for embryogenesis and the subsequent stages of life, offering a context in which epigenetic marks are introduced to the genome.

From the beginning of life, there has been tension between calcium and lipid homeostasis, alleviated by the formation of calcium channels by exploiting those self-same lipids, yielding a common evolutionary strategy. The subsequent rise in atmospheric carbon dioxide, generating carbonic acid when dissolved in water, caused increased calcium leaching from rock. Calcium is essential for all metabolism and it is through calcium-based mechanisms that the inception of life is marked with a calcium spark kindled by sperm fertilization of the ovum, a process that marks the processes of life until the time of death; perhaps the aura chronicled by those who have experienced near-death is that very same calcium spark.

A cohesive, mechanistically integrated view of physiology has long been sought. L.L. Whyte described it as "unitary biology," but the concept lacked a scientifically causal basis, so it remained philosophy. But with the discovery in 1978 of growth factor signaling as the mechanistic basis for molecular embryology, his vision of a singularity may now be realized.

Throughout this book the contrast between conventional descriptive physiology and the deep mechanistic insight gained by referring back to the epistatic balance between calcium and lipids, mediated through homeostasis, has been highlighted. It is characteristic of the self-organizing, self-referential nature described for the origin of life itself. Using this organizing principle avoids the perennial pitfalls of teleology, conversely providing a way of resolving such seeming dichotomies as genotype and phenotype, emergence and contingence, cells and vast multicellular organisms. Insight to the fundamental interrelationship between calcium and lipid homeostasis was first chronicled in our book *Evolutionary Biology, Cell-Cell Communication, and Complex Disease*. Further investigation will solidify the utility of focusing on the advent and roles of cholesterol in eukaryotic evolution, extending from unicellular to multicellular organisms, and provide novel insights to the true nature of the evolutionary continuum in an unprecedented, predictive, and reproducible manner.

This understanding of the how and why of evolution provides the unprecedented basis for a "central theory of biology," which is long overdue. Many have given up on the notion of a predictive model for biology like those for chemistry or physics. This is due to the fact that biology remains descriptive, and that describing a mechanism is not the same as determining causation based on first principles, like quantum mechanics and relativity theory. This is surprising in the wake of the publication of the human genome, which is only 20% of its expected size. That alone should have generated criticism for the prevailing way in which biology is seen, as a *fait accompli*, characterized by correlations and associations. John Ioannidis has declared that "most published biomedical research findings are false." This may be because we are using a descriptive framework, which will not allow for prediction.

The current Chapter addresses the value added in seeing physiology and evolution from their origins. Chapter 14, entitled "The predictive value of the cellular approach to evolution," recapitulates the concept that by starting from the cellular origins of life the seemingly complex, indecipherable physiologic principles can be understood and expanded to all of physiology.

Suggested Reading

Bohm D. (1980) *Wholeness and the Implicate Order*. Routledge & Kegan Paul, London.

Dummett M. (1976) Is logic empirical? In Lewis HD (ed.), *Contemporary British Philosophy*, 4th series. Allen & Unwin, London, pp. 45–68. Reprinted in Dummett M. (1978) *Truth and other Enigmas*. Duckworth, London, pp. 269–89.

Putnam H. (1968) Is logic empirical? In: Cohen RS, Wartofsky MW (eds), *Boston Studies in the Philosophy of Science*, vol. 5. D. Reidel, Dordrecht, pp. 216–41. Reprinted as "The logic of quantum mechanics" in *Mathematics, Matter and Method*, 1975, pp. 174–97.

Reichenbach H. (1944) *Philosophic Foundations of Quantum Mechanics*. University of California Press. Reprinted by Dover, 1998.

Torday JS, Rehan VK. (2012) *Evolutionary Biology, Cell-Cell Communication, and Complex Disease*. Wiley-Blackwell, Hoboken, NJ.

Whyte LL. (1949) *The Unitary Principle in Biology and Physics*. The Cresset Press, London.

14

The Predictive Value of the Cellular Approach to Evolution

The concept that the visceral organs are homologs that evolved from the plasmalemmae of unicellular organisms as the core precept of this book is illuminating. Since the writings of Galen and Harvey we have been exposed to a top-down perspective on physiology akin to plumbing and electrical wiring, rather than to the more insightful biologic principles that fostered Claude Bernard's concept of the *milieu intérieur*, and Walter Cannon's homeostasis. In this age of the human genome, we need a perspective for physiology that is predictive, integrated, and robust, not one where the whole is less than the sum of its parts – less in the sense that its counterintuitive nature is a waste of intellectual energy. The cellular-molecular, ontogenetic, phylogenetic, forward-directed approach that we have promoted in this book offers just such an opportunity. To that end, we provide some examples, as follows.

Observations of Pre-Adaptation in the Evolution Literature are Pervasive, but Why?

As one reads the evolutionary biology literature, observations of pre-adaptations are constantly encountered. Perhaps that is because we are looking at the process of evolution from its ends instead of its means. *A priori*, if one follows pre-adaptation to its logical extension, it culminates in the unicellular state, which is the origin of metazoans, both ontogenetically and phylogenetically. By looking at the process based on this recurrent observation, one sees the processes of ontogeny and phylogeny in reverse. Now, by moving in the forward direction, bearing in mind that the evolutionary changes occurred in the context of the ever-changing environment, the causal relationships become clear, as we have shown for the evolution of the lung (Figure 14.1): by regressing the genes that have determined structure and function during lung ontogeny and phylogeny in the face of major changes in the environment – ocean salinity, the drying-up of the oceans, fluctuations in atmospheric oxygen, as Cartesian coordinates – one can see the adaptive strategy of internal selection due to physical forces, mediated by physiologic stress, starting with the advent of the peroxisome as balancing selection against calcium dyshomeostasis. The lung may be the optimal example, or cipher, for such evolutionary changes in vertebrate visceral physiology because of the powerful selection pressure for its evolution during the water-to-land transition (WLT) – there were no alternatives, it was either adapt or go extinct.

The lesson learned from that event becomes even more self-evident when thinking about the specific implications of the two gene duplications and one mutation of

Evolution, the Logic of Biology, First Edition. John S. Torday and Virender K. Rehan.
© 2017 John Wiley & Sons, Inc. Published 2017 by John Wiley & Sons, Inc.

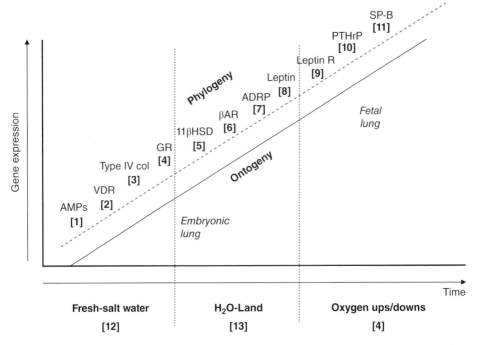

Figure 14.1 The effect of environmental change on the evolution of the lung. The genes involved in the development and phylogeny of the lung are aligned according to both processes, and are regressed against major environmental upheavals, as follows: [1] antimicrobial peptides; [2] vitamin D receptor; [3] type IV collagen; [4] glucocorticoid receptor; [5] 11β-hydroxysteroid dehydrogenase; [6] β-adrenergic receptor; [7] adipocyte differentiation related protein; [8] leptin; [9] leptin receptor; [10] parathyroid hormone-related protein; [11] surfactant protein B; [12] increased salinity; [13] water-to-land transition; [14] atmospheric oxygen swings up and down during the Phanerozoic eon.

receptors known to have occurred during that transition: the parathyroid hormone-related protein receptor (PTHrPR), the β-adrenergic receptor (βAR), and the glucocorticoid receptor (GR). The first of these may have duplicated because it is pleiotropic for air breathing, skeletal support, and the skin as a barrier, all of which were obviously necessary for terrestrial adaptation. And experimentally, if you delete the PTHrP gene from a developing mouse, it results in structural and functional deficits in the lung (no alveoli), bone (failure to calcify), and skin (immature barrier), consistent with all of the aforementioned phenotypes.

The literature would have us think that these genetic mutations occurred by chance alone in association with the WLT because they are not thought of within both an ecologic and a biologic context, instead resorting to the random mutation and selection dictated by Darwinian evolution. But that is far from the case, since vascular shear in response to physiologic stress can cause genetic mutations and duplications. That interrelationship is of particular interest regarding the microvascular beds of visceral organs, because the shearing effects would occur within the structural-functional contexts of specific biologic constraints caused by specific factors in the environment, the internal physiologic "niches" of cells and tissues serving as functions formed by evolution going all the way back to the unicellular origins of vertebrate evolution. And as such, stresses would tend to select for specific adaptive changes over evolutionary time (or the lineage would go extinct) – what Darwinian evolutionists refer to euphemistically as "survival of the fittest," though now with specific, mechanistically testable hypotheses.

Genetic remodeling of the alveolar bed for stretch-regulated PTHrP signaling would have had dual physiologic adaptational advantages, initially by stimulating alveolar surfactant production, relieving the inevitable episodic stress of alveolar insufficiency resulting in hypoxia during the process of evolution. That would have been followed by alveolar demand for PTHrP acting both to generate more alveoli, and as a potent vaso-dilator, accommodating the concomitant increase in alveolar microvascular blood flow in the short-run, and as an angiogenic factor, laying down additional capillaries over the long haul.

Another gene duplication that occurred during the WLT was for the βAR, which ulti-mately provided local pulmonary regulation of alveolar capillary blood pressure, neces-sitated by the constraints imposed by the systemic blood pressure on the alveolar capillary system – think of the capillaries as "circuit breakers." That new physiologic trait may have evolved as a result of the coevolution of PTHrP signaling in both the anterior pituitary and in the adrenal cortex, increasing adrenocorticotropic hormone (ACTH) and glucocorticoid production, respectively, in adaptation to terrestrial physiologic stress. That increased responsiveness to physiologic stress by the pituitary-adrenal axis (PAA) would have amplified adrenaline (epinephrine) production, since the corticoids produced in the adrenal cortex flow out to the circulation through the adrenal medulla, where they physiologically stimulate the rate-limiting step in adrenaline production, phenylethanolamine-N-methyltransferase (PNMT). Moreover, the increased PTHrP flowing through the medulla may actually have fostered the medullary vascular arcade, since PTHrP is angiogenic. This terrestrial adaptation mediated by the PAA may have been brought on by the inevitable episodic pulmonary insufficiencies that would have occurred during the process of lung evolution in adaptation to land, causing intermit-tent periods of hypoxia, the most potent agonist for the physiologic stress reaction. And that adaptation to hypoxia would have been further reinforced by the subsequent fluc-tuations in atmospheric oxygen over the course of the last 500 million years, varying between 15 and 35% during the Phanerozoic eon. The resultant over-expression of adrenaline would have transiently alleviated the physico-chemical constraint on the alveolar wall by stimulating surfactant secretion via the βAR, lowering alveolar surface tension in combination with transiently increasing blood flow due to PTHrP's potent vasodilatory effect. It should be emphasized that all of these pre-adapted physiologic traits were recruited in service to optimized air breathing, and ultimately were selected for by the concerted effects of internal selection in combination with natural selection.

Ultimately, the well-recognized phylogenetic increase in alveolar capillary βAR density, potentially due to the stress-stimulated effect of glucocorticoids on βAR expression, alle-viated the constraint on the circulatory system by allowing for independent regulation of the pulmonary and systemic blood pressures, accommodating the ever-increasing meta-bolic demand for increased lung surface area to facilitate gas exchange based on this allostatic mechanism.

Interestingly, the glucocorticoid receptor also evolved from the mineralocorticoid receptor during this same window of time, due to the addition of two amino acid resi-dues to the mineralocorticoid receptor. The evolution of glucocorticoid signaling from the mineralocorticoid signaling mechanism would have decreased the mineralocorticoid contribution to elevated blood pressure caused by increased gravitational effects due to land habitation, which was constraining the evolution of lung surface area, as indicated above; the concomitant positive selection for glucocorticoids as agonists for βAR expression complemented the evolution of the local alveolar blood pressure regulation in service to increased oxygenation.

As "proof of principle" for the relevance of βARs to land adaptation, deletion of that gene in mice results in simplification of the newborn heart from four chambers to two, which is homologous with a fish heart. This should not be surprising, since the lung and heart evolved in tandem from the one-chambered worm heart, to the two-chambered fish heart, to the three-chambered frog heart, to the four-chambered mammalian heart. The coordination of lung and heart evolution by the PTHrP and βAR gene duplications through the complementary interactions described above would have allowed land habitation and further evolutionary adaptation in response to tandem increases in both gas-exchange and blood pressure regulation under physiologic stress conditions.

This synergistic relationship between PTHrP signaling and βAR signaling may have fostered the formation of the glomerulus from the primitive glomus found in fish kidneys phylogenetically. This evolutionary mechanism is consistent with contemporary mammalian physiology since PTHrP signaling mediates glomerular filtration, and βARs regulate urinary output under stress.

There is no physical fossil evidence for the above-cited visceral organ adaptations during the WLT, but we do know that, based on the skeletal fossil record, there were at least five attempts by vertebrates to adapt to living on land. Such attempts to escape the drying-up of oceans, lakes, rivers, and streams would have been mediated by the increased gravitational effect on land, causing the remodeling and calcification of bone according to Wolff's law – the theory developed by the German anatomist and surgeon Julius Wolff (1836–1902) in the nineteenth century, that bone in a healthy person or animal will adapt to the loads under which it is placed. If loading on a particular bone increases, the bone will remodel itself over time to become stronger to resist the deleterious effects of loading. Increased tension on bone causes increased PTHrP expression locally within the bone, allowing for tissue-specific, conditionally appropriate remodeling and calcification. Such skeletal changes would have been accompanied by changes in the visceral organs, particularly in response to positive selection for PTHrP and βAR signaling, which would have promoted the development and phylogeny of the lung, skin, bone, and kidney, as described above.

Host Defense as a Level of Selection: Lessons from a Frog Experiment

We have implicated the hormone leptin in lung development since it is a product of adipocytes, and lung alveolar lipofibroblasts are homologs. Erica Crespi and Robert Denver had shown experimentally that leptin treatment stimulates *Xenopus* limb development, gaining our attention since the implication of a metabolic hormone in the development of "locomotion" represents two "legs" of the vertebrate evolutionary "stool," respiration being the third "leg." Crespi and Denver had detected the leptin receptor in the frog lung, inferring a hypothetical effect of leptin on *Xenopus* lung development. Upon treatment of the tadpole lung with leptin, we found that it had all of the developmental characteristic effects seen in the mammalian lung – thinning of the alveolar wall, increased surfactant synthesis, and decreased basement membrane thickness. Which was counterintuitive since the frog is a buccal breather, forcing the air down into its lungs, in the absence of a diaphragm. And frogs have faveoli, which are 1000 times larger than alveoli, obviating the need for surfactant to prevent atelectasis since they do not generate high surface tension at the air–liquid interface like alveoli do. On the other hand, we observed that leptin stimulated frog lung expression of surfactant protein A (SP-A), suggesting that perhaps its mechanism of action was relevant to the evolution of the lung from the gut, necessitating increased host defense molecules like SP-A and SP-D. These antimicrobial peptides subsequently facilitated surface tension reduction in combination with surfactant phospholipids to accommodate the alveoli becoming progressively smaller phylogenetically to increase gas-exchange.

PTHrP and Hypothalamic-Pituitary-Adrenal Regulation of Physiologic Stress

The Role of PTHrP Expression in Pituitary/ACTH Regulation

Extensive experimental evidence from our laboratory has demonstrated the central role of PTHrP in normal lung development, beginning with the embryonic deletion of PTHrP causing failed lung development due to failure to form alveoli. In various insult models of lung disease – oxotrauma, barotrauma, infection – we have documented the decrease in PTHrP expression in all of these instances. Moreover, infants who develop bronchopulmonary dysplasia (BPD) are PTHrP deficient based on measurement of the molecule in bronchoalveolar lavage.

The Role of PTHrP Expression in Adrenal Cortex/Corticoid Synthesis

More recently, it has been discovered that PTHrP is expressed in both the pituitary, where it stimulates ACTH, and in the adrenal cortex, where it mediates the ACTH stimulation of corticoid synthesis. This pathway would amplify physiologic stress (see Figure 1.3), increasing adrenaline production by the adrenal medulla. This mechanism may have evolved during the water-to-land transition, during which the lung would episodically have been unable to effectively generate adequate amounts of oxygen to meet the rising metabolic demand of land adaptation, causing hypoxia. Hypoxia is the most potent physiologic agonist known; by stimulating adrenaline production, which causes lung alveolar surfactant secretion, it would have transiently alleviated the stress on the lung alveoli.

The Upregulation of Adrenaline

Such a mechanism may refer all the way back to the era when our rodent-like ancestor had to be nimble in order to avoid being crushed or eaten by predators. Over time the stress on the microvasculature of the lung, pituitary, and adrenal cortex may have "remodeled" all of these structures. This would have included the adrenal medulla, known to have evolved a complex vascular arcade in mammals, acting like an "echo chamber" to enhance adrenaline production in response to stress, both baseline and regulated states.

The Evolution of Peroxisome Biology as a Prime Example of the Utility of "Ancestral Health"

Agonists to peroxisome proliferator activated receptor gamma (PPARγ) have been found either to prevent or effectively treat a wide variety of inflammatory diseases, ranging from the lung to the kidney, liver, and brain. Such findings have been serendipitous, based on clinical observation of patients being treated with thiazolidinediones for type II diabetes. The exception is the use of such compounds in the treatment of BPD, a chronic lung disease of the newborn. Such studies in our laboratory were a rational approach, predicated on the paracrine determination of normal lung development, which is truncated in infants born prematurely. In that circumstance, the failure to develop lipofibroblasts in the alveolar wall is the developmental and evolutionary cause of BPD – lipofibroblasts probably evolved in the alveolar wall to protect it from oxidant injury. The production of leptin by these cells, in turn, facilitated the cell–cell interactions that led to stretch- and hormone-regulated lung development. PPARγ is the molecular determinant of lipofibroblast differentiation, and as such has facilitated lung evolution. It has turned out to be an effective treatment for lung immaturity, showing none of the deleterious effects of glucocorticoids, including the absence of sex-specificity that has been problematic since the inception of antenatal corticosteroid use.

The Elimination of Space and Time from the Analysis of Evolution

Since Linnaeus instituted the discipline in biology of naming plants and animals using binomial nomenclature, the overt activity of biologists has largely been the description of life in space and time. But that enterprise, necessary as it was to determine the breadth and width of the biology, is languishing in the name of progress. Now that we have the complete genomes for humans, mice, fish, birds, worms, and so forth, it is time to address the central questions of how and why the biota evolved. To date, all attempts to do so have failed, and we contend that that is because we are still stuck in the descriptive modality, whether at the genetic or phenotypic level, and everything in between. As we have mentioned elsewhere in this book, there is a fundamental problem in this regard because it is the responsibility of the evolutionists to answer such questions, yet they have eschewed cell biology due to a "historical accident." Over the past decade or so, we have been examining the value added in reintroducing cell biology into the question of how and why visceral organs have evolved. By starting from the putative origins of the cell, and moving forward in space and time, it has become clear that there were specific cellular adaptational changes that have facilitated eukaryotic evolution:

- the spontaneous formation of primitive cells from lipids in water;
- the reduction in entropy within the cell;
- the advent of cholesterol and its incorporation into the eukaryotic cell membrane;
- the entraining of external physical factors (oxygen, nitrogen, minerals) by the endomembrane system;
- the development of homeostatic control of metabolism;
- cell–cell signaling for metazoan evolution.

Once these processes were seen across ontogeny and phylogeny in the context of specific physiologic traits – lung, skin, kidney, bone, brain – it became apparent that there were "first principles of physiology" that have constituted the origins of complex physiology, in contrast to Harvey's version of physiology as a loose association of parts. *Reducing ontogeny/phylogeny to one process through cell biology eliminates "time." The recognition of the cell as the unit of selection eliminates "space."* That perspective has eliminated the overt complexity of physiology as it appears, reducing it to simple unicellular structure/function relationships, as reflected by our perennial return to the unicellular state in the zygote.

The Unicellular Plasmalemma as the Homolog of Metazoan Visceral Organs

By definition, physiology is mechanistically integrated, starting with the unicellular state. This approach is foundational in being consistent with the origins of cellular life, and integrates genotype and phenotype by definition. The existence of a mechanism for determining the fidelity of gene mutations through epigenetics is facilitated by the return of metazoans to their unicellular origins as a means of monitoring, policing, and "editing" such mutations within the context of homeostasis as the criterion for selection, both ontogenetically and phylogenetically, as proposed by Ernst Haeckel in his biogenetic law. Such a systematic, diachronic mechanism would, for example, explain how and why evolution ensures both stability and change. Such a developmental mechanism is far more economical than Darwinian reproductive selection is, for example.

Cholesterol in the Plasmalemma as a Catalyst for Vertebrate Evolution

Starting with the notion that the visceral organs and brain coevolved from the plasmalemma of unicellular organisms, how did such homologies foster integrated physiology? Starting with the insertion of cholesterol in the plasmalemma as the "trigger" for eukaryotic evolution, we have previously speculated that cholesterol caused the vertical integration of physiology from unicellular to multicellular organisms, starting with the cell membrane fostering locomotion, respiration, and metabolism. This triumvirate gave rise to the lipid raft – lipid densities in the cell membrane where cell surface receptors are located – further fostering multicellular organism physiology by providing the structural basis for cell–cell signaling, followed by the endocrine system.

There is a commonality for lipid homeostatic mechanisms in the lung, skin, and brain, all of which increased during the water-to-land transition. Mechanistically, lysosomal activity is common to all three as well, and there are pathophysiologic overlaps between skin and brain through lipodystrophies. There is also a pathway connecting melanocyte receptor 1 (MC1R) to skin and brain, though that linkage is very convoluted. And there are also interconnections through MC2-5R. Since MC1R has been implicated in alopecia, the thought arose that perhaps upregulation of MC1R in skin caused hair loss, and in the central nervous system (CNS) it enhanced brain development.

The Ever-Transcendent Unicellular State

We can trace the arc of the evolution of complex physiologic principles from primitive cells lying at the interface between water and land (i.e., sea foam) to the human brain. The detailed process is easier said than done, yet all of the structural-functional links have formed in service to calcium flux, mediated by the lipids that form a continuum from cholesterol in primitive eukaryotes, to the myelination of neurons. Underpinning all that is the foundational principle of life – fomented by negentropy generated by chemiosmosis and endomembranes, sustained by homeostasis – a squishy organic ball of its own making and devices. This is akin to the Greek metaphor of life, the ouroboros – the snake catching its own tail, self-organizing and self-perpetuating. Viewed from this vantage point, the notion that the unicellular state of the life cycle is the primary site for selection pressure becomes tenable: the life cycle is not stagnant, it has a vectorial direction and magnitude of change, either moving upward or downward as it evolves or devolves towards extinction. If so, what is the initiating event, since this process must be inhomogeneous? Conventionally, we think of the adult form as the determinant of such a mechanism, but that is a facile anthropocentric viewpoint, like geocentrism, and we know where that ended up. Alternatively, we should consider the unicellular zygote, particularly in view of new evidence that epigenetic marks on the egg and sperm are not eliminated during meiosis, as had long been held to be the case. What is the significance of such epigenetic marks, and what determines which ones are retained, and which ones are eliminated? Such considerations are not trivial since only about 1–2% of inherited diseases in humans are Mendelian, leaving a huge vacuum in the outer space of heritable diseases that may be filled by epigenetics. And perhaps this is why we are destined to recapitulate phylogeny during the process of ontogeny. Haeckel's biogenetic law was rejected long ago for lack of experimental evidence that embryogenesis faithfully hit all the phenotypic milestones of phylogeny, but that was before we learned about the cellular-molecular changes that underpin such morphogenetic changes. Such data are referred to in the evolutionary literature as "ghost lineages," but in molecular embryogenesis they are how and why structure and function develop as a continuous process. That knowledge also alludes to the possibility that we need to mechanistically recapitulate phylogeny in order to ensure that any newly acquired epigenetic mutations are in compliance with the homeostatic and allostatic mechanisms that they might affect in introducing them into the organism's gene pool. But in so saying, the unicellular state is the overall determinant and arbiter of this process – the unicellular state rules! And if so, we as a species need to reassess our priorities among our cousins, plant and animal alike.

Importantly, this evolutionary model of physiology also lends itself to thinking about health and disease as a continuum, rather than as a mutually exclusive dichotomy, revealing the true nature of aging, for example, as an integral part of the life cycle, not merely as a consequence of cumulative pathology. Ultimately, this epistemologic change in our view of biology and medicine would form the basis for bioethics based on logic, rather than on anecdote and subjective *a posteriori* reasoning. This approach thus provides a platform for rational, ethical healthcare policies, and for effective societal resource allocation.

Maybe this will lead to bioethics based on first principles rather than on self-serving actions that emanate from our anthropocentric attitudes. We have thrived as a species up until now by using resources that were not exclusively ours. And maybe we need to reconsider Haeckel's dictum in formulating a central theory of biology. And in so doing, reconsider internal selection as an extension of natural selection.

The Predictive Value of Determining Evolutionary Mechanisms from their Origins Instead of their Consequences

Literally all of the evolution literature is based on the descriptive basis for biology and medicine. As a result of this reversed epistemology, the wrong conclusions are arrived at, to put it bluntly. Given that evolution began from the unicellular state, what else would one expect? Yet to our knowledge the problem has never been examined from that perspective. In so doing, by invoking the principles of developmental embryology, starting with the zygote, the mechanisms of growth, development, and homeostasis are arrived at through cell–cell interactions, providing an understanding of how and why such structures evolved, particularly when looked at phylogenetically – why the swim bladder evolved into the lung, the heart evolved auricles and ventricles, the neuroendocrine and endocrine systems evolved to support allostasis.

By starting from the origins of life as derivative of the physical environment, generating negentropy, sustained by homeostatic mechanisms, one can see the continuum from the inert to the vital. Nowhere is this more evident than in the literature on consciousness and mind. It runs the gamut from those who see homology between chemical bonds and biologic "bonds" to the top-down rationalization of mind as the aggregate of the various structures of the brain. It is highly likely that consciousness and mind, like all other evolved structures and functions, evolved as a result of the recombination of pre-existing traits. And such traits were not necessarily neurologic in nature; indeed, they probably were not since (i) they have not been determined, and (ii) the central nervous system of vertebrates evolved from the vermiform skin (skin brain), making it highly likely that there was some other atavistic trait that fostered consciousness and mind. There are other characteristics of humans, such as hairlessness, upright posture, and protracted childhood that may lie at the basis for consciousness and mind.

Humans are Integral with Nature

The traditional perspective for physiology, as portrayed by Galen and Harvey, is like Lego® blocks, with one biochemical process linked to another, until an entire biochemical structure is revealed. In contrast to that *post facto* narrative, a predictive approach can be asserted – there actually are founding first principles for physiology that originated in and emanate from the unicellular stage of life. Einstein's insight to relativity theory emerged from a dream in which he traveled in tandem with a light beam, seeing it as an integral particle and wave. Similarly, viewing physiology as a continuum from unicellular to multicellular organisms provides fundamental insight to ontogeny and phylogeny as an integral whole, directly linking the external physical environment to the internal environment of physiology, and even extending beyond, to the metaphysical realm, bearing in mind that the calcium waves that mediate consciousness in paramecia and in our brains are one and the same mechanism.

Life probably began much like the sea foam that can be found on any shoreline, since similar lipids naturally form primitive "cells" when vigorously agitated in water. Algae, for example, are as much as 73% lipid. Such primitive cells provided a protected space for catalytic reactions that decreased and stabilized the internal energy state within the cell, and from which life could emerge. Crucially, that cellular space permits the circumvention of the second law of thermodynamics. (The entropy of an isolated system such as a unicell never decreases since such systems always decay toward thermodynamic equilibrium as a state of maximum entropy.) That violation of physical law is the essential property of life as self-organizing, and self-perpetuating, always in flux, staying apace with, and yet continually separable from a stressful, ever-changing external environment.

Even from the inception of life, rising calcium levels in the ocean have driven a perpetual balancing selection for calcium homeostasis, mediated by lipid metabolism. Metaphorically, the Greeks called this phenomenon ouroboros (Figure 14.2), an ancient symbol depicting a serpent eating its own tail. The ouroboros embodies self-reflexivity or cyclicity, especially in the sense of something constantly recreating itself. Just like the mythologic phoenix, it operates in cycles that begin anew as soon as they end. Critically, the basic cell permits the internalization of factors in the environment that would otherwise have destroyed it – oxygen, minerals, heavy metals, microgravitational effects, and even bacteria – all facilitated by an internal membrane system that compartmentalized those factors within the cell to make them useful. These membrane interfaces are the biologic imperative that separates life from non-life – "Good fences make good neighbors."

The Advent of Multicellularity

Unicellular organisms dominated the Earth for the first 4 billion years of its existence. Far from static, these organisms were constantly adapting. From them, the simplest plants evolved first, producing oxygen and carbon dioxide that modified the nitrogen-filled atmosphere. The rising levels of atmospheric carbon dioxide, largely generated by cyanobacteria, acidified the oceans by forming carbonic acid, progressively leaching more and more calcium from rock into the ocean waters, eventually forcing a proliferation of life from sea to land. The existence of a protected space within primitive "cells" allowed for the formation of the endomembrane system, giving rise to chemiosmosis, or the generation of bioenergy through the partitioning of ions within the cell, much

Figure 14.2 Ouroboros, an ancient symbol depicting a serpent eating its own tail.

like a battery. Early in this progression, the otherwise toxic ambient calcium concentrations within primitive cells had to be lowered by the formation of calcium channels, composed of lipids embedded within the cell membrane, and the complementary formation of the endoplasmic reticulum, an internal membrane system for the compartmentalization of intracellular calcium. Ultimately, the advent of cholesterol synthesis facilitated the incorporation of cholesterol into the cell membrane of eukaryotes, differentiating them (our ancestors) from prokaryotes (bacteria), which are devoid of cholesterol. This process was contingent on an enriched oxygen atmosphere, since it takes eleven oxygen molecules to synthesize one cholesterol molecule. The cholesterol-containing cell membrane thins out, critically increasing oxygen transport, enhancing motility through increased cytoplasmic streaming; this is also conducive to endocytosis, or cell eating. All of these processes are the primary characteristics of vertebrate evolution. At some point in this progression of cellular complexity, impelled by oxygen promoting metabolic drive, the evolving physiologic load on the system resulted in endoplasmic reticulum stress, periodically causing the release of toxic calcium into the cytoplasm of the cell. The counterbalancing, or epistatic mechanism was the "invention" of the peroxisome, an organelle that utilizes lipids to buffer excess calcium. That mechanism became homeostatically fixed, further promoting the movement of ions into and out of the cell. Importantly, the internalization of the external environment by this mechanism reciprocally conveyed functional biologic information about the external surroundings, and promoted intracellular communication – what Claude Bernard referred to as the *milieu intérieur*. Walter B. Cannon later formulated the concept that biologic systems are designed to "trigger physiological responses to maintain the constancy of the internal environment in face of disturbances of external surroundings," which he termed homeostasis. He emphasized the need for reassembling the data being amassed for the components of biologic systems into the context of whole organism function. Hence, in 1991, Ewald Weibel, Richard Taylor, and Hans Hoppeler tested their theory of "symmorphosis," the idea that physiology has evolved to optimize the economy of biologic function; interestingly, the one exception to this theory was the lung, which they discovered was "over-engineered," but more about that later. Harold Morowitz is a proponent of the concept that the energy that flows through a system also helps organize that system. Geoffrey West, James Brown, and Brian Enquist have derived a general model for allometry (the study of the relationship of body size to shape, anatomy, physiology, and behavior). They proposed a mathematical model demonstrating that metabolism complies with the 3/4 power law for metabolic rates (i.e., the rate of energy use in mammals increases with mass with a 3/4 exponent). Back in 1945, Norman Horowitz hypothesized that all of biochemistry could be reduced to hierarchical networks, or "shells." Based on these decades of study, investigators acknowledge that there are fundamental rules of physiology, but they do not address how and why these rules have evolved.

As eukaryotes thrived, they experienced increasing pressure for metabolic efficiency in competition with their prokaryotic cousins. Via endocytosis they ingested bacteria, which were assimilated as mitochondria, providing more bioenergy to the cell for homeostasis. Eventually, eukaryotic metabolic cooperativity between cells gave rise to multicellular organisms, which were effectively able to compete with prokaryotes. As Simon Conway Morris has archly noted, "First there were bacteria, now there is New York." Bacteria can through such behavioral traits as quorum sensing and through biofilm formation, behave even at this primitive stage as a pseudo-multicellular organism.

The subsequent counter-balancing evolution of cellular growth factors and their signal-mediating receptors in our vertebrate ancestors facilitated cell–cell signaling, forming the basis for metazoan evolution. It is this same process that is recapitulated each time the organism undergoes embryogenesis.

This cellular focus on the process of evolution serves a number of purposes. First, it regards the mechanism of evolution from its unicellular origins as the epitome of the integrated genotype and phenotype. This provides a means of thinking about how and why multicellular organisms evolved, starting with the unicellular cell membrane as the common origin for all evolved complex traits. Further, it offers a discrete direction for experimentally determining the constituents of evolution based on the ontogeny and phylogeny of cellular processes. For example, it is commonplace for evolution scientists to emphasize the fact that any given evolved trait had its antecedents in an earlier phylogenetic species as a pre-adapted, or exapted trait. These ancestral traits can then subsequently be cobbled together to form a novel structure and/or function. Inescapably, if followed to its logical conclusion, all metazoan traits must have evolved from their unicellular origins.

Evolution, Cellular-Style

Moving forward in biologic space and time, how might such complex biologic traits have come about? Physiologic stress must have been the primary force behind such a generative process, transduced by changes in the homeostatic control mechanisms of cellular communication. When physiologic stress occurs in any complex vertebrate, it increases blood pressure, causing vascular wall shear stress, particularly in the microvascular beds of visceral organs. Such shear stress generates reactive oxygen species (ROS), specifically at points of greatest vascular wall friction. ROS are known to damage DNA, RNA, and proteins, and to do so particularly at those sites most affected by the prevailing stress. This can result in context-specific gene mutations, and even gene duplications, all of which can profoundly affect the process of evolution. So it should be borne in mind that such genetic changes are occurring within the integrated structural-functional context of that tissue and organ. However, understanding the biochemical processes undergirding the genetic ones equips a profound and testable mechanism for understanding the entire aggregate of genetic changes as both modifications of prior genetic lineages, and yet "fit enough to survive" in their new form.

Over evolutionary time, such varying modifications of structure and function would iteratively have altered various internal organs. These divergences would either successfully conform to the conditions at hand, or failing to do so, cause yet another round of damage-repair. Either an existential solution was found, or the lineage became extinct; either way, such physiologic changes would have translated into both phylogenetic and ontogenetic evolution. Such an evolutionary process need not be unidirectional. In the forward direction, developmental mechanisms recapitulate phylogenetic structures and functions, culminating in homeostatically controlled processes. And in the reverse direction, the best illustration lies with the genetic changes that occur under conditions of chronic disease, usually characterized by simplification of structure and function. For example, all scarring mechanisms are typified by reversion of fibroblasts to their primordial signaling pathway. This sustains the integrity of the tissue or organ through the formation of scar tissue, albeit suboptimally, yet allowing the organism to reproduce before being overwhelmed by the ongoing injury-repair.

The Cellular Approach to Evolution is Predictive

The reduction of the process of evolution to cell biology has an important scientific feature – it is predictive. For example, it may answer the perennially unsolved question as to why organisms return to their unicellular origins during their life cycles. Perhaps, as Samuel Butler surmised, "a hen is just an egg's way of making another egg" after all. It is worth considering the proposal that since all complex organisms originated from the unicellular state, a return to the unicellular state is necessary in order to ensure the fidelity of any given mutation with all of the subsequently evolved homeostatic mechanisms, from its origins during phylogeny, through all the elaborating permutations and mutational combinations of that trait during the process of evolution. One way of thinking about this concept is to consider that perhaps Haeckel's biogenetic law is correct after all – that ontogeny actually does recapitulate phylogeny. His theory has been dismissed for lack of evidence for intermediary steps in phylogeny occurring during embryonic development, like gill slits and tails. However, that was during an era when the cellular-molecular mechanisms of development were unknown. A testament to the existence of such molecular lapses is the term "ghost lineage," which fills such gaps in the fossil record with euphemisms. We now know that there are such cellular-molecular physiologic changes over evolutionary time that are not expressed in bone, but are equally as important, if not more so, in other organ systems. In all likelihood, ontogeny must recapitulate phylogeny in order to vouchsafe the integrity of all of the homeostatic mechanisms that each and every gene supports in facilitating evolutionary development. Without such a "fail-safe" mechanism for the foundational principles of life, there would be inevitable drift away from our "first principles," putting the core process of evolution in response to environmental changes itself at risk of extinction. Stephen J. Gould famously wondered whether an evolutionary "tape" replayed would recapitulate itself? In this construct, the answer would resoundingly be "no" at least qualitatively, since all of the same components – bacteria, oxygen, minerals, heavy metals – are still present, and it would be expected that first principles would still remain as they are, but the sequence of environmental events would be radically different.

One implication of this perspective on evolution, starting from the unicellular state phylogenetically, and being recapitulated ontogenetically, is the likelihood that it is the unicellular state that is actually the object of selection. The multicellular state, which Gould and Richard Lewontin referred to as "spandrels," is merely a biologic "probe" for monitoring the environment between unicellular stages in order to register and facilitate adaptive changes. This consideration can be based on both *a priori* and empiric data. Regarding the former, emerging evidence for epigenetic inheritance demonstrates that the environment can cause heritable changes in the genome, but they only take effect phenotypically in successive generations. This would suggest that it is actually the germ cells of the offspring that are being selected for. The starvation model of metabolic syndrome may illustrate this experimentally. Maternal diet can cause obesity, hypertension, and diabetes in the offspring. But they also mature sexually at an earlier stage due to the excess amount of body fat. Though seemingly incongruous, this may represent the primary strategy to accelerate the genetic transfer of information to the next generation (positive selection), effectively overarching the expected paucity of food. The concomitant obesity, hypertension, and diabetes are unfortunate side effects of this otherwise adaptive process in the adults. Under these circumstances, it can be surmised that it is the germ cells that are being selected for; in other words, the adults are disposable, as Dawkins has opined.

Hologenomic evolution theory provides yet another mechanism for selection emerging from the unicellular state. According to that theory, all complex organisms actually represent a vast collaborative of linked, co-dependent, cooperative, and competitive localized environments and ecologies functioning as a unitary organism toward the external environment. These co-linked ecologies are comprised of both the innate cells of that organism, and all of the microbial life that is cohabitant with it. The singular function of these ecologies is to maintain the homeostatic preferences of their constituent cells. In this theory, evolutionary development is the further expression of cooperation, competition, and connections between the cellular constituents in each of those linked ecologies in successive iterations as they successfully sustain themselves against a hostile external genetic environment. Ontogeny would then recapitulate phylogeny since the integrity of the linked environments that constitute a fully developed organism can only be maintained by reiterating those environmental ecologies in succession toward their full expression in the organism as a whole.

Another way to think about the notion of the unicellular state as the one being selected for is to focus on calcium signaling as the initiating event for all of biology. There is experimental evidence that increases in carbon dioxide during the Phanerozoic eon have caused acidification of the oceans, causing leaching of calcium from the ocean floor. The rise in calcium levels can be causally linked to the evolution of the biota, and is intimately involved with nearly all biologic processes. For example, fertilization of the ovum by sperm induces a wave of calcium, which triggers embryogenesis. The same sorts of processes continue throughout the life cycle, until the organism dies. There seems to be a disproportionate investment in the zygote from a purely biologic perspective. However, given the prevalence of calcium signaling at every stage, on the one hand, and the participation of the gonadocytes in epigenetic inheritance on the other, the reality of the vectorial trajectory of the life cycle becomes apparent – it cannot be static, it must move either toward or away from change.

By using the cellular-molecular ontogenetic and phylogenetic approach described above for the water-to-land transition as a major impetus for evolution, a similar approach can be used moving both forward and backward from that critically important phase of vertebrate evolution. In so doing, the gaps between unicellular and multicellular genotypes and phenotypes can realistically be filled in systematically. But it should be borne in mind that until experimentation is done, these linkages remain hypothetical. Importantly though, there are now model organisms and molecular tools to test these hypotheses, finally looking at evolution in the direction in which it occurred, from the earliest iteration forward. This approach will yield *a priori* knowledge about the first principles of physiology, and how they have evolved to generate form and function from their unicellular origins.

Anthropic Principle Redux: We are not *in* this Environment, we are *of* it

The realization that there are first principles in physiology, as predicted by the cellular-molecular approach to evolution, is important because of its impact on how we think of ourselves as individual humans and as a species, and our relationship to other species. Once it is recognized and understood that we, as our own unique species, have evolved from unicellular organisms, and that this is the case for all of the other organisms on Earth, including plant life, the intense and intimate interrelationships between all of us must be embraced. This kind of thinking has previously been considered in the form of genes that are common to plants and animals alike, but not as part of a larger and even more elemental process of evolution from the physical firmament. This perspective is on a par with the reorientation of humans to their surroundings once it was acknowledged that the Sun, not the Earth, was the center of the Solar System. That shift in thought gave rise to the Age of Enlightenment! Perhaps in our present age, such a frame-shift will provide insight to black matter, string theory, and multiverses.

In retrospect, it should have come as no surprise that we have misapprehended our own physiology. Many discoveries in biomedicine are serendipitous, medicine is post-dictive, and the Human Genome Project has not yet yielded any of its predicted breakthroughs. However, moving forward, knowing what we now do, we should countenance our own existence as part of the wider environment – that we are not merely in this world, but literally of this world – with an intimacy that we had never previously imagined.

This unicellular-centric vantage point is heretical, but like the shift from geocentrism to heliocentrism, our species would be vastly improved by recognizing this persistent, systematic error in self-perception. We are not the pinnacle of biologic existence, and we would be better stewards of the land and our planet if we realized it. We have learned that we must share resources with all of our biologic relatives. Perhaps through a fundamental, scientifically testable, and demonstrable understanding of what we are and how we came to be so, more of us will behave more consistently with Nature's needs instead of subordinating them to our own narcissistic whims. As we become deeply aware of our true place in the biologic realm, such as we are already witnessing through our increasing recognition of the immense microbial array of fellow travelers as our microbiome, we may find a more ecumenical approach to life than we have been practicing for the last 5000 years.

Bioethics Based on Evolutionary Ontology and Epistemology, not Descriptive Phenotypes and Genes

By definition, a fundamental change in the way we perceive ourselves as a species would demand a commensurate change in our ethical behavior. Such thoughts are reminiscent of a comment in a recent biography of the British philosopher Derek Parfit in *The New Yorker* magazine, entitled "How to be Good," in which he puzzles over the inherent paradox between empathy and Darwinian survival of the fittest. These two concepts would seem to be irreconcilable, yet that is only because the latter is based on a false premise. Darwin's great success was in making the subject of evolution user friendly by providing a narrative that was simple and direct. Pleasing as it may be, it is at best, entirely incomplete. Think of it like the transition from Newtonian mechanics to relativity theory. As much is learned about the unicellular world with its surprising mechanisms and capacities, new pathways must be imagined. It is clear that we as humans are hologenomes, and all other complex creatures are too. In fact, there are no exceptions. The reasons for this can only be understood properly through a journey from the "Big Bang" of the cell forward, with all of the cell's faculties and strictures. By concentrating on cellular dynamics, an entirely coherent path is mapped. Tennyson's line about "Nature, red in tooth and claw" is only the tip of what the iceberg of evolution really constitutes. As pointed out above, we evolved from unicellular organisms through cooperation, co-dependence, collaboration, and competition. These are all archetypical cellular capacities. Would we not then ourselves, as an example of cellular reiteration, have just those self-same and self-similar behaviors?

Summary

By looking at the process of evolution from its unicellular origins, the causal relationships between genotype and phenotype are revealed, as are many other aspects of biology and medicine. These have hitherto remained anecdotal and counterintuitive because the prevailing descriptive, top-down portrayal of physiology under Darwinism is tautologic and teleologic. In opposition to that, the cellular-molecular, bottom-up approach is conducive to prediction, which is the most powerful test of any scientific concept. Though there is not a great deal of experimental evidence for the intermediate steps between unicellular and multicellular organisms compared to what is known of ontogeny and phylogeny of metazoans, it is hoped that the perspectives expressed in this chapter will encourage more such fundamental physiologic experimentation in the future.

The present Chapter reprises the concept that by starting from the cellular origins of life the underlying principles of seemingly complex, indecipherable physiologic principles can be understood and expanded to all of physiology. Chapter 15, entitled "Homeostasis as the mechanism of evolution," provides a mechanistic integration for the how and why of evolution. Beginning with the protocell, homeostasis acts as the integrating principle on a scale-free basis.

15

Homeostasis as the Mechanism of Evolution

Introduction

It has been helpful in understanding biologic purpose to use teleology, but harmful in thinking about its mechanistic origins, because the evolved trait is a combinatorial of non-purposed historic traits. This is, in essence, François Jacob's "tinkering" within boundaries. For example, certain gene duplications occurred during the vertebrate water-to-land transition – the parathyroid hormone-related protein (PTHrP) receptor, the β-adrenergic receptor, and the glucocorticoid receptor. All three were key to the physiologic changes necessary for vertebrate adaptation to a terrestrial environment – skeletal, pulmonary, renal, dermal, and vascular. The repurposing of these genes was the consequence of both the past and present conditions, allowing for future emergent biologic traits.

We have taken a unique approach to understanding the processes involved in evolution based on the cell as the principal level of selection. This perspective makes the assumption that the cell evolved from the physical fundament some 4.5 billion years ago to spontaneously form liposomes, providing a protected space within which catalysis generated bioenergy, allowing for a reduction in entropy, promulgated by homeostatic mechanisms. That perspective subsumes an across-time and -space process that holistically connects the past, present, and future of the organism, rather than the conventional quasi-static snapshot view of homeostasis maintaining the status quo much like your home thermostat. The primary reason that this point of view has not been put forward is because of the general acceptance of teleological thinking due to the descriptive, non-mechanistic nature of biology. We would like to offer a causal, deterministic, and predictive way of thinking about biology that would supersede teleology.

Homeostasis is Dynamic, not Static

In earlier chapters, we alluded to the fact that the source for evolutionary change is contained within the unicell, and that processes of evolution must be seen in the broad historic context as ontogeny and phylogeny. Nowhere is the paradoxical distinction between traditionally descriptive biology and mechanistically robust and interactive evolution more evident than in the way we conceive of homeostasis as static, when in

Evolution, the Logic of Biology, First Edition. John S. Torday and Virender K. Rehan.
© 2017 John Wiley & Sons, Inc. Published 2017 by John Wiley & Sons, Inc.

reality it is highly dynamic. Homeostatic regulators of physiology are constantly in flux, monitoring the environment, and always resetting themselves, on the one hand, or on the other hand providing the reference values for evolutionary change if necessary for survival in an ever-changing environment. Whereas the former perspective is a consequence of contemporary descriptive biology, the latter is best seen in the field of developmental physiology, particularly when it is truncated in the preterm infant, or reversed, as in the case of chronic diseases. For it is the growth factor-receptor signaling mechanisms of development, regeneration, and repair that underlie all of these processes, providing a way of seeing the continuum by which structure and function change over the course of ontogeny and phylogeny, and reach equipoise to maintain, sustain, and perpetuate physiologic stability.

Ontogeny and phylogeny as one continuous mechanism of lung homeostasis turns out to be a unique and empowering insight to the fundamental mechanism of evolution – how homeostasis can act simultaneously as both a stabilizing agent and as the mechanism for evolutionary change. For example, in *Evolutionary Theory: the Unfinished Synthesis*, R.G.B. Reid pointed out the paradoxical relationship between homeostasis and evolution, though he failed to invoke either a developmental or phylogenetic dimension. The bottom line is that biologists commit a systematic error in describing the different phases of the life cycle without considering the mechanistic interrelationships between them, which must logically exist, but the data have been siloed within the various subdisciplines of biology. It is this fractious nature of descriptive biology that is hindering our understanding of what evolution actually constitutes.

The Historic Concept of Homeostasis, from Bernard to Cannon

Homeostasis is defined as the property of a system in which variables are regulated so that internal conditions remain stable and relatively constant. Examples of homeostasis include the regulation of temperature and the balance between acidity and alkalinity. It is a process that maintains the stability of the organism's internal environment in response to fluctuations in external conditions.

The conceptualization of homeostasis as the *milieu intérieur* was first described by Claude Bernard in his book *An Introduction to the Study of Experimental Medicine* in 1865. The term homeostasis was coined by Walter Bradford Cannon in his *Organization for Physiological Homeostasis* in 1926. Conrad Waddington preferred the more dynamic term homeorhesis. Although the term was originally used to refer to processes within living organisms, it is frequently applied to automatic control systems. Homeostasis requires a sensor that is sensitive and specific to the condition being regulated, an effector mechanism that can vary in response to that condition, and a negative feedback connection between the two.

All living organisms depend on maintaining a complex set of interacting metabolic chemical reactions. From the simplest unicellular organisms to the most complex plants and animals, internal processes operate to keep the conditions within tight limits to allow these reactions to proceed. Homeostatic processes act at the level of the *cell*, the *tissue*, and the *organ*, as well as for the organism as a whole, referred to as allostasis.

Negative Feedback

All homeostatic control mechanisms have at least three interdependent components for the variable being regulated: The receptor is the sensing component that
monitors and responds to changes in the environment. When the receptor senses a
stimulus, it signals information to the nucleus, which sets the range at which a variable is maintained. The nucleus determines an appropriate response to the stimulus. The nucleus then sends signals to an effector, which can be other cells, tissues,
organs, or other structures that receive signals for homeostasis. After receiving the
signal, a change occurs to correct the deviation by depressing or damping it with
negative feedback.

Negative feedback mechanisms consist of reducing the output or gain of any organ or
system back to its normal range of functioning. A good example of this is regulating
blood pressure. Blood vessels can sense the resistance to blood flow when blood pressure increases. The blood vessels act as receptors and they relay the message to the
brain. The brain then sends a message to the *heart* and blood vessels, both of which are
effectors. The heart rate will decrease as the blood vessels increase in diameter (known
as vasodilation). This change would cause the blood pressure to fall back to its normal
range. The opposite would happen when blood pressure decreases, which would cause
vasoconstriction.

Another important example is seen when the body is deprived of food. In response,
the body will then reset the metabolic set-point to a lower value. This allows the
body to continue to function at a slower metabolic rate even though the body is
starving. Therefore, people depriving themselves of food while trying to lose weight
find it easy to shed weight initially and much harder to lose more thereafter. This is
due to the body readjusting itself to a lower metabolic set-point to allow the body to
survive with its lower supply of energy. Exercise can change this effect by increasing
the metabolic demand.

Another good example of a negative feedback mechanism is thermoregulation.
The hypothalamus, which monitors body temperature, is capable of determining
even the slightest variations in body temperature. Response to such variation
could be stimulation of the glands that produce sweat to reduce body temperature
due to evaporative cooling, or signaling various muscles to shiver to increase body
temperature.

Homeostatic Imbalance

Pathologic states cause disturbances in homeostasis symptomatic of disease. Similarly,
as the organism ages, the efficiency in the control of physiologic systems erodes due to
breakdown in cell–cell signaling. These inefficiencies gradually result in an unstable
internal environment that increases the risk of illness, and leads to the physical changes
associated with aging.

Certain homeostatic imbalances, such as a high core temperature, a high concentration of salt in the blood, or a low concentration of oxygen, can generate homeostatic
emotions (such as warmth, thirst, or breathlessness), which motivate behavior aimed at
restoring homeostasis.

Cart and Horse Diachronic Perspective

> It is not enough to see the horse pulling a cart past the window as the good working horse it is today; the picture must also include the minute fertilized egg, the embryo in its mother's womb, and the broken-down old nag it will eventually become.
>
> *C.H. Waddington*

There is a fundamental epistemologic causation problem in evolution theory – the perspective Waddington suggests in the above quote is that it is necessary to see the entire process of life as a continuum in order to understand the underlying evolutionary principle. Reducing vertebrate physiologic evolution to the cellular, mechanistic level has allowed us to examine ontogeny and phylogeny across space and time as one continuous process; that the short- and long-term histories of the organism are, in fact, one and the same, consistent with what Waddington is expressing – that life is a continuum. In retrospect, ours was an important breakthrough because it demonstrated the fallacy in looking at the processes of the life cycle independently of one another, as is the case when looked at only in its present form. But by looking at the process from its cell-molecular mechanistic basis, one can see it as a continuous process of adaptation to atmospheric oxygen accommodating the metabolic demand of vertebrate evolution.

This conceptual breakthrough has encouraged us to rethink Ernst Haeckel's biogenetic law that ontogeny recapitulates phylogeny in light of our newfound ability to conceive of the life cycle as a continuum rather than as individual components without any mechanistic connectivity. By suggesting that ontogeny and phylogeny were interrelated, Haeckel inferred that they had common properties, providing important insight to the mechanism of evolution. But it is only at the cellular-molecular level that the structural-functional homologies can be seen, data that were inaccessible to Haeckel. The ability of the life form to sustain and "reinvent" itself has allowed vertebrates to adapt to an ever-changing environment over eons through the process of embryogenesis. The most logical integrating mechanism that transcends such divergent scales of time and space for adaptation is homeostasis. And the recapitulation of phylogeny ontogenetically may act to constrain evolutionary changes that are internally consistent with homeostatic control at key stages of embryologic development referring all the way back to the unicellular state.

In turn, that raises the basic ontologic, epistemologic question as to what homeostasis has evolved in support of. We have suggested that reduced, sustained, perpetuated negentropy is the underlying driving force behind evolution, a property of life that requires homeostatic control at its core. By reducing evolution to the evolution of homeostasis, a fundamental change in our understanding of the causal nature of this process emerges. Chance mutation and natural selection are epiphenomena, whereas adaptation of the internal environment of the organism to the external environment of the physical realm in service to homeostasis provides a testable, predictable model of evolution.

Gerhard Fankhauser's classic paper on the effects of cell size on newt development exemplifies the advantage of a mechanistic over a descriptive view of the function of homeostasis. Polyploid embryos had fewer but larger cells, which counterintuitively had no effect on tissue or body size based on descriptive biology. For example, the size of the kidney ducts was unaffected by the number of epithelial cells surrounding them.

This finding even baffled the great Einstein, prompting him to state that "It looks as if the importance of the cell as ruling element of the whole had been overestimated previously. What the real determinant of form and organization is seems obscure." The authors then questioned what the real determinant of form and organization is. Yet, if one hypothesizes that homeostasis is the underlying selection pressure for solute exchange over the surface area of the duct, the lack of overall structural change now makes sense.

With the luxury of hindsight, we have literally and figuratively been putting the phenotypic-genotypic "cart" before the homeostatic "horse," allowing our thought processes to be diverted by descriptive, top-down, *a posteriori* biology. We must begin thinking along cellular-molecular lines regarding evolution if we are to make advances in biology and medicine, or we will plod along as the alchemists did until chemistry, the periodic table, and quantum theory set us on the road to predictive physics.

Homeostatic Regulation as Downward Causation

Downward causation is defined as a causal relationship from higher levels of a system to lower-level parts of that system: for example, mental events acting to cause physical outputs. The term was coined in 1974 by the philosopher and social scientist Donald T. Campbell. In his paper "A theory of biological relativity: no privileged level of causation," the author develops the concept of "downward causation," schematically depicted in Figure 15.1. He concludes from this analysis that there is no hierarchical level of selection in biologic systems. In our opinion, this is a classic example of the systematic error we make in thinking about life forms only from the perspective of their adult state as the end-result of evolution, in contrast to Waddington's "cart and horse" metaphor, which encourages us to think beyond the present circumstance to the continuum of life,

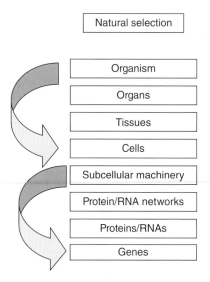

Figure 15.1 Downward causation. Teleologically, there is no privileged level of causality in biologic systems.

including the next generation as the fertilized egg in the mother's womb, referred to now as the Barker hypothesis. As a result, Dennis Noble, the developmental cardiologist, has similarly concluded teleologically that there is no privileged level of causality in biologic systems. Yet we maintain that there is, but it can only be seen by running the evolutionary "tape" forward from unicellular state to unicellular state over the entire course of the life cycle of the organism.

Contrary to popular belief, it is now known that the epigenetic marks acquired during the life cycle of the organism are not all eliminated during meiosis. Moreover, there is accumulating evidence that such epigenetic marks are consequential because they can be inherited and have biologic effects. Therefore, the salient question is what controls this process and how does it affect evolution. Our laboratory has been studying the effect of maternal smoking on the transgenerational inheritance of the asthma phenotype, first documented epidemiologically as "the grandmother effect," namely, that cigarette smoke exposure is more strongly associated with whether your grandmother smoked than whether your mother smoked. The nicotine contained in cigarette smoke induces specific epigenetic changes in the upper airway of the lung and the gonads of the offspring for at least three generations. These findings beg the question as to the level of selection because newly acquired epigenetic mutations only affect the offspring, not the adults.

Downward causation is biased toward vertically integrated evolutionary biology, top-down or bottom-up. In contrast to that, we have been advocating for a "middle-out," cell–cell signaling approach based on the cellular-molecular mechanism of embryogenesis (Figure 15.2). The vertical integrating perspective is consistent with conventional Darwinian "descent with modification" and natural selection, whereas the middle-out approach across spatio-temporal ontogeny and phylogeny is more in concert with environmental forces such as the sun (above) and gravity (below). Seen as a vectorial product of these forces, evolution would be propelled horizontally from generation to generation, always gaining information from the environment in the process.

Perhaps the reason we go through the life cycle from zygote to zygote is to acquire genetically heritable information from the environment and selectively integrate it into our genome, or not. The "filtering" mechanisms are those of ontogeny and phylogeny, providing the short-term and long-term "histories" of the organism as a means of determining the homeostatic relevance of the acquired mutations. Homeostasis is integral to morphogenesis since the growth factor signaling mechanisms of embryogenesis become homeostatic mechanisms in the offspring. As such they also can discriminate between

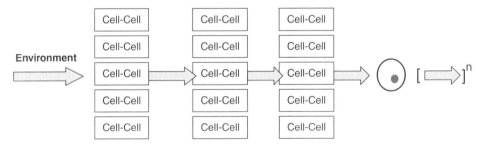

Figure 15.2 Evolution as cell–cell signaling. Environmental "stress" affects cell–cell communication mechanisms that determine homeostatic control, resulting in genetic "mutations" that modify structure and function evolutionarily.

adaptive and maladaptive genetic mutations that affect homeostasis, either indirectly through the developmental process, or directly through the regulatory mechanisms of physiology.

Growth Factor Signaling is Common to Development, Homeostasis, and Regeneration

Embryogenesis is determined by paracrine growth factor cell–cell signaling, forming the spatio-temporal patterns that provide form and function of tissues and organs. The lung is the best characterized organ because of its critical importance to survival at the time of birth in humans (see Figure 15.1).

In order to form an effective, diffusible interface for gas exchange with the circulation, the lung endoderm and mesoderm undergo extensive branching morphogenesis and alveolization closely associated with angiogenesis and vasculogenesis. It is becoming clear that many of the key factors determining the process of branching morphogenesis, particularly of the respiratory organs, are highly conserved throughout evolution. Synthesis of information from deletional mutations in *Drosophila* and mouse indicates that members of the sonic hedgehog/patched/smoothened/Gli/FGF/FGFR/sprouty pathway are functionally conserved and extremely important in determining respiratory organogenesis through mesenchymal–epithelial inductive signaling, which induces epithelial proliferation, chemotaxis, and organ-specific gene expression. Transcriptional factors, including Nkx-2.1, HNF family forkhead homologs, GATA family zinc finger factors, pou and hox, helix-loop-helix (HLH) factors, Id factors, and glucocorticoid and retinoic acid receptors mediate and integrate the developmental genetic instruction of lung morphogenesis and cell lineage determination. Signaling by the insulin-like growth factor (IGF), epidermal growth factor (EGF), and transforming growth factor β (TGF-β)/ bone morphogenetic protein (BMP) pathways, extracellular matrix components, and integrin signaling pathways also directs lung morphogenesis as well as proximo-distal lung epithelial cell lineage differentiation. Soluble factors secreted by lung mesenchyme comprise a "complete" inducer of lung morphogenesis. In general, peptide growth factors signaling through cognate receptors with tyrosine kinase intracellular signaling domains such as fibroblast growth factor receptor (FGFR), EGF receptor (EGFR), IGF receptor (IGFR), platelet-derived growth factor receptor (PDGFR), and c-Met stimulate lung morphogenesis. On the other hand, cognate receptors with serine/threonine kinase intracellular signaling domains, such as the TGF-β receptor family, are inhibitory, although BMP4 and BMPR also play key inductive roles. Pulmonary neuroendocrine (PNE) cells differentiate earliest in gestation from among multipotential lung epithelial cells. MASH1-null mutant mice do not develop PNE cells. Proximal and distal airway epithelial phenotypes differentiate under distinct transcriptional control mechanisms.

It is becoming clear that angiogenesis and vasculogenesis of the pulmonary circulation and capillary network are closely linked with and may be necessary for lung epithelial morphogenesis. Like epithelial morphogenesis, pulmonary vascularization is subject to balance between positive and negative factors. Angiogenic and vasculogenic factors include vascular endothelial growth factor (VEGF), which signals through cognate receptors, while novel anti-angiogenic factors include endothelial monocyte-activating peptide II (EMAP II).

Homeostasis, Agent for Change During the Water-to-Land Transition

Specific gene duplications occurred during the vertebrate water–land transition, facilitating the physiologic adaptation to terrestrial life. In fish, parathyroid hormone-related protein receptors (PTHrPRs), β-adrenergic receptors (βARs), and glucocorticoid receptors (GCRs) all mediate key physiologic functions for adaptation to water. In land vertebrates, these genes were "amplified" due to their duplications for respiratory, skeletal, kidney, and skin barrier adaptations. The genetic amplification of these specific genes was more than just Darwinian chance mutations and selection; they were existential for either adapting or becoming extinct. Why they duplicated is answered by analyzing their developmental and functional roles in the lung, neuroendocrine system, and metabolism. How is speculative, but it would seem to have occurred as a result of microvasuclar shear stress causing remodeling of these specific tissues and organs.

Homeostasis as the Result of Developmental Mechanisms

As one peruses the evolutionary biology literature, observations of pre-adaptations come up over and over again. Perhaps that is because we are looking at the process of evolution from its ends instead of its means. If one follows pre-adaptation to its logical extension, it culminates in the unicellular state, which is the historic origin of metazoans. By looking at the process as a series of pre-adaptations, one sees the processes of ontogeny and phylogeny in reverse. Instead, by looking at phylogeny in the forward direction, thinking about the cellular-molecular evolutionary changes in the context of the ever-changing environment, the causal relationships become ever more clear, as we have shown for the evolution of the lung: by regressing the genes that have determined structure and function during lung ontogeny and phylogeny against major changes in the environment – ocean salinity, the drying-up of the oceans, atmospheric oxygen – as Cartesian coordinates, one can see the adaptive strategy of internal selection due to physical forces, mediated by physiologic stress, starting with the advent of the peroxisome as balancing selection against calcium dyshomeostasis. The lung may be the optimal algorithm for such evolutionary changes in vertebrate visceral physiologic evolution because of the powerful selection pressure for its evolution during the water–land transition – there were no alternatives, it was either adapt or become extinct.

The lesson learned from that event becomes even more self-evident when thinking about the specific implications of the three gene duplications that occurred during that transition (see Figure 1.3): the PTHrPR, the βAR, and the GR [= step 5]. The PTHrPR may have duplicated primarily because it promotes bone remodeling, and the vertebrate skeleton is known to have evolved on at least five separate occasions based on the fossil record, providing ample opportunity for the coevolution of the visceral organs necessary for land physiologic adaptation. But PTHrP signaling is also important for air breathing and for the skin as a barrier, both of which were also necessary for terrestrial adaptation. Experimentally, if you delete the PTHrP gene of a developing mouse, it results in developmental deficits in the lung (no alveoli), bone (failure to calcify), and skin (immature barrier), consistent with all of the aforementioned phenotypes.

The literature would lead us to think that these gene duplications occurred by chance alone based on Darwinian evolution theory, which is far from the case. In fact, to consider the biologic context in which the gene duplications occurred during the vertebrate water–land transition is highly illuminating. Regarding the biologic mechanism for such gene duplications, the physiologic stress of the water–land transition would have caused vascular shear, generating radical oxygen species known to cause genetic mutations and duplications. That causal interrelationship would have been particularly relevant to the organs most affected by such adaptational stresses because the shearing effects would have predominantly occurred within those tissues and organs most heavily affected (or the lineage would have gone extinct) – what Darwinian evolutionists refer to euphemistically as "survival of the fittest," though now with the opportunity for formulating specific, mechanistically testable, and refutable hypotheses.

Genetic remodeling of the alveolar bed for stretch-regulated PTHrP signaling would have had dual physiologic adaptational advantages, initially by stimulating alveolar surfactant production, relieving the inevitable episodic stress of alveolar insufficiency, resulting in hypoxia during the process of evolution. That would have been followed over evolutionary time by PTHrP acting both to generate more highly evolved alveoli, and, as a potent vasodilator, accommodating the concomitant increase in alveolar microvascular blood flow.

PTHrP and Hypothalamic-Pituitary-Adrenal Regulation of Physiologic Stress

Extensive experimental evidence from our laboratory has shown the central role of PTHrP in normal lung development, beginning with the genetic deletion of PTHrP causing impaired lung development due to failure to form alveoli. In various insult models of lung disease – oxotrauma, barotrauma, infection – we have documented the decrease in PTHrP expression in all of these instances leading to lung simplification indicative of "reverse evolution." Moreover, infants who develop bronchopulmonary dysplasia (BPD), the chronic lung disease of the newborn, are PTHrP deficient based on measurement of the protein in bronchoalveolar lavage.

The Role of PTHrP Expression in Adrenal Corticoid Synthesis

More recently, it has been found that PTHrP is expressed in the pituitary, where it stimulates adrenocorticotropic hormone (ACTH), and in the adrenal cortex, where it enhances ACTH stimulation of corticoid synthesis. This pathway would amplify the hypothalamic-pituitary-adrenal (HPA) response to physiologic stress, increasing adrenaline (epinephrine) production by the adrenal medulla. This mechanism may have evolved during the water–land transition, wherein the lung would periodically have been unable to effectively generate adequate amounts of systemic oxygen, causing hypoxia. Hypoxia is the most potent physiologic agonist for the HPA known; by stimulating adrenaline production, which stimulates alveolar surfactant secretion, PTHrP expression would have transiently alleviated the atelectatic stress on the lung by making the alveoli more distensible.

Such a mechanism may refer all the way back to the rodent-like ancestor of placental mammals, which had to be nimble in order to avoid being crushed and/or eaten by predators. Over time, the stress on the microvasculature of the lung, pituitary, and adrenal cortex may have "remodeled" all of these structures. This includes the adrenal medulla, known to have evolved a complex microvascular arcade in mammals, acting like an "echo chamber" to enhance adrenaline production in response to stress, both baseline and regulated states.

Homeostatic Regulation is Diachronic

Understanding the interrelationship between homeostasis and embryogenesis lies in recognizing the across-space and -time, diachronic nature of the overall mechanism as evolution (see Figure 1.3). During embryogenesis, paracrine growth factor signaling to cognate receptors on cells of different germline origins determines the physiologic structure and function of the offspring, including the homeostatic set-points of the tissues and organs of the body. After birth, those set-points may be challenged during postnatal life, inherently being maintained by many of the same signaling principles used for embryogenesis. If the limits of homeostasis are tested, growth factor signaling mechanisms may revert to their ancestral forms in order to sustain the organism, some-times causing fibrosis as the structural default mode that grants the organism the ability to reproduce under suboptimal physiologic conditions – what a Darwinian would call survival of the fittest. Under extreme conditions, such as the five known mass extinc-tions, or the water–land transition, physiologic stress has caused pragmatic remodeling of organs in order to adapt. Those members of the species best able to mount such an adaptive strategy transmit such homeostatically adaptive genes to their offspring, generating a heritable phenotype in the process. Hence, the relationship between homeostasis and phenotypic change is a continuum, mediated by growth factor signaling properties that are mechanistically common to both.

This intimate relationship between environmental physiologic stress, homeostasis, and remodeling goes all the way back to the inception of life, when protocells formed from micelles, generating an internal environment (*milieu intérieur*) using both the plasmalemma for endo- and exocytosis, and the endomembrane system to compart-mentalize physiologic properties. The resulting generation of negentropy, sustained by chemiosmosis under homeostatic control, forms the basis for the vital principle. The ability to recapitulate this process from one generation to the next, acquiring new "knowledge" through reproduction and epigenetics, allows the system to perpetuate itself indefinitely.

Allostasis as Integrated Homeostasis

The concept of allostasis is yet another example of how our seeming inability to see homeostasis in its fullest form, diachronically, as the integrating mechanism underlying all of biology, has distracted us from seeing the continuum of biologic processes. Bruce McEwen and John Wingfield define allostasis as a process that supports homeostasis, the "setpoints" and other boundaries of control that must change...the physiological and/or life history stages that must change to achieve stability. Allostasis clarifies an inherent ambiguity in the term "homeostasis" and distinguishes between the systems that are essential for life ("homeostasis") and those that maintain these systems in balance ("allostasis") as environment and life history stage changes.

This misperception of a supra-homeostatic control system independent of homeostasis is the consequence of failing yet again to recognize the primacy of the cell in mediating the evolutionary principle. If in fact life is a continuum that is focused on the unicellular state, then homeostasis functions at all levels of biology, independent of scale. So the properties of allostasis are the expression of the same homeostatic principles at a higher level. The examples used by McEwen and Wingfield – blood pressure, metabolism, pH, complex patterns of bird migration – are all derived from homeostatic regulation of the unicellular state that has evolved in support of multicellular organisms. The latter example of migratory birds was used by Ernst Mayr to exemplify the fundamental difference between proximate and ultimate causation in evolutionary biology, which drove a wedge between those interested in structure-function relationships (proximate) and the process of evolution itself (ultimate), generating volumes of descriptive data, undermining any attempt to understand how and why evolution has occurred based on principles of cell biology. But in this day and age of molecular mechanisms, that distinction between proximate and ultimate causation in evolution is antiquated. We now know that changes in the wavelength of ambient light affect the pineal, altering neuroendocrine hormones that regulate reproduction. That is the foundation for bird migration, seen as one continuous process instead of independent non-mechanistic associated events.

In our book titled *Evolutionary Biology, Cell-Cell Communication and Complex Disease*, we formulated the life span of the organism as a continuous series of ligand–receptor interactions from morphogenesis to the maintenance of physiologic homeostasis, to the loss of homeostatic control mechanisms during aging, culminating in death. Seen in this light, allostasis takes on a very different set of characteristics, stress having short-term effects that are physiologically beneficial for the reproductive strategy; but over the long-haul, such adaptive responses can have deleterious effects that occur as unintended consequences of the optimization of the primary homeostatic mechanisms involved. In other words, acceleration of development would bring on precocious aging and death as a continuous mechanism selecting for the unicellular state. By conventionally focusing on the pathologic aspects of allostatic load, we get a skewed view of its function. For example, we have yet to understand the role of premature adrenarche – the production of androgens by the adrenal cortex at the onset of puberty. We know that it is associated with intrauterine growth retardation and overweight children, particularly as it relates to altered metabolism *in utero*. Yet from a cellular mechanistic perspective, precocious puberty due to low food abundance during development *in utero* is adaptive, accelerating sexual maturation in the offspring, accelerating the option for the next generation to live in an environment of greater food abundance. But from a synchronic, pathophysiologic perspective, this condition appears to be maladaptive because it gives rise to metabolic syndrome – obesity, hypertension, heart attack, and stroke. So these are diametrically opposite perspectives on the significance of homeostasis during the life history of the organism.

Furthermore, McEwen and Wingfield characterize allostasis as "protection vs damage," yet again only seeing the immediate consequences of metabolic control on physiology when the true agenda – the acquisition of newly acquired epigenetic mutations – is actually of an intergenerational nature, providing a very different interpretation of the observed phenomenon. We have to see the organism as both its past and its future, not just as it appears in its present condition, in order to understand epigenetic inheritance. Here again, the interpretation of protection versus damage in the context of adaptive and maladaptive is very different from that perspective, when what appears as damage is actually the consequence of an adaptive response from the perspective of optimized reproduction.

Conclusion

So, in closing, ignorance of the fundamental "first principles of physiology" has led us astray. Conversely, by focusing on the unicellular state as the primary level of selection, we gain insight to such ontologic and epistemologic principles as the life cycle, embryogenesis, life history, and homeostasis. Such deep understanding is of critical importance to our effective utilization of genomic information.

Both Pierre Duhem and W.O.V. Quine have pointed out the limitations of science as "underdetermination." Quine said that knowledge is "a man-made fabric which impinges on experience only along the edges," and then went on to say that "a conflict with experience at the periphery occasions readjustments in the interior of the field." If we knew what our physiologic "experience" actually constituted, perhaps we could avoid the pitfalls of such scientific subjectivity.

This Chapter has provided a mechanistic integration for the how and why of evolution. Beginning with the protocell, homeostasis acts as the integrating principle on a scale-free basis. Chapter 16, entitled "On the evolution of development," takes the dogma of development and shows how it becomes part of the continuum of evolution using the principles provided in the previous chapters.

Suggested Readings

Kirschner M, Gerhart J, Mitchison T. (2000) Molecular "vitalism". *Cell* 100:79–88.
Reid GB. (1985) *Evolutionary Biology: the Unfinished Synthesis*. Cornell University Press, Ithaca, NY.

16

On the Evolution of Development

Introduction

Calcium homeostasis counterbalanced by lipid homeostasis seems to underlie all of cellular evolution. Right from the moment of fertilization there is a calcium burst in the zygote that sustains us until the penultimate moment when we die. Throughout life, as sentient beings we experience calcium-dependent events such as creative bursts, procreative bursts, the runner's high, the near-death experience as the light seen during this phenomenon being the return of the calcium "spark" of life, snatched from the jaws of death.

It is hypothetically possible that the zygote is the principal state of evolutionary selection; this is counterintuitive based on descriptive biology, but is the logical conclusion reached from a cellular-molecular approach, working backwards in ontogeny and phylogeny to our origins in unicellular life. And during the course of the life cycle, epigenetic inheritance occurring via the germ cell, bypassing the adults, would suggest the primacy of the unicellular state. Does this skewed vision of the fundamental nature of biology provide more answers than questions? It is certainly worthy of consideration, given the dogmatic, uncontestable understanding of how complex physiology has evolved, not to mention its lack of predictive power for the scientific practice of medicine.

Life began within lipid droplets called liposomes as a dynamic equilibrium of balancing selection between lipids and calcium to sustain negentropy within the protocell, circumventing the second law of thermodynamics. Since this constitutes life, evolution merely acting to modify biology in order to maintain calcium-lipid homeostasis under the ever-changing conditions in the environment, is it any wonder we go back to those first principles during the life cycle. Or perhaps we never left the unicellular state in the first place; in other words, the multicellular state is an agent for perpetuating the first principles, used to monitor the ever-changing environment, gleaning information, subsequently having it filtered by the embryo as a mechanism for stability or change, as the case may be. That perspective is radically different from seeing the adult as the principal state of being, in contrast to the unicell mechanistically in sync with what actually transpires in the environment. Like quantum theory, we are aware that we are made up of atoms, but we do not think in those terms on a moment-to-moment basis other than to contemplate our existence from time to time. If we were to embrace the idea that all the biota evolved from the same cellular construct, like kinship theory, which says that empathy is a function of how closely related we are, we would be far more universally compassionate as a species!

Evolution, the Logic of Biology, First Edition. John S. Torday and Virender K. Rehan.
© 2017 John Wiley & Sons, Inc. Published 2017 by John Wiley & Sons, Inc.

There have been several breakthrough moments in human history, such as Archimedes' realization of buoyancy (Eureka!), Magellan demonstrating that the world is round, and Copernicus pointing the way to heliocentrism. And there were several attempts to construct a predictive periodic table of elements, yet only one was successful. Dmitri Mendeleev finally came up with the correct initial condition for periodicity of the elements, that of atomic number as the organizing principle. Similarly, there have been innumerable theories for the existence of life, ranging from creationism to Darwinism, but all of them are predicated on descriptive, materialistic biology. Only once it is realized that the initial conditions of life are autonomous, self-organizing, self-referential cells, circumventing the second law of thermodynamics, can the meaning of life actually be understood. Like equating mass and energy ($E = mc^2$), that idea is a game changer. And because it is predictive of otherwise counterintuitive phenomena in biology, it seems to be correct. For example, did you ever wonder why life goes in reiterative cycles from zygote to adult, and back to the zygote?

The evolutionary "arc" of complex physiologic principles can be traced using the calcium "spark," from primitive cells lying at the interface between water and land – sea foam – to the human brain. Easier said than done, yet all structural-functional links have formed in service to calcium flux, mediated by the lipids that form a continuum from cholesterol in primitive eukaryotes, to the myelination of neurons. And bear in mind that the homeostatic mechanisms involved are highly conserved at every functional level – cell, tissue, organ, systemic physiology. Undergirding all that is the foundational principle of Life, fomented by negentropy, generated by chemiosmosis and endomembranes, sustained by homeostasis – a squishy, compliant, organic automaton of its own making and devices. This recalls the Greek metaphor of life, the ouroboros, the snake catching its own tail, self-organizing and self-perpetuating. Viewed from this vantage point, the notion that the unicellular state of the life cycle is the primary site for selection pressure becomes tenable: The life cycle is not stagnant, it has a vectorial direction and magnitude of change, either moving upward or downward as it evolves to adapt, or devolves toward extinction. If so, what is the initiating event, since this process must be inhomogeneous? Conventionally, we think of the adult form as the apotheosis of such a mechanism, but that is a narcissistic, anthropocentric viewpoint, like geocentrism, and we know how that ended up.

Alternatively, we should consider the unicellular zygote, particularly in view of new evidence that epigenetic marks on the egg and sperm are not eliminated during meiosis, as had long been thought. What is the functional significance of such epigenetic marks, and what determines which ones are retained and which ones are eliminated? Such considerations are not trivial, since only ~1–3% of inherited human diseases are Mendelian, leaving a huge void in the constellation of heritable diseases that may be filled by epigenetics. And perhaps this is why we are destined to recapitulate phylogeny during the process of ontogeny. Ernst Haeckel's biogenetic law was rejected long ago for lack of experimental evidence that embryogenesis faithfully recapitulated all the phenotypic milestones of phylogeny, but that was before the discovery of the cellular-molecular signaling mechanisms that underpin such morphogenetic changes. Such data are referred to in the evolutionary biology literature as "ghost lineages," but in molecular embryogenesis they are how and why structure and function develop as a continuous process. That knowledge also alludes to the possibility that we may need to mechanistically recapitulate phylogeny in order to ensure that any newly acquired epigenetic mutations are in compliance with the homeostatic and allostatic mechanisms that they might affect in introducing them into the organism's gene pool. In so saying, the unicellular state is ultimately the overall determinant and arbiter of this process – the unicellular state dominates. And if so, we as a species need to reassess our priorities among our cousins, plant and animal alike. Maybe this will lead to bioethical considerations based on first principles rather than on self-serving actions that emanate from our jingoist attitudes. We have thrived on this planet as a species until now by using resources that were not truly only ours, though that's how we have behaved.

So maybe we need to reconsider Haeckel's dictum in formulating a central theory of biology. And in so doing, reconsider internal selection as a mechanistic extension of natural selection.

Until recently there was a widely unacknowledged blind spot created by the absence of cell biology from evolution theory. The only explanation for this seeming oversight has come from Betty Smocovitis, who noted in her book *Unifying Biology* that there was a parting of the ways between the embryologists and the evolutionists back at the end of the nineteenth century, resulting in the absence of a cellular approach to evolution. That has been exacerbated in the interim by evolutionists lacking training in cell biology, and the resultant cultural breakdown between cell biology and evolution theory, exemplified by the comment by Stephen J. Gould that internal selection is tantamount to cancer. And even the more recent advent of evolutionary developmental biology merely places these two disciplines in close proximity to one another, but does not take advantage of the mechanistic synergy between them. But recriminations aside, what is the residual of this abyss between cell biology and evolution? I will roll out a cellular perspective on the process of evolution, and discuss its benefits for your consideration.

Embryologic development is the only process we have for determining the mechanisms for morphogenesis of tissues and organs. The major breakthrough in this field was the discovery that embryonic development is dependent on cell–cell signaling mediated by soluble growth factors and their cognate receptors, signaling for the growth and differentiation of the cells that ultimately determine form, function and homeostasis. But what if Hans Spemann had been able to determine the nature of his "organizer" back in the nineteenth century? How would that have affected evolution theory? I have conceived of a way of tracing the evolution of the lung, all the way back to its origins in the unicellular plasmalemma, affording a way of then looking in the forward direction to determine how and why the lung evolved from that simple unicellular structure for gas exchange, as follows.

The mammalian lung develops from the foregut starting on the 9th day of development in a mouse embryo. The trachea forms from the esophagus, and the major conducting airways are subsequently formed, followed by the alveoli, all through reciprocating, sequential interactions between the endodermal and mesodermal germ layers, ultimately giving rise to more than 40 different cell-types. The key to understanding both lung development and evolution is the formation of the alveolar epithelium, which produces lung surfactant, a soapy material that prevents the collapse of the alveoli on deflation. By comparing gene regulatory networks both across phyla and during development, the sequence of events by which structures and their functions evolved can be determined.

In a recent publication, we have shown how the lung may have evolved from the swim bladder of fish based on the parathyroid hormone-related protein (PTHrP) signaling pathway, a pathway necessary for both lung homeostasis and development. PTHrP signaling predicts the magnitude and direction of lung maturation, and may also predict the phylogenetic changes in the vertebrate lung, characterized by decreasing alveolar diameter, accompanied by the thinning and buttressing of the alveolar wall.

PTHrP is expressed throughout vertebrate phylogeny, beginning with its expression in the fish swim bladder as an adaptation to gravity as buoyancy; microgravity downregulates the expression of PTHrP by alveolar type II epithelial cells, and by the bones of rats exposed to $0 \times g$, suggesting that PTHrP signaling has evolved in adaptation to gravity. PTHrP signaling is upregulated by stretching alveolar type II cells and interstitial fibroblasts, whereas over-distension downregulates PTHrP and PTHrP receptor expression, further suggesting a deep evolutionary adaptation since these genes evolved independently over biologic time. Both surfactant homeostasis and alveolar capillary perfusion are under PTHrP control, indicating that alveolarization and ventilation/perfusion matching, the physiologic principle of the alveolus, may have evolved under the influence of PTHrP signaling.

PTHrP is a highly evolutionarily conserved, stretch-regulated gene that is unusual among the paracrine growth factors that have been identified to mediate lung development because:

1) The PTHrP gene deletion is stage-specific, and results in failure to form alveoli, the major lung adaptation for gas-exchange in land vertebrates.
2) Unlike other such growth factors that emanate from the mesoderm and bind to the endoderm, PTHrP is unusual in being expressed in the endoderm and binding to the mesoderm, providing a reciprocating mechanism for morphogenesis.
3) Only PTHrP has been shown to act pleiotropically to integrate surfactant synthesis and alveolar capillary perfusion, mediating the on-demand surfactant mechanism of alveolar homeostasis.

In contrast to this, others have focused on the importance of the epithelial-mesenchymal trophic unit, and on the importance of the fibroblasts of the "scaffold" that act as "sentinels" to regulate local inflammatory responses. However, PTHrP signaling from the epithelium to the mesoderm is highly significant. The earliest developmental signals for alveolar development originate from the endoderm, and we have demonstrated the dependence of the fibroblast phenotype on epithelially derived PTHrP for development, homeostasis, and repair. All of these features of PTHrP biology justify its use as an archetype for our proposed model of lung evolution. We have schematized this integrated approach for lung developmental and comparative biology, homeostasis, and repair in Figure 5.1.

Ontogeny and Homeostasis

Stimulation of PTHrP and its receptor by alveolar wall distension coordinates the physiologic increase in surfactant production with alveolar capillary blood flow, maximizing the efficiency of gas exchange across the alveolar wall, referred to conventionally based on descriptive biology as ventilation/perfusion (V/Q) matching. V/Q matching is the net result of the evolutionary integration of cell-molecular interactions by which the lung and pulmonary vasculature have functionally adapted to the progressive increase in metabolic demand for oxygen as vertebrates evolved to accommodate land life. The structural adaptation for gas exchange is threefold:

1) the decrease in alveolar diameter;
2) the thinning of the alveolar wall; and
3) the maximal increase in total surface area.

These structural adaptations have resulted from the phylogenetic amplification of the PTHrP signaling pathway. PTHrP signaling through its receptor is coordinately stimulated by stretching the alveolar parenchyma. The binding of PTHrP to its receptor activates the cyclic adenosine monophosphate (cAMP)-dependent protein kinase A (PKA) signaling pathway. Stimulation of this signaling pathway results in the differentiation of the alveolar interstitial lipofibroblast, characterized by increased expression of adipocyte differentiation related protein (ADRP) and leptin. ADRP is necessary for the trafficking of substrate for surfactant production, and leptin stimulates the differentiation of the alveolar type II cell. PTHrP thus affects the cellular composition of the alveolar interstitium in at least three ways that are synergistic with one another:

1) Inhibition of fibroblast growth and stimulation of apoptosis, causing septal thinning.
2) Stimulation of epithelial type II cell differentiation by leptin, which can inhibit epithelial cell growth.
3) Leptin may upregulate type IV collagen synthesis, reinforcing the alveolar wall.

Type IV collagen likely evolved in the water-to-land transition as a natural water barrier, since its evolved amino acid composition is hydrophobic.

Ontogeny and Phylogeny

Cell–cell interactions between primordial lung endoderm and mesoderm cause the differentiation of those germ layers into over 40 different cell-types. We know a great deal about the growth factor signaling that determines these processes, and the downstream signals that alter nuclear readout. And because a great deal of effort has been put into understanding the consequences of preterm birth, we also know how these mechanisms lead to homeostasis, or fail to do so, in which case the phenotype for chronic lung disease informs us of the mechanism of lung fibrosis.

Embryonic lung development is subdivided into branching morphogenesis and alveolarization, the former being "hard-wired," the latter being highly plastic. Deleting the PTHrP gene results in failed alveolarization, implying the relevance of PTHrP to lung evolution, since alveolarization was the primary mechanism for vertebrate lung evolution. Note, for example, that the lung specifically evolved from the swim bladder of physostomous fish, which have a pneumatic duct connecting their esophagus to the swim bladder, a homolog of the trachea, but have no alveoli; however, they do form two subdivisions of the swim bladder, which may be homologs of alveoli. Because PTHrP and its receptor are highly conserved, and stretch-regulated, linking the endoderm and mesoderm to the vasculature, we are compelled to investigate its overall role in lung phylogeny and evolution.

The combined effects of features 1–3 in the previous section would lead to natural selection for progressive, concomitant decreases in both alveolar diameter and alveolar wall thickness through ontogeny and phylogeny, optimally increasing the gas-exchange surface area-to-blood volume ratio of the lung. PTHrP shuts off myofibroblast differentiation by inhibiting the glioblastoma gene *Gli*, the first molecular step in the mesodermal

Wingless/int (Wnt) pathway, and by inactivating β-catenin, followed by the activation of LEF-1/TCP, C/EBPα, and peroxisome proliferator activated receptor gamma (PPARγ). The downstream targets for PPARγ are adipogenic regulatory genes such as ADRP and leptin. PTHrP induces the lipofibroblast phenotype, first described by Dennis Vaccaro and Jerome Brody. This cell-type is expressed in the lungs of a wide variety of species, including both newborn and adult humans. Lipofibroblasts are found next to type II cells in the adepithelial interstitium, and are characterized by neutral lipid inclusions wrapped in ADRP, which actively mediates the uptake and trafficking of lipid from the lipofibroblast to the type II cell for surfactant phospholipid synthesis, protecting the alveolar acinus against oxidant injury. The concomitant inhibitory effects of PTHrP on both fibroblast and type II cell growth, in combination with PTHrP augmentation of surfactant production, would have the net effect of distending and "stenting" the thinning alveolar wall, synergizing with the upregulation of PTHrP, and physiologically stabilizing what otherwise would be an unstable structure that would tend to collapse.

Myofibroblast Transdifferentiation as Evolution in Reverse

Lung development prepares the fetus for birth and physiologic homeostasis. Surfactant production in particular is crucial for effective gas exchange. Based on this integrated functional linkage between lung development and homeostasis, we have generated data demonstrating that the underlying mechanisms of repair may recapitulate ontogeny. If lung fibroblasts are deprived of PTHrP, their structure changes: First, the PTHrP receptor is downregulated, as are its downstream targets ADRP and leptin; the decline in the lipofibroblast phenotype is mirrored by the gain of the myofibroblast phenotype, characteristic of fibrosis.

During the process of fetal lung development, the mesodermal fibroblasts are characterized by Wnt/β-catenin signaling that determines the splanchnic mesodermal fibroblast. We have shown that during alveolarization the formation of lung fluid actively upregulates the PTHrP signaling pathway in the endoderm by distending the alveolar wall, causing the downregulation of the Wnt/β-catenin pathway, leading to the differentiation of the lipofibroblast. These cells dominate the alveolar acinus during fetal lung development, but are highly apoptotic in the postnatal lung, giving rise to the alveolar septa. A functional hallmark of this paracrine determination of the mesodermal cell-types is the failure of the fibroblasts to terminally differentiate.

Phylogenetically, the swim bladder and frog lung interstitium are characterized by myofibroblasts; lipofibroblasts do not appear during phylogeny until reptiles and mammals. The recapitulation of myofibroblasts during lung injury is consistent with the similarities between lung ontogeny and phylogeny, and with the molecular mechanisms of fibroblast transdifferentiation described above, and may, therefore, represent the process of lung evolution in reverse.

A wide variety of factors can inhibit the normal paracrine induction of the lipofibroblast, and promote myofibroblast proliferation and fibrosis, including prematurity, barotrauma, oxotrauma, nicotine, and infection. In all of these instances, injury of the epithelial type II cell can cause downregulation of PTHrP, so that the mesodermal fibroblasts default to the myofibroblast phenotype. Myofibroblasts cannot promote the growth and differentiation of the alveolar type II cell for alveolarization; moreover, they produce angiotensin II, which further damages the type II cell population.

The PTHrP receptor is present on the surfaces of adepithelial fibroblasts. Stretching of the alveolus by fluid or air coordinately upregulates both PTHrP ligand and PTHrP receptor activity, promoting surfactant production by the type II cell, and lipofibroblast neutral lipid uptake, protecting both of them against oxidant injury. PTHrP receptor binding stimulates cAMP-dependent PKA expression, which determines the lipofibroblast phenotype. Treatment of the transdifferentiating myofibroblast, either *in vitro* or *in vivo*, with PPARγ agonists blocks the transdifferentiation of the myofibroblast, preventing fibrotic injury.

The Roles of PPARγ in Ontogeny and Repair

PTHrP induces lipofibroblast differentiation via the PKA pathway, which blocks Wnt signaling by inhibiting both Gli and glycogen synthase kinase (GSK) 3β, and upregulates the lipofibroblast phenotype, PTHrP receptor, ADRP, leptin, and triglyceride uptake by stimulating PPARγ expression.

On the basis of the minimalist idea that development culminates in homeostasis, disruption of homeostasis may lead back to earlier developmental and evolutionary motifs. This occurs in various lung diseases, and by focusing on the continuum from development and evolution to homeostasis, we can select treatments that are more consistent with promoting cellular physiologic reintegration than merely stopping inflammation. For example, bronchopulmonary dysplasia (BPD) can be induced by over-distending an otherwise healthy but immature newborn baboon lung. Destabilizing the homeostatic balance of the alveolus by knocking out surfactant protein genes B, C, or D leads to alveolar remodeling that is either grossly flawed (B) or less than optimal (C, D) physiologically. Interfering with epithelial-mesenchymal signaling blocks lung development, usually resulting in alveolar simplification (or "reverse evolution"). Conversely, replacing missing developmental elements can re-establish lung development, homeostasis, and structure.

Repair recapitulates ontogeny because it is programmed to express the cross-talk between epithelium and mesoderm through evolution. This model is based on three key principles:

1) the cross-talk between epithelium and mesoderm is necessary for homeostasis;
2) damage to the epithelium impedes the cross-talk, leading to loss of homeostasis and readaptation through myofibroblast proliferation;
3) normal physiology will either be faithfully re-established, or cell/tissue remodeling/altered lung function may occur, and/or fibrosis will persist, leading to chronic lung disease.

The cell-molecular injury affecting epithelial-mesenchymal cross-talk recapitulates ontogeny and phylogeny (in reverse), providing effective diagnostic and therapeutic targets.

The Overall Relevance of Lung Evolution to Physiologic Evolution

The premise of this chapter is that the "first principles of physiology" (FPPs) do exist and are knowable. We just need to be artful enough to identify, define, and validate them. Our working hypothesis is that such FPPs were reprised and co-opted during the transition of vertebrates from water to land, beginning with the acquisition of cholesterol by unicellular yeast-like eukaryotes, facilitating unicellular evolution over the course of the first 4.5 billion years of the Earth's history in service to the reduction in intracellular entropy, far from equilibrium – that feat is what initiated life, which has been perpetuated by chemiosmosis and homeostasis from that day to this. The iterative process of sustaining and perpetuating negentropy in service to life forms ultimately gave rise to the metazoan homologs of the gut, lung, kidney, skin, bone, and brain. Central to this working hypothesis is that homeostatic control is flexible, or "plastic," allowing for inheritance of a range of set-points, referred to as reaction norms, rather than just one genetically fixed state. It is important to note that this perspective is 180 degrees out of phase with traditional Darwinian evolution, which is exclusively based on random mutation and natural selection. Yet such plasticity is totally in keeping with contemporary ideas as the Barker hypothesis for the fetal origins of adult disease, the role of epigenetics, and what we know of the variation in growth factor determination of morphogenesis and homeostasis. *Tiktaalik*, the fossilized fish-tetrapod transitional organism first reported by Neil Shubin in 2004, provides a heuristic template for the vertebrate water-to-land transition. To make that transition, *Tiktaalik* had to have been "pre-adapted" for respiration as well as for those traits of the kidney, skin, gut, bone, and brain amenable to land life. Thought of in the context of fish physiology as the antecedent for such a critical transition, the swim bladder, importantly, has been definitively shown to be structurally, functionally (as a gas exchanger), and genomically homologous to the tetrapod lung. Both the swim bladder and lung are outpouchings of the gut, mediating the uptake and release of atmospheric oxygen and carbon dioxide. Furthermore, among the most highly expressed genes in the zebrafish swim bladder during development is PTHrP, whose signaling receptor underwent a gene duplication event during the phylogenetic transition from fish to amphibians. That event made atmospheric gas exchange for the water-to-land transition feasible, since PTHrP promotes the formation of lung alveoli. If one deletes the PTHrP gene in mice, the offspring die at birth due to the absence of alveoli. PTHrP is expressed in the epithelial cells of both the swim bladder of fish and the lung alveoli of land vertebrates. In alveoli, PTHrP stimulates the production of surfactant, which maintains alveolar structure and function by reducing surface tension; in the absence of surfactant, the alveoli will collapse, rendering them dysfunctional.

PTHrP: Determinant of Lung Cell-Molecular Evolutionary Homeostasis

PTHrP is a small peptide that is secreted by alveolar type II cells in response to lung inspiration distending the walls of the alveoli. PTHrP acts locally via its cell surface receptor on the adjacent mesoderm, inducing specialized connective tissue fibroblasts to become lipofibroblasts. The lipofibroblasts are critical to understanding the evolution of the lung for two reasons: (i) they protect the alveolus against oxidant injury by actively recruiting and storing neutral lipids from the alveolar microcirculation, acting as antioxidants; and (ii) the stored neutral lipids are actively mobilized from the lipofibroblasts to the alveolar type II cells for surfactant synthesis through the mechanically coordinated biochemical effects of PTHrP, leptin, and prostaglandin E_2, which act via their cognate receptors residing on the apposing surfaces of neighboring epithelial type II cells and lipofibroblasts. Ultimately, PTHrP regulates alveolar epithelial calcium homeostasis, reprising its evolutionary history in maintaining calcium/lipid homeostatic balance as follows: calcium concentrations in the alveolar aqueous, protein-containing aqueous hypophase regulates the formation and dissolution of tubular myelin. That in turn determines the surface tension-reducing effect of the tubular myelin on the alveolus; tubular myelin is a lipid-β-defensin complex homologous with the lipid-β-defensin barrier generated by the epithelium of the skin to form the stratum corneum, preventing fluid leakage and protecting against microbial infection. Since the skin is the most primitive organ of land vertebrate gas exchange, it may have provided a molecular homeostatic co-option for further evolution of the lung.

It can be calculated that such a mechanism, dependent upon coordinated interactions between the endoderm and mesoderm for the existential regulation of surfactant, would have taken 9×10^{16} years to have occurred by chance, which is seven orders of magnitude longer than the estimated 5×10^{9} year existence of the Earth. Alternatively, the exaptation of the FPPs, starting with cholesterol facilitating gas exchange in unicellular organisms, iteratively co-opted over the course of vertebrate evolution ontogenetically and phylogenetically, would have only taken the last 500 million years to have transpired.

ADRP as a Deep Homology that Interconnects Evolved Functional Homologies, or "Oh, the Places You'll Go!" – Dr Seuss

The key molecule that mediates neutral lipid trafficking between the alveolar microcirculation, lipofibroblast, and epithelial type II cell is adipocyte differentiation related protein (ADRP). It is a member of the perilipin-ADRP-TIP47, or PAT, family of intracellular lipid cargo proteins that mediate lipid uptake, storage, and secretion in a wide variety of cells, tissues, and organs, ranging from fat cells to endothelium, liver, and steroidogenic endocrine organs. PAT proteins are expressed in many organisms, ranging from mammals to slime molds and fungi. ADRP was first discovered to be involved in early adipocyte differentiation and subsequently shown to be necessary for the uptake and storage of intracellular lipid droplets when overexpressed in Chinese hamster ovary cells, which do not naturally express ADRP.

In the lung, ADRP in lipofibroblasts is physiologically stimulated by stretching alveolar type II cells, which produce PTHrP. The PTHrP then binds to its cognate receptors on lipofibroblasts, stimulating PPARγ, which subsequently upregulates ADRP. This mechanism may have initially evolved to protect the alveolar wall against hyperoxia, since the rising atmospheric oxygen tension over the course of the Phanerozoic eon caused the differentiation of myofibroblasts into lipofibroblasts. This mechanism may subsequently have been co-opted to regulate surfactant synthesis during the vertebrate water-to-land transition, consistent with the phylogenetic adaptation of the alveolus from the swim bladder of fish to the highly adapted lungs of mammals and birds. This phenomenon is of particular interest in the context of exploiting such functional molecular homologies when one considers the homologies between the alveolar lipofibroblast and endocrine steroidogenesis. For example, oxygen in the atmosphere did not increase linearly from 0 to 21%; rather, it has gone up and down episodically, ranging between 15 and 35% over the past 500 million years. Bearing in mind that hypoxia is the most stressful of all physiologic agonists, alternating hyperoxia and hypoxia would have put huge physiologic constraints on both the evolving lung and endocrine systems. Perhaps fortuitously, the vertebrate pulmonary and endocrine systems were pre-adapted for such circumstances through the action of PAT genes; thus, the well-recognized effects of the adrenocortical system on lung development and homeostasis can be seen as part of a logical evolutionary progression of positive external and internal selection mechanisms.

This is not a tautologic "Just So Story," since, for example, the same morphogenetic mechanisms occur during both ontogeny and phylogeny, and we witness the reversal of this evolutionary process in chronic lung diseases, in which there is "simplification" of the alveolar bed, resulting in a frog-like structure in emphysema, for example. In this vein, experimentally, Valérie Besnard *et al.* found that when they deleted a gene necessary for the synthesis of cholesterol, the most primitive of lung surfactants, specifically by mouse lung alveolar type II epithelial cells, the lung was unaffected. Upon microscopic examination, the lung developmentally "compensated" for the poor quality of the cholesterol-less lung surfactant by the hyperproliferation of the lipofibroblast population in the alveoli, suggesting that these cells have an evolutionary capacity to facilitate surfactant production, both ontogenetically and phylogenetically. This rational cell-molecular approach to understanding how and why the lung evolved can be carried one step further, since catecholamine/β-adrenergic receptor signaling was essential for the regulation of blood pressure in the lung independent of the systemic circulation, facilitating a further increase in the surface area of the evolving lung. Our ancestors were able to survive the whipsawing physiologic effects of alternating hyperoxia and hypoxia by structurally and functionally adapting their pulmonary and endocrine systems (see below). Here again, as in the case of PTHrP signaling, the β-adrenergic receptor also underwent a gene duplication during the fish–amphibian transition that allowed for a further increase in lung surface area to support metabolic demand. At this phase in vertebrate evolution, the glucocorticoid receptor is documented to have evolved from the mineralocorticoid receptor, perhaps as a counterbalancing selection for the blood pressure-elevating effect of mineralocorticoids, alleviating the additional increase in blood pressure due to life on land versus water. The shunting of mineralocorticoid activity to glucocorticoids was achieved by the addition of two amino acid residues to the mineralocorticoid receptor, likely due to the cumulative effects of vascular shear stress. That epistatic mechanism would have been synergized by concomitant glucocorticoid stimulation of β-adrenergic receptor expression, further alleviating the blood pressure constraint imposed by the systemic circulation on the lung microvasculature.

The emergence of the physiologic glucocorticoid mechanism may have been further facilitated by the presence of pentacyclic triterpenoids in land vegetation, a product of rancidification unique to the land vegetation. These compounds inhibit 11β-hydroxysteroid dehydrogenase type II (11β-HSD2), which inactivates cortisol's blood pressure-stimulating activity, causing positive selection pressure for the tissue-specific expression of 11β-HSD1,2 in a wide variety of glucocorticoid target organs, including the lung, thereby permitting more efficient, local tissue-specific activation and inactivation of cortisol. Reinforcing this hypothesis, when pituitary adrenocorticotropic hormone (ACTH) stimulates glucocorticoid production by the adrenal cortex, the hormone passes through the intra-adrenal portal vascular system of the medulla, providing it with uniquely high local concentrations of glucocorticoids. These high concentrations are needed to induce the medullary enzyme phenylethanolamine-N-methyltransferase (PNMT), which controls the rate-limiting step in catecholamine synthesis, thus coordinately upregulating both of the primary adrenal stress hormones for a maximally adapted "fight or flight" response. As a further proof of this mechanism, in fish the adrenal cortex and medulla are independent structures, further attesting to the active selection pressure for robust adrenaline (epinephrine) production in response to physiologic stress.

PTHrP and the Evolution of Kidney Cell-Molecular Homeostasis

Akin to its role as a stretch-regulated gene product for alveolar homeostasis, PTHrP is also integral to renal physiology (see Figure 9.3). In the glomerulus, PTHrP is produced by the epithelially derived podocytes that envelop the glomerular capillaries, maintaining the function of the mesangium, a stretch-sensitive fibroblast-derived structure that determines systemic fluid volume and electrolyte homeostasis by regulating glomerular filtration. As an aside, the functional molecular homology between the lung and kidney should not be surprising, since both structures contribute to the formation of amniotic fluid during embryonic development. It should also be borne in mind that the glomerulus also makes its appearance during the phylogenetic transition from fish to amphibians, and subsequently to reptiles, mammals, and birds.

The fish kidney is comprised of a primitive microvascular structure, the glomus. Elevated blood pressure during the water–land transition may have promoted capillary formation within that structure since PTHrP is angiogenic. The net result may have been the emergence of the PTHrP-regulated glomerulus in terrestrial vertebrates, emerging from the constitutively pre-adapted glomus.

PTHrP and the Evolution of Skin Cell-Molecular Homeostasis

PTHrP is essential for the development of skin, mediated by paracrine interactions between melanocytes and keratinocytes, the epithelium generating the stratum corneum as a barrier for water and microorganisms, essential for preventing desiccation in terrestrial vertebrates. It is noteworthy that the alveolar type II epithelial cells and the skin epithelium of the stratum corneum exhibit a functional homology at the cell-molecular level, packaging lipids together with host defense peptides, secreting them in the form of lamellar bodies to generate lipid-based barriers against water loss (from the inside out), and host invasion (from the outside in) in both structures.

The evolutionary significance of the homology between lung and skin as barriers is further exemplified by the pathophysiology of asthma. Patients with asthma are also afflicted with the skin disease atopic dermatitis, or atopy. Both of these phenotypes are common to humans and dogs, and have been mechanistically linked through β-defensin polymorphisms, which mediate innate host defense in both skin and lung. In dogs, β-defensins in the skin determine coat color, which serves a variety of adaptive advantages, ranging from protective coloration to their associated reproductive strategies. The β-defensin CD103 has also been shown to cause atopic dermatitis in dogs, and possibly asthma, since it is also found in dog airway epithelial cells. Therefore, hierarchically, host defense and reproduction take evolutionarily adaptive precedence over wheezing due to asthma. A similar interrelationship between β-defensins and asthma has been documented in a cohort of Chinese children.

The Goodpasture Syndrome: Lesson in Evolution

Vertebrates transitioned from water to land approximately 300 million years ago, causing selection pressure for type IV collagen, which acts to physically maintain epithelial integrity throughout the body, including the walls of the alveoli and glomeruli. Since the extracellular matrix forms during the processes of cellular growth and differentiation, it is highly likely that modifications of the basement membrane occurred early in the evolutionary adaptation to land. Molecular evolutionary studies of Goodpasture syndrome, characterized by the formation of antibodies to type IV collagen, sometimes causing death due to lung and kidney failure, have revealed that the 3α Goodpasture isoform of type IV collagen evolved during the phylogenetic transition from fish to amphibians due to positive selection pressure for specific amino acid substitutions that rendered the molecule more hydrophobic and negatively charged, providing a natural barrier against water loss. Basement membrane extracts from *Caenorhabditis elegans*, *Drosophila melanogaster*, and *Danio rerio* do not bind Goodpasture autoantibodies, while frog, chicken, mouse, and human basement membranes do bind such autoantibodies. The type IV collagen isoform characteristic of Goodpasture syndrome is not present in worms (*C. elegans*) or flies (*D. melanogaster*), and is first detected phylogenetically in fish (*D. rerio*). Three-dimensional molecular modeling of the human Goodpasture type IV collagen isomer suggests that evolutionary alteration of electrostatic charge and polarity due to the emergence of critical serine, aspartic acid, and lysine amino acid residues, accompanied by the loss of asparagine and glutamine, contributed to the emergence of the Goodpasture epitopes, as the protein evolved from over the ensuing 450 million years.

The functional difference between the duplication of the PTHrP receptor, the β-adrenergic receptor, and the glucocorticoid receptor and that of the Goodpasture type IV collagen, all evolving during the water-land transition, is of interest. Although all of these evolutionary adaptations were the result of gene mutations brought on by vascular shear stress, only the matrix protein causes pathology. We speculate that the receptor-mediated mechanisms were constrained by the adaptive signaling history of the physiologic "niches" in which they appeared, whereas the type IV collagen lacked limiting biofeedback.

Internal and External Selection, PTHrP, and the Water-to-Land Transition

Another way to think about the co-option of cell-molecular mechanisms of evolution is as serial interactions between internal and external selection pressures. Such external environmental constraints to the transition from water to land as air breathing, gravitational effects on blood pressure, and desiccation were all hypothetically adapted through a common, internal cell-molecular pathway for development and homeostasis – PTHrP and its cognate G protein-coupled receptor. This model is also predictive, since PTHrP is a potent vasodilator, and an angiogenic factor (promotes capillary formation), potentially explaining why the fish kidney glomus, as a microvascular structural-functional derivative of the renal artery, may have evolved glomeruli in the transition from fish to amphibians.

The significance of PTHrP in the vertebrate transition from water to land may be as follows: Such organisms must have been selected for their ability to spontaneously overexpress PTHrP signaling, initially for lung evolution from the swim bladder, specifically in physostomous fish like zebra fish, which possess a tracheal homolog, the pneumatic duct that connects the esophagus and swim bladder for gas filling and emptying. At the cell-molecular level, the smooth muscle that forms both the pneumatic duct and trachea is determined by fibroblast growth factor 10 (FGF10) expression.

Independent evidence for such positive selection for the lung comes from the comparative study of physiologic systems by Ewald Weibel and his colleagues. They had hypothesized that physiologic traits were efficiently adapted to their functions, which they termed symmorphosis. Their results were consistent with this hypothesis, with the exception of the lung, which they found to be "over-engineered." This may be due to the necessity for PTHrP–PTHrP receptor signaling plasticity as a *sine qua non* for air breathing.

The PTHrP-mediated mechanisms in the kidney and skin hypothetically followed suit, since they would have protected against desiccation during the terrestrial adaptation, both being dependent on PTHrP for their development and homeostasis on land. Calcification of bone in response to increased gravitational force on land would have further facilitated adaptation to terrestrial life; Wolff's law states that bone will conform to the physical load under which it is placed. PTHrP is a gravity-sensitive paracrine hormone that is integral to bone development and homeostasis, determining bone calcium uptake and incorporation into cartilaginous structures, facilitating the adaptation of terrestrial organisms to environmental gravitational forces. This scenario of an iterative process for the acquisition of traits that facilitated the water-to-land transition is consistent with data showing that vertebrates attempted the water-to-land transition at least five times. It is reasonable to assume that the visceral organs also had to have evolved in adaptation to land habitation.

Based on parsimony, one can propose that these processes were all realized as a result of the PTHrP receptor gene duplication event that occurred during the water-to-land transition, beginning with the lung, by necessity, and that those organisms that could upregulate their PTHrP–PTHrP receptor signaling through ligand-receptor-mediated paracrine mechanisms evolved as the forebears of contemporary land vertebrates. In contrast, lineages that were unable to accomplish this feat became extinct. This perspective is supported by our demonstration of the correlation between the cell-molecular genetic motifs common to ontogeny and phylogeny of the lung and major environmental epochs (see Figure 15.1). Note the apparently seamless alternations between internal and external selection mechanisms in association with major ecologic stresses; we postulate that there are no gaps between these genetic adaptations because the data are derived from the surviving land vertebrates; conversely, those lineages of the species that failed to adapt died off, and thus are not accounted for in this analysis.

In further support of this concept, it is noteworthy that many chronic lung diseases are typified by simplification, or "reverse evolution." This mechanism is due to the loss of signaling between the epithelial and mesenchymal compartments of the alveoli, leading to increased diameter of the alveoli, which seemingly revert to earlier ontogenetic-phylogenetic stages in lung evolution. At the cellular level, this is characterized by the atavistic expression of both the myofibroblast, determined by the Wnt pathway, which is re-expressed due to loss of PTHrP signaling from neighboring epithelial cells. Moreover, experimentally providing the spatio-temporal signals that generated the mammalian alveolus ontogenetically and phylogenetically – PTHrP, leptin, and PPARγ – recapitulates the evolution of the lung, providing solid evidence for this mechanism.

Cellular Growth Factors, the Universal Language of Biology

This approach to a fundamental, *a priori* understanding of vertebrate physiology, not as a top-down descriptive process, but as a series of exaptations originating from the cell membrane of unicellular organisms, will ultimately lead to an understanding of the FPPs based on their evolutionary origins. The actualization of such FPPs would have numerous advantages, primarily a predictive model for physiology and medicine, as well as a functional merging of biology, chemistry, and physics into a common algorithm for the natural sciences. Such a perspective would allow us to de-emphasize the human signature from our anthropocentric view of our physical environment, on the scale of the Copernican recentering of the solar system on the Sun, which clarified our perception of our Universe, and that of other universes.

What Predictions Derive from a Cellular Approach to Evolution

Starting with the premise that ontogeny is the only biologic process we know of that generates structures and functions, we should exploit this process to understand evolution, since it does so throughout the phylogenetic history of the organism. By focusing on cell–cell interactions, particularly those mediated by soluble growth factors and their cognate receptors, one can deconvolute the evolution of the lung, functionally tracking it back to the swim bladder of physostomous fish; unlike physoclistous fish, this category of bony fish has a tracheal homolog that connects the esophagus to the swim bladder, determined by FGF10 signaling in common with the lung. The developing zebrafish swim bladder has been shown to express all of the genes necessary for lung development.

Since the adaptation of fish to land was contingent on efficient atmospheric gas exchange, the lung can be seen as the cellular-molecular template for the evolution of other physiologic adaptations to land life. By systematically tracing the functional molecular homologies between the lung, adrenals, skin, kidney, gut, bone, and brain across developmental, phylogenetic, homeostatic, and pathophysiologic space and time, the FPPs can be determined. Once such relationships are traced back to unicellular organisms, the underlying physiologic principles can be used to replay the evolutionary tape, and predict and prevent homeostatic failure as disease. Perhaps as E.O. Wilson has suggested, the reduction of biology to ones and zeros can offer the opportunity to merge biology, chemistry, and physics into one common user- friendly algorithm as a "periodic table of nature".

The Zygote as the Primary Level of Selection for Vertebrate Evolution

Another prediction of the cellular-molecular integrated approach to evolution is the primacy of the unicellular state as the founding and ongoing mechanism of evolution. It provides the functional bauplan for the biota as the reference point for sustaining life on Earth. In support of that notion, a thought experiment is in order. What if there were organisms that deviated from the ubiquitous use of nucleotides for biologic "memory"; I would submit that they are extinct because the absence of such a mechanism for informing the organism of its past evolutionary course puts it at a competitive disadvantage.

Like the realization of our physical place in the Universe by shifting to a heliocentric perspective, ushering in the Age of Enlightenment, our recognition that it is the unicellular state that is being selected for, and that all other aspects of the life cycle are in service to it, would enlighten us as to our biologic place in the Universe. That "paradigm shift" would inform the relationship between hominids and other biota as a continuum with the physical realm, and permit us to make rational *a priori* decisions about bioethical questions based on FPPs, rather than "guessing" the right course of action after the fact, as we have been doing since time immemorial. In short, the cell-centric perspective would herald a new Age of Enlightenment, and none too soon, given the exorbitant cost of healthcare, the pollution of our environment, and climate change.

The multicellular state, which Gould and Richard Lewontin called "spandrels," is merely a biologic agent for monitoring the environment between unicellular stages in order to register and facilitate adaptive changes. Samuel Butler speculated that perhaps "A hen is only an egg's way of making another egg," begging the question of hierarchy in the life cycle. This consideration can be based on both *a priori* and empiric data. Regarding the former, emerging evidence for epigenetic inheritance demonstrates that the environment can cause heritable changes in the genome, but they only take effect phenotypically in successive generations. This would suggest that it is actually the germlines of the offspring that are being selected for. The starvation model for the metabolic syndrome illustrates this experimentally. Maternal food deprivation can cause obesity, hypertension, and diabetes in the offspring. But the offspring also enter puberty precociously due to premature adrenarche, due to the ability of the adrenal gland to produce dehydroepiandrosterone. Though seemingly incongruous, this may represent the primary strategy to accelerate the genome transfer to the next generation (positive selection), effectively foreshortening the exposure to a food-deprived environment, risking death. The concomitant obesity, hypertension, and diabetes are unfortunate side effects of this otherwise adaptive process in the adults. Under these circumstances, it can be surmised that it is the germ cells that are being selected for; in other words, the adults are disposable, as Richard Dawkins has opined.

Hologenomic evolution theory provides yet another mechanism for selection emerging from the unicellular state. According to that theory, all complex organisms actually represent a collaborative of linked, co-dependent, cooperative, and competitive localized environments and ecologies functioning as a unitary organism. These linked ecologies are comprised of both the innate cells of the organism, and all of the microbial life that is cohabitant with it. The singular function of these ecologies is to maintain the homeostatic preferences of the constituent cells. According to this theory, evolutionary development is the further expression of cooperation, competition, and connections between the cellular constituents in each of those linked ecologies in successive generations as they successfully sustain themselves against a hostile external environment. Ontogeny would then recapitulate phylogeny since the integrity of the linked environments that constitute a fully developed organism can only be maintained by reiterating those environmental ecologies in succession toward their full expression in the organism as a whole.

Another way to think about the notion of the unicellular state as the one being selected for is to focus on calcium signaling as the initiating event for all of biology. There is experimental evidence that increases in carbon dioxide during the Phanerozoic eon caused acidification of the oceans, with consequent leaching of calcium from the ocean floor. The rise in calcium levels can be causally linked to the evolution of the biota, and is intimately involved with nearly all biologic processes. For example, fertilization of the ovum by sperm induces a wave of calcium, which triggers embryogenesis. The same sorts of processes continue throughout the life cycle, until the organism dies. There seems to be a disproportionate investment in the zygote from a purely biologic perspective. However, given the prevalence of calcium signaling at every stage, on the one hand, and the participation of the gonadocytes in epigenetic inheritance, on the other, the reality of the vectorial trajectory of the life cycle becomes apparent – it cannot be static, it must move either toward or away from change.

By using the cellular-molecular ontogenetic and phylogenetic approach described above for the water–land transition as a major impetus for evolution, a similar approach can be used moving both forward and backward developmentally and phylogenetically from that critically important phase of vertebrate evolution. In so doing, the gaps between unicellular and multicellular genotypes and phenotypes can realistically be filled systematically. The caveat is that until experimentation is done, these linkages remain hypothetical. Importantly though, there are now model organisms and molecular tools to test these hypotheses, finally looking at evolution in the direction in which it occurred, from the earliest iteration forward. This approach will yield *a priori* knowledge about the first principles of physiology, and how they have evolved to generate form and function from their unicellular origins.

Bioethics Based on Evolutionary Ontology and Epistemology, not Descriptive Phenotypes and Genes

By definition, a fundamental change in the way we perceive ourselves as a species would cause a commensurate change in our ethical behavior. Such thoughts are reminiscent of a comment in a recent biography of the British philosopher Derek Parfit in the *New Yorker* magazine, entitled "How to be Good," in which he puzzles over the inherent paradox between empathy and Darwinian survival of the fittest. These two concepts would seem to be irreconcilable, yet that is only because the latter is based on a false premise. Darwin's great success was in making the subject of evolution user friendly by providing a narrative that was simple and direct. Pleasing as it may be, it is at best, entirely incomplete. Think of it like the transition from Newtonian mechanics to relativity theory. As much is learned about the unicellular world with its surprising mechanisms and capacities, new pathways must be imagined. It is clear that we as humans are hologenomes, and all the other complex creatures are too. In fact, there are no exceptions. The reasons for this can only be understood properly through a journey forward from the "Big Bang" of the cell, with all its faculties and strictures. By concentrating on cellular dynamics, an entirely coherent path is opened up. Tennyson's line about "Nature, red in tooth and claw" is only the tip of what the iceberg of evolution really constitutes. As pointed out above, we evolved from unicellular organisms through cooperation, co-dependence, collaboration, and competition. These are all archetypical cellular capacities. Would we not then ourselves, as an example of cellular reiteration, have just those self-same and self-similar behaviors?

Summary

By looking at the process of evolution from its unicellular origins, the causal relationships between genotype and phenotype are revealed, as are many other aspects of biology and medicine that have remained anecdotal and counterintuitive. That is because the prevailing descriptive, top-down portrayal of physiology under Darwinism is tautologic. In opposition to that, the cellular-molecular, bottom-up approach is conducive to prediction, which is the most powerful test of any scientific concept. Though there is not a great deal of experimental evidence for the intermediate steps between unicellular and multicellular organisms compared to what is known of the ontogeny and phylogeny of metazoans, it is hoped that the perspectives expressed in this chapter will encourage more such fundamental physiologic experimentation in the future.

The current Chapter takes the dogma of development and shows how it becomes part of the continuum of evolution using the principles provided in the previous chapters. Chapter 17, entitled "A central theory of biology," provides the first comprehensive perspective on the "first principles of biology." By utilizing the unique view provided by cell biology as the common denominator for ontogeny and phylogeny, biology can be seen as having a logic.

17

A Central Theory of Biology

Introduction

The underlying unity of nature has been sought ever since the time of the ancient Greek philosophers. More recently, Lancelot L. Whyte formulated a way of thinking about unitary biology, but it lacked scientific support. Others like David Bohm and Herb Benson have offered ways of generating unity, acknowledging the underlying problem of our own self-perception. The present approach offers a scientific basis for viewing biology as primarily being unicellular in nature, multicellularity being an epiphenomenon. This conceptualization is scale-free and predictive, offering a "central theory of biology."

Vertebrate evolution is a chronicle of the adaptation to oxygen for ever-increasing metabolism. One assumes that in their contemporary forms, vertebrates evolved in direct response to metabolic drive, but this process is far more interactive than just evolution being "fueled" by oxygen; the cellular mechanisms by which oxygen is integrated into the biologic cellular mechanisms of ontogeny and phylogeny behave like a cipher. Seen longitudinally, evolution is a functionally linked continuum of emergent and contingent processes resulting from the recombination and permutation of genetic traits that were first expressed in unicellular organisms. That is a very different image, more like doing a crossword puzzle, and the answer pops out of the matrix.

Conventionally, evolutionary biology is teleological, undermining its mission in explaining the processes involved. Instead, identifying mechanisms that were exapted from seemingly unrelated ancestral traits is particularly valuable in avoiding such circular "Just So Stories."

In this context, the events surrounding the water-to-land transition that fostered vertebrate adaptation to land are instructive, and are highly relevant to human physiology. Moreover, because they provide insight to the emergent and contingent mechanisms underlying endothermy/homeothermy in mammals and birds, they can be reverse-engineered to determine the intermediate physiologic steps in land vertebrate evolution.

Water-to-Land Transition as the Platform for Vertebrate Evolution

Alfred Romer hypothesized that land vertebrates emerged from water some 400 mya in response to the desiccating effect of rising levels of carbon dioxide in the atmosphere, drying up bodies of water globally. Indeed, based on the fossil record, vertebrates

Evolution, the Logic of Biology, First Edition. John S. Torday and Virender K. Rehan.

breached land on at least five occasions, indicating the magnitude and direction of the selection pressure to "gain ground." But it is remarkable that no attention has been paid to the self-evident, concomitant evolution of the visceral organs involved during this key transitional period, other than to document the phylogenetic differences between fish, amphibians, reptiles, mammals, and birds. This becomes particularly disturbing when you see the commonalities between the genes involved in this process.

The disconnect between such phenotypic observations and the underlying mechanisms of evolution is due to the systematic emphasis placed on random mutation and population selection by conventional Darwinian evolutionists. In contrast to this dogmatic approach, we have pointed out the value added in determining the cellular-molecular adaptation to oxygenation in forming the mammalian lung through the specific cell–cell interactions known to determine its embryogenesis. Such cellular developmental mechanisms are mediated by soluble growth factors and their receptors, acting iteratively in response to alternating external and internal selection pressures to generate form and function based on homeostatic principles. The history of such cellular-molecular interrelationships can be traced as far back as the unicellular state by following the pathways formed by lipids in accommodating calcium homeostasis, and their consequent effects on oxygen uptake by cells, tissues, and organs. Through this *a priori* understanding of the fundament of evolution, the pitfalls of teleology and tautology can be avoided, and instead a predictive model of evolutionary biology can be formulated, as follows.

PTHrP Signaling is Essential for Understanding the Evolution of the Lung

The key empiric observation for understanding the evolution of the mammalian lung was the discovery that parathyroid hormone-related protein (PTHrP) (Figure 17.1), is necessary for the formation of alveoli, the gas exchange units that have fostered the evolution of the lung from the swim bladder of fish. When the PTHrP gene is deleted from the developing mouse embryo, the lung does not form alveoli. PTHrP is synthesized and secreted by the alveolar epithelial type II cells, and binds to the neighboring

(a) (b)

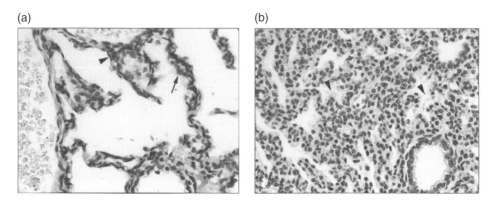

Figure 17.1 Parathyroid hormone-related protein (PTHrP) knockout lung. **a)** Normal lung with alveoli (arrow heads) in PTHrP +/+ mice. **b)** Deletion of the PTHrP gene in the developing mouse embryo (PTHrP –/–) results in failed alveolar formation at birth.

lung fibroblast via the G-protein-coupled PTHrP receptor (PTHrPR). This triggers the intracellular protein kinase A pathway, inducing the lipofibroblast phenotype. These cells protect the lung against oxidant injury by actively accumulating and storing neutral lipids. Lipofibroblasts subsequently evolved the capacity to actively provide neutral lipid substrate for lung surfactant phospholipid synthesis. Paracrine signaling from the lipofibroblast to the alveolar type II cell is mediated by the locally acting paracrine hormone leptin, which stimulates lung surfactant synthesis by the alveolar type II cells. These mutually interactive cell–cell interactions facilitate the molecular cross-talk between PTHrP and leptin for the mechanically regulated production of surfactant, since PTHrP, leptin, and their respective cell-surface receptors are all coordinately stretch-regulated genes. The neutral lipid trafficking process is orchestrated by adipocyte differentiation related protein (ADRP), which mediates the uptake, storage, and transit of neutral lipid from the lipofibroblast to the alveolar type II cell.

Once these cellular-molecular aspects of the functionally integrated mechanism for homeostatic regulation of lung surfactant were reconstructed, it was evident that they must have resulted from selection pressure for specific cellular functions since chance mutations alone would have taken longer than the existence of the Earth itself to have mediated such changes ($>9 \times 10^{18}$ years). Therefore, understanding the functional interrelationships between the molecular mechanisms and their phenotypes lay in how the lung surfactant subserves the alveoli both ontogenetically and phylogenetically. That is to say, the overarching process of lung evolution is characterized by the progressive decrease in alveolar diameter, which facilitates gas exchange by increasing the surface area-to-blood volume ratio between the alveolus and the alveolar capillaries that transfer oxygen to the peripheral tissues and organs.

The Physics of Lung Evolution

Laplace's law dictates that surface tension is inversely proportional to the diameter of a sphere such as the alveolus. We know from extensive phylogenetic studies by Christopher Daniels and Sandra Orgeig that the composition of the surfactant, and therefore its surface tension-reducing capacity, has changed progressively to compensate for the increasing surface tension caused by the evolutionary decrease in alveolar diameter. But that begs the question as to what cellular-molecular mechanisms facilitated such accommodations. Given that epithelial–mesenchymal interactions are responsible for alveolar morphogenesis, culminating in surfactant-mediated alveolar homeostasis, the logical hypothesis was that the epithelial and mesenchymal cells generating the alveoli evolved under selection pressure to modify the composition and production of the surfactant, fostering both the phylogenetic and ontogenetic decreases in alveolar diameter.

The mammalian lung evolved from the fish swim bladder, which uses gases to regulate buoyancy for feeding and other bodily functions. The swim bladder of physostomous fish is an outpouching of the esophagus, connected to the alimentary tract by the pneumatic duct, which is homologous with the trachea. For example, at the cell-molecular level both the pneumatic duct and trachea are formed from smooth muscle controlled by the interaction between Hedgehog protein and fibroblast growth factor 10 (FGF10). Furthermore, the swim bladder is lined by gas gland epithelial cells that synthesize and secrete cholesterol, the most primitive form of lung surfactant. Moreover, PTHrP is among the most highly expressed genes in zebrafish swim bladder development. Therefore, the functional homology between the swim bladder and lung can be discerned as the utilization of lipid to facilitate gas exchange. Utilizing cholesterol, the most primitive surfactant, to lubricate the inner surface of the swim bladder facilitates buoyancy for feeding on algae, which are as high as 68% lipid. This gas exchange mechanism is functionally homologous with the mammalian lung, utilizing surfactant phospholipids to facilitate gas exchange for efficient metabolism. This is essentially how François Jacob famously described evolution – as "tinkering." However, until now this process has been seen as the chance result of Darwinian mutation and selection, whereas in the present model structure and function have evolved from pre-existing cellular-molecular traits, determined by homeostatic changes in growth factor-mediated cell–cell communication.

Functional Homology Between Membrane Lipids and Oxygenation

These cellular-molecular homologies raise the question as to what atavistic unicellular trait or traits might have formed the basis for the functional interrelationships between membrane lipids and oxygenation. Early in the evolution of unicellular organisms, oxidant stress caused endoplasmic reticulum stress, resulting in the release of potentially toxic levels of stored calcium into the cytoplasm. Christian de Duve hypothesized the advent of the peroxisome in unicellular organisms to compensate for this potential calcium toxicity; the peroxisome is an organelle that utilizes lipids to protect against such excesses in intracellular calcium. This ancient relationship between the peroxisome, endoplasmic reticulum stress, and calcium homeostasis may underpin the ubiquitous effects of peroxisome proliferator activated receptor gamma (PPARγ) in preventing and treating a wide variety of inflammatory diseases. And this same receptor is crucial to longevity in laboratory mice, lifespan being determined by the same cell–cell communications that evolved to maintain calcium-lipid homeostasis. PPARγ is the nuclear transcription factor that determines the adipocyte phenotype, which protects against oxidant injury. When PPARγ is inhibited, the adipocyte re-expresses its atavistic muscle phenotype, characterized by alpha smooth muscle actin (αSMA). The contrast between the adipocyte and muscle phenotypes reprises the seminal role of cholesterol in facilitating the evolution of eukaryotes. As for the evolutionary origins of this relationship, Barbara Wold's research has shown that cultured muscle cells will spontaneously differentiate into adipocytes in 21% oxygen (room air), but not if cultured in 6% oxygen, bespeaking the role of atmospheric oxygen in the origins of the adipocyte phenotype.

Another functional indication for the role of lipid-calcium epistasis in evolution is the homology between the lung and skin (Figure 17.2). Both organs synthesize and secrete lipid-containing lamellar bodies in combination with host defense peptides to form watertight, antimicrobial "barriers." In the case of the skin, the stratum corneum secretes such an extracellular lipid-antimicrobial barrier. In the case of the alveolus, the alveolar type II cell secretes the surfactant film, termed tubular myelin, a lipid-protein complex composed of phospholipids and surfactant protein A (an antimicrobial peptide), similar in composition and structure to the lipid barrier formed by the stratum corneum.

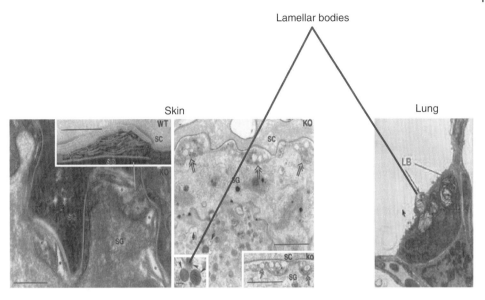

Figure 17.2 Molecular homology between skin and lung. Both the epithelial layer of the skin (*left panel*) and the alveolar type II cell of the lung (*right*) form lamellar bodies (LBs) composed of lipid and antimicrobial peptides (circled structures). In both cases, the LBs generate barriers against fluid loss.

So there is a fundamental homology between lipids, antimicrobial peptides, and barrier function exhibited by both the lung and skin. These structural-functional homologies refer as far back as the unicellular state, in which the cell membranes of eukaryotes were populated by cholesterol. In turn, the advent of cell membrane cholesterol promoted gas exchange, motility, and metabolism, the major evolutionary characteristics of all vertebrates. And since we now have experimental evidence that the unicellular form expresses the complete "toolkit" for multicellular organisms, it is feasible that the lipid-oxygen-barrier homology between the lung and skin evolved from the plasmalemma of unicellular organisms. Experimentally, manipulation of cell membrane cholesterol has shown that increasing the cholesterol content is cytoprotective, whereas loss of membrane cholesterol can cause cell death.

Atmospheric Oxygen, Physiologic Stress, Gene Duplication, and Lung Evolution

As indicated at the outset, the hypothesis to be tested is that visceral organ changes during the water-to-land transition were caused by physiologic stress. Based on the adaptive changes cited above, consider the consequences of episodic fluctuations in environmental oxygen, initially protected against by sterol hopanoids found in prokaryotic bacteria. Mechanistically, oxygen stimulates the sterol regulatory element binding protein (SREBP)/Scap family of enzymes that regulate sterol biosynthesis in prokaryotes and eukaryotes alike, reflecting the depth of this evolved trait. Konrad Bloch had hypothesized that the synthesis of cholesterol was due to the increased availability of atmospheric oxygen, since it takes six molecules of oxygen to make one molecule of cholesterol; however, bacteria do not produce cholesterol, so the oxygen-sterol connection must have some other origin.

For example, David Deamer has written extensively on the role of polycyclic hydrocarbons, omnipresent throughout the Universe, in the origins of life. Aromatic molecules delivered to the young Earth during the heavy bombardment phase in the early history of our Solar System were likely to be among the most abundant and stable organic compounds available. The "aromatic world" theory suggests that aromatic molecules might function as container elements, energy transduction elements, and templating genetic components for early life forms. These molecules can experimentally stabilize fatty acid vesicles much like cholesterol does in contemporary cell membranes, and can foster the biosynthesis of nucleotides.

During the Phanerozoic eon, much larger fluctuations in atmospheric oxygen, ranging between 12% and 35%, are widely recognized to have caused dramatic increases in animal body size; what has not been addressed previously are the physiologic consequences of the concomitant episodic decreases in oxygen that followed the increases, documented by Robert Berner *et al.* The effect of hypoxia, the most potent physiologic stressor known, is mediated by the hypothalamic-pituitary-adrenal axis in vertebrates. Pituitary adrenocorticotropic hormone (ACTH) stimulating corticoid production by the adrenal cortex subsequently stimulates catecholamine production by the downstream adrenal medulla. This physiologic mechanism is of evolutionary significance because catecholamines cause surfactant secretion from the lung alveoli, which would acutely have alleviated the hypoxic stress on the lung by further reducing surface tension, consequently increasing the distension of the alveolar wall. In turn, that would have stimulated alveolar type II cells to produce PTHrP, coordinately increasing both alveolarization and alveolar vascular perfusion. PTHrP is both a potent vasodilator and an angiogenic factor, thus comprehensively promoting the physiologic increase in gas exchange surface area over the course of evolutionary time.

Most importantly, the PTHrPR duplicated during the water-to-land transition, amplifying the PTHrP signaling pathway, thus validating this hypothetical evolutionary mechanism based on empiric evidence. One might wonder why the PTHrPR gene duplicated at this critical juncture in vertebrate evolution. As mentioned above, the visceral adaptive changes occurred in concert with at least five independent skeletal changes in order to breach land. The success of this path may specifically relate to the PTHrP signaling pathway, which directly affects bone formation and remodeling. Bone will re-conform structurally in response to physical force, referred to as Wolff's law. The only known mechanism for this effect is mediated by PTHrP, a gravisensor that regulates calcium uptake and accumulation by bone locally.

Duplication of the β-Adrenergic Receptor and the Glucocorticoid Receptor Genes

In further support of this hypothetical mechanism for physiologic adaptation, the other two gene duplications known to have occurred during the water-to-land transition were the β-adrenergic receptor (βAR) and the glucocorticoid receptor (GR), both of which facilitated vertebrate land adaptation. The increase in βARs alleviated the constraint on pulmonary blood pressure independent of systemic blood pressure. The GR evolved from the mineralocorticoid receptor (MR), likely due to the constraint of the orthostatic increase in blood pressure due to the increased force of gravity on land-adapting vertebrates; this was exacerbated by the effect of stress on mineralocorticoid stimulation of blood pressure, now offset by diverting some MR expression to the GR. This, combined with the synergistic effect of adrenocortical glucocorticoid production on adrenomedullary βAR production, synergized integrated physiology.

Increased PTHrP signaling in soft tissues such as the lung during the water-to-land transition would initially have promoted positive selection for those members of the species adapting to land having higher levels of PTHrP to facilitate bone adaptation. Moreover, physiologic stress is known to cause microvascular capillary shear stress, which causes genetic mutations, including gene duplications. Such an effect, particularly on the nascent pulmonary microvasculature, was critical for land adaptation, increased breathing causing stress on the lung microvasulature in particular.

Evolution of Endothermy/Homeothermy as Evidence for the Effect of Stress on Vertebrate Physiologic Evolution

One can easily dispute whether these physiologic adaptations were causal since there is no "hard" fossil evidence for this sequence of events, though the functional relationships are consistent with their contemporary roles in ontogenetically forming and phylogenetically maintaining homeostasis *a posteriori*. There is also an *a priori* scenario for the subsequent evolution of these integrated physiologic traits that is internally consistent with their ontogeny and phylogeny through the advent of endothermy/homeothermy. Since a non-teleologic explanation for the evolution of endothermy/homeothermy has not previously been formulated, by exploiting the above-mentioned gene duplications, a mechanism that entails such pre-existing physiologic traits that may conditionally have given rise to endothermy/homeothermy is proposed. In the scenario cited above for the selective advantage of catecholamines alleviating the constraint on air breathing, catecholamines would secondarily have caused the secretion of fatty acids from peripheral fat cells, consequently increasing metabolism and body temperature.

In tandem with the effect of intermittent hypoxia on catecholamine release of fatty acids from fat cells, it also has been shown to stimulate leptin secretion by adipocytes. Leptin, in turn, has been shown to increase the basal metabolic rate of ectothermic fence lizards.

The increase in body temperature would have interacted synergistically with the evolved mammalian lung surfactant, composed of saturated phosphatidylcholine, which at 37 °C functions 300% more actively to reduce surface tension than at 25 °C. This effect is due to the elevated phase transition temperature of saturated phosphatidylcholine (41 °C), the temperature at which the lung surfactant film collapses, no longer acting to reduce surface tension. The selection pressure for the coevolution of saturated phosphatidylcholine production by the alveoli and endothermy/homeothermy may have been due to the pleiotropic effects of catecholamines, stimulating both surfactant secretion by the alveoli, and coordinately increasing the unsaturated fatty acid composition of peripheral cell membranes, thereby increasing oxygen uptake by increasing membrane fluidity. The progressive phylogenetic increase in the percentage of saturated phosphatidylcholine in lung surfactant is indicative of the constitutive change in adaptation to endothermy/homeothermy. These fundamental changes in lipid composition in service to metabolism are exaptations of the events that initiated eukaryotic evolution. Considering the severe conditions generated by Romer's gap, during which vertebrates were virtually wiped off the face of the Earth, it should not be surprising that such deep homologies were recruited during this critical phase of vertebrate evolution.

Hibernation as Reverse Evolution

The causal nature of the interrelationship between physiologic stress, catecholamines, and endothermy/homeothermy is validated by the reverse effects of hibernation or torpor on lung surfactant lipid composition and cell membrane fatty acid composition. Under such conditionally low stress conditions, decreased catecholamine production results in both increased surfactant cholesterol, rendering lung surfactant less surface active, and decreased unsaturated fatty acid content of cell membranes, adaptively reducing oxygen uptake.

There is a phylogenetic precedent for lung surfactant facultatively accommodating ambient temperature. For example, in a study by Lester Lau and Kevin Keough it was found that maintaining map turtles at different ambient temperatures adaptively altered the composition of their lung surfactant. Ultimately, the ability to optimize lung alveolar physiology at various environmental temperatures may have been the precursor to endothermy/homeothermy. Experimental evidence for the causal interrelationships between body temperature, surfactant composition, and catecholamine regulation of surfactant secretion supports this hypothesis.

The cellular accommodation of environmental temperature by lipids is hypothetically an exaptation for the fundamental enabling effects of cholesterol at the origins of eukaryotic evolution. That this is not merely an association is corroborated by the evolution of the alveolar lipofibroblast in mammals. These adipocyte homologs provide a ready source of substrate for increased surfactant phospholipid production under physiologic demand for oxygen via the stretch-regulated mechanism described above. As further evidence for this hypothetical evolutionary mechanism, when cholesterol synthesis by alveolar type II cells is experimentally inhibited in the developing mouse lung alveolar type II cell by deleting the *Scap* gene, the lung compensates by increasing the number of lipofibroblasts. This compensatory mechanism is apparently due to the observed increase in PPARγ expression by these cells, likely due to endoplasmic reticulum stress, reprising how peroxisomes evolved in the first place.

As further evidence in support of the hypothesized role of hypoxia-induced endothermy, other significant mammalian-specific changes that occurred during vertebrate evolution are functionally consistent with this mechanism. First, PTHrP appears in both the mammalian pituitary and adrenal cortex, thus amplifying the fight-or-flight mechanism (Figure 17.3). Furthermore, Richard Wurtman has discovered that there are complex vascular arcades in the mammalian adrenal medulla, which act to amplify the production of catecholamines under stress conditions. Glucocorticoids produced in the adrenal cortex pass through the adrenal medulla, where they stimulate the rate-limiting step in catecholamine biosynthesis, phenylethanolamine-N-methyltransferase, thus enhancing its production for the stress reaction. This expansion of the medullary

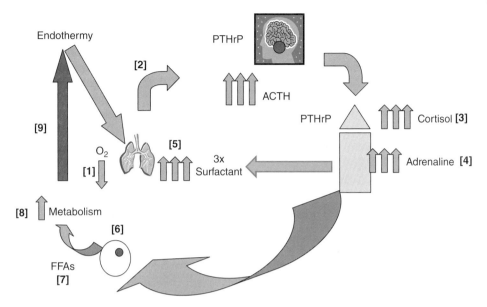

Figure 17.3 Evolution of endothermy. Periodic episodes of hypoxia [1] caused physiologic stress [2], stimulating cortisol production by the adrenal cortex [3], followed by adrenaline (epinephrine) production by the adrenal gland [4], releasing the constraint on the lung by stimulating alveolar surfactant secretion [5]. Ultimately, distension of the alveoli stimulated parathyroid hormone production, generating more alveoli. In tandem, adrenaline causes lipolysis of fat cells [6], releasing free fatty acids (FFAs) to the circulation [7], stimulating metabolism [8], increasing body temperature, and resulting in endothermy [9]. ACTH, adrenocorticotropic hormone; PTHrP, parathyroid hormone-related protein.

microvasculature may itself have been caused by the adrenocortical secretion of PTHrP, which is directly angiogenic. Speculatively, the combined effects of PTHrP on the adrenal cortex and medulla may have fostered the structural integration of the independent cortical and chromaffin tissues of fish in transition to the amphibian corticomedullary configuration.

It is also feasible that this complex cascade of physiologic stress-mediated cellular mechanisms gave rise to the kidney glomerulus, which is largely absent in fish, but is ubiquitous in amphibians, reptiles, mammals, and birds. PTHrP is the mediator of fluid and electrolyte balance in the glomerulus, being secreted by the podocytes lining this compartment, binding to its receptor on the mesangium, which regulates the amounts of fluid and electrolytes entering the kidney tubules. As is the case for the lung, the distension of the glomerulus is sensed by the podocyte, which then transduces that signal for fluid and electrolyte balance via PTHrP signaling. Here again is a functional homology between seemingly structurally and functionally disparate tissues and organs, based on descriptive biology, representing the pleiotropic distribution of the same cellular-molecular trait for both breathing and urination. This trait may also have evolved under the influence of increased catecholamine production due to physiologic stress, since it inhibits loss of water and salt from the kidney in adaptation to land.

In further support of this complex scenario for the evolution of land vertebrate physiology, it has been observed that the genome decreased by about 80–90% after the

Cambrian extinction. The advent of endothermy may explain this phenomenon because ectotherms require multiple isoforms for the same metabolic enzyme in order to function at variable ambient temperatures, whereas the uniform body temperature of endotherms only requires one metabolic isoform to function optimally. Since metabolic genes account for 17% of the human genome, representing a fraction of the number of metabolic genes expressed by ectotherms, this reduction in metabolic enzyme heterogeneity may have contributed to the dramatic decrease in post-Cambrian genomic size.

Hibernation and the Predictive Power of the Cellular-Molecular Approach to Evolution

The evolutionary physiologic interrelationship between stress, metabolism, and endothermy may underlie the effect of meditation on hypometabolism. It has long been known that yogis have the capacity to regulate their metabolism at will, and formal study of this phenomenon has validated it scientifically. Functionally linking to ever-deeper principles of physiologic evolution through meditation and biofeedback may prove to be of wider benefit in healing, both conventional and self-healing alike.

Conclusion

By focusing on the necessity and utility of lipids in initiating and facilitating the evolution of eukaryotes, a cohesive evolutionary strategy becomes evident. In fostering metabolism, gas exchange, locomotion, and endocytosis/exocytosis, cholesterol in the cell membrane of unicellular eukaryotes formed the basis for what was to come. The basic difference between prokaryotes and eukaryotes is the soft, compliant cell membrane of the latter, interacting with the external environment, adapting to it by internalizing it using the endomembrane system as an extension of the cell membrane. This iterative process was set in motion in competition with prokaryotes, which can emulate multicellular behaviors like biofilm formation and quorum sensing. All of the examples cited in this chapter – peroxisomes, water-to-land transition, lipofibroblasts, endothermy – are functional fractals of the originating principle of lipids in service to the evolution of eukaryotes.

Following the course of vertebrate physiology from its unicellular origins instead of its overt phenotypic appearances and functional associations provides a robust, predictive picture of how and why complex physiology evolved from unicellular organisms. This approach lends itself to a deeper understanding of such fundamentals as the First Principles of Physiology. From these emerge the reasons for life cycles and why all organisms always return to the unicellular state, pleiotropy, and homeostasis. A coherent rationale is provided for embryogenesis and the subsequent stages of life, offering a context in which epigenetic marks are introduced to the genome.

From the beginning of life, there has been tension between calcium and lipid homeostasis, alleviated by the formation of calcium channels by exploiting those self-same lipids, yielding a common evolutionary strategy. The subsequent rise in atmospheric carbon dioxide, generating carbonic acid when dissolved in water, caused increased calcium leaching from rock. Calcium is essential for all metabolism and it is through calcium-based mechanisms that the inception of life is marked with a calcium spark kindled by sperm fertilization of the ovum, a process that characterizes the processes of life until the time of death; perhaps the aura chronicled in near-death experiences is that very same calcium spark.

A cohesive, mechanistically integrated view of physiology has long been sought. L.L. Whyte described it as "unitary biology," but the concept lacked a scientifically causal basis, so it remained philosophy. But with the advent of growth factor signaling as the mechanistic basis for molecular embryology in 1978, his vision of a singularity may now be realized.

Throughout this chapter, the contrast between conventional descriptive physiology and the deep mechanistic insight gained by referring back to the epistatic balance between calcium and lipids, mediated through homeostasis, has been highlighted. It is emblematic of the self-organizing, self-referential nature described for the origin of life itself. Using this organizing principle avoids the perennial pitfalls of teleology, conversely providing a way of resolving such seeming dichotomies as genotype and phenotype, emergence and contingence, cells and vast multicellular organisms. Insight to the fundamental interrelationship between calcium and lipid homeostasis was first chronicled by us in *Evolutionary Biology, Cell-Cell Communication and Complex Disease.* Further research will solidify the utility of focusing on the advent and roles of cholesterol in eukaryotic evolution, extending from unicellular to multicellular organisms, and provide novel insights to the true nature of the evolutionary continuum in an unprecedented, predictive, and reproducible manner.

This understanding of the how and why of evolution provides the unprecedented basis for a "central theory of biology," which is long overdue. Many have given up on the notion of a predictive model for biology akin to those for chemistry or physics. This is largely due to the failure to realize that biology remains descriptive, and that describing a mechanism is not the same as actually determining causation based on founding principles, like quantum mechanics and relativity theory. This is surprising in the wake of the publication of the human genome, which is only 20% of the predicted size. That alone should have generated criticism of the prevailing way in which biology is seen as a fait accompli, characterized by correlations and associations. John Ioannidis has declared that "most published research findings are false." This may be because we are using a descriptive framework, which will not allow for prediction.

This Chapter has provided the first comprehensive perspective on the "first principles of biology." By utilizing the unique view provided by cell biology as the common denominator for ontogeny and phylogeny, biology can be seen as having a logic. Chapter 18, entitled "Implications of evolutionary physiology for astrobiology," demonstrates how the principles of physiology and evolution on Earth can provide a logical way of thinking about extraterrestrial life.

18

Implications of Evolutionary Physiology for Astrobiology

Introduction

Over the course of the last two centuries, strides have been made in the description of evolutionary development and phylogeny. However, the ontologic and epistemologic underpinnings of physiology have relatively recently been questioned given the teleologic nature of the conventional Darwinian narrative. An alternative perspective to that general narrative can now be offered based on cellular patterns of communication and cell–cell signaling mechanisms for morphogenesis. This complex structured communication is enabled by signaling molecules and growth factors, but also through other important non-molecular means that directly influence unicellular and multicellular metabolism, growth, and development. When evolutionary development is focused on cellular communication, a different perspective unfolds that uniquely depicts the evolutionary progression toward more complex organisms. In contrast to the more rigid Darwinian frame traditionally imposed on the ontogeny and phylogeny of physiologic traits, the cellular communicative-molecular approach affords a logically progressive evolutionary narrative based on basic cell properties and reciprocating interactions between the environment and the organism that extend beyond selection. According to this model, the cell is a fundamental fractal unit of activity that reciprocally communicates with and reacts to its environment iteratively. This specific path has implications that pertain to how life elsewhere in our universe could evolve, and how communication with such life might be affected.

If the processes that enable the cell are willingly considered as the further basis for the evolution of life on our planet, then the physical conditions on other planets must be scrutinized, as best we can, on the basis of compatibility with those conducive conditions that have been so far apprehended as present within our own sphere. Therefore, a deep examination of the cellular nature of life on Earth as it reiterates at every scope and scale should be expected to provide insight into the nature of that search, or the manner in which communication with other forms of life might be conducted. If this is our frame, then any syllogistic assumption that life is solely carbon-based, has bilateral symmetry, and expresses the same biologic imperatives that organisms on Earth have for food, water, shelter, and reproduction must be reconsidered, as not all cells have similar requirements in equal measures. However, even given that set of variations, any rational approach to the challenging problem of the enactment of life on Earth or how we might conceive a means of recognizing and communicating with extraterrestrial life requires a thorough recapitulation of what is now known about the cellular-molecular processes leading from the unicell to our own vertebrate evolution.

In the Beginning

Life has existed on Earth for billions of years, starting with primitive cells that evolved into unicellular organisms over the course of the last 3 billion years of Earth's existence. The evolution of complex biologic organisms began with the symbiotic relationship between prokaryotes and eukaryotes that is estimated to have begun between 1.6 and 1.2 billion years ago. It is conceivable that life is even more ancient, flourishing and being extinguished multiple times under the anaerobic conditions that are believed to have been present in earlier times. However, there can be no doubt that all genetic aggregates, and specifically in the cellular realm, are impelled to seek both individual advantages and collaborative partnerships. Whether this latter impulse proceeds by engulfment, or by other means such as the "inside out" blebbing mechanism that has recently been proposed and is available to some cell types, is as yet unknown. In either case, endosymbiotic relationships are generally believed to have given rise to mitochondria, and the resulting diversity of unicellular organisms led to their metabolic cooperativity, mediated by ligand–receptor interactions and cell–cell signaling. However, at all steps, the crucial aspects of these cell–cell interactions on the path to multicellularity are their amplification through cell–cell communication as a means of reciprocal interactivity with the environment. It is this process that is the actuating driver for evolutionary development. This crucial intercommunication serves as a self-reinforcing set of mechanisms between nucleobases, proteins, lipids, and calcium under the boundary conditions of a semi-permeable membrane. Such a membrane must be present for evolutionary development to occur. It is the reciprocal combination of those organic entities with a limiting boundary membrane (Figure 18.1) that creates those properties that enact cellular life as a series of interrelated reciprocal reactions among the constituent entities. The indisputable product of this amalgam is cellular awareness as a basic property of the cell. Explicit examples of that capacity in the microbial realm are quorum sensing and biofilm formation, which arise through the reciprocating functions of cellular awareness and intercellular communications.

Living things, and cells as their basic constituents, are aware and communicate as two elemental properties. Therefore, information transfer must be regarded as the essential component of life. Certainly, the purpose of information transfer is to permit aware entities to either avoid or align with others. Cells exhibit both actions. Awareness is a complex set that has several components: ipseity (awareness of self), quiddity (awareness of otherness), and awareness of self and similar. This last property may be so fundamental to our physical systems that it is even apparent within polymer chemistry. It is as yet unclear exactly where any explicit epicenter of awareness might lie within any cell. It may not be separately imbued in any specific component of the cell, such as DNA or proteome or lipid membrane. Most likely, awareness is a reciprocal property between all the elements that make a functioning cell, and then results in an ability to receive information and communicate it. Sentience without the instinct or means to communicate it is lifeless.

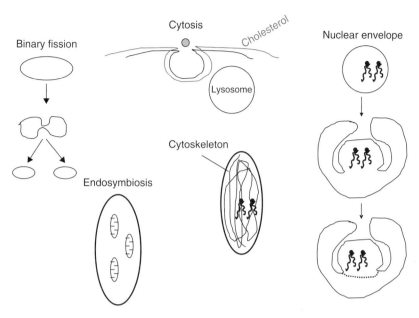

Figure 18.1 Role of membranes in evolution. Both the cell membrane and the endomembrane system were key to eukaryotic cell evolution. Various physiologic traits conferred by membranes are depicted – binary fission, endosymbiosis, endocytosis, nuclear envelope – all within the framework of the cytoskeleton. Cholesterol, fostered by atmospheric oxygen, was critical for the evolution of eukaryotes, as discussed in the text.

The concept of self-organization has been seen as one means of accounting for ourselves. Yet, this also is a disembodied process in and of itself if separated from any fundamental property of awareness. Without such, it remains just chemistry. However, once it is accepted that awareness exists at every scope and scale within biologic entities, by whatever means, a path that directs how life has furthered itself on this planet can be based upon the cognitive cell as a self-reinforcing and communicative unit capable of reciprocal interactions with other cells, and with its greater environment. Therefore, there are powerful motives to view the unicell as fundament in all its multifarious manifestations. The cell, with its intrinsic properties once achieved, then becomes a unit that can be employed as a fractal enlargement of basic properties toward more complex organisms.

Since all life is defined by communication (Figure 18.2), reproduction too must be regarded as just another means of communication. This immediately places communication beyond any simple Darwinian selection mechanism, since cellular purposes might extend beyond mere reproductive frequency. Evolutionary development then travels along mechanisms that mirror cellular capacities, and importantly are based on the patterns and limits of intercellular communications. Information is transferred beyond lineal reproduction along robust paths that are only now being defined, and can be singularly influenced by environmental factors. The emerging science of epigenetics is such a means. Therefore, any breakdown of cellular communication can then be productively seen as failed homeostatic signaling, creating a pathway to pathology on the one hand, and repair on the other. In this manner, the cellular drive toward complexity

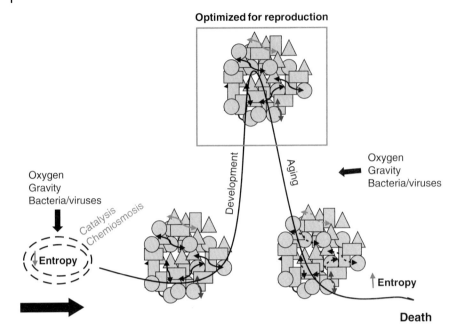

Figure 18.2 All of life as cell communication. Life begins (*far left*) as a reduction in entropy in response to oxygen, gravity, and bacteria within protocells. Cell–cell communication is the basis for metazoan development, leading to optimized reproduction. Aging (*far right*) is the breakdown in cell communication due to oxygen, gravity, and bacteria, terminating in death as increased entropy.

and novelty guides development toward ontogeny and phylogeny, which are then best understood as manifestations of patterns of signaling and communication that emanate hierarchically from fundamental cellular capacities. The Darwinian quest to understand evolution through macroorganisms and their traits rests upon the examination of a terminus rather than any predictive model that would properly understand points of initiation. And derivatively, the quest to discern the origins of life on Earth, or the search for extraterrestrial life, thereby reduces to our improved understanding of cell–cell communications on our planet.

How do Cellular Mechanisms Drive the Evolution of Physiology?

The essential component of understanding evolutionary development is gained through insights into signaling pathways. As a prime example (see Figure 5.1), we have exploited the evolution of the lung at the cellular-molecular level as a ploy to reverse-engineer the organ of gas exchange back to its unicellular origins. The premise that eukaryotic calcium homeostasis is counterbalanced by lipids at multiple levels, both within and between cells (i.e., lipid rafts) allows an understanding of how complex physiology translates to the seemingly simple homeostatic regulation practiced by unicellular organisms. Similar means can be used for other metabolic pathways within cells. At all levels, from the first protocell to eukaryotes and all complex organisms, similar reciprocating and reiterative patterns are reinforced to enable the diversity of forms and faculties that can be observed. The nature of that end-point has only recently been apprehended. All complex creatures are collaborations of deep and extensive cellular ecologies as hologenomes. All complex life on Earth is just that. There are no exceptions. Whether enacted as unicellular aggregates in complex biofilms, or as multicellular organisms, the same basic impulses of awareness and communication yield collaborative forms as iterations of cellular processes critically dependent upon lipids and calcium homeostasis.

Why Return to the Unicellular State During the Life Cycle?

It is a crucial aspect of life cycles that all multicellular organisms return to the unicellular zygotic state. There must be purpose in this as it is the very basis upon which all life thrives. Therefore, it can be advanced that re-experiencing the unicellular state through meiosis is necessary to maintain the balance of the critical cell–cell signaling mechanisms that have accumulated epigenetic experience in the organism's multicellular form. This is largely predicated on burgeoning evidence that epigenetic inheritance is far more prevalent than had previously been thought. The fact that epigenetic "marks" accumulated by eggs and sperm during the life cycle of the organism are not expunged during meiosis, but are selected for (or against), is highly significant because it implies that this form of inheritance essentially bypasses the parental DNA, only affecting the progeny. However, the unicell extends beyond mere selection. It is empowered by awareness as a purposeful means to achieve cellular aims to meet and adapt to complex environments. The requirement of returning to the unicellular state during the life cycle of all complex multicellular organisms would then be a necessary mechanism for the regulated filtering of epigenetic marks, ensuring that the cycle of cellular reproduction proceeds via fidelity to basic signaling and communication pathways. Some epigenetic experiences are accepted and others expunged.

If it is hypothesized that everything advances from the cell, then how might physiology that maintains life have unfolded? The key to that understanding resides within the evolution of vertebrates from their unicellular origins. This path requires a focus on lipids as part of the reciprocating mechanism of cellular communication and development. Cholesterol (Figure 18.3) arose as a result of increasing amounts of oxygen in the atmosphere. The cell membrane of unicellular organisms made life possible by establishing the boundary between the inner workings of the organism and its physical surroundings. The plasmalemma acts to mediate the flow of material and information in and out of the intracellular environment, or the *milieu intérieur* as Claude Bernard phrased it. The lipid-protein bilayer is a highly interactive structure that determines the metabolism, respiration, and locomotion of the cell – the three fundaments of vertebrate evolution. The physical and chemical environments may be biologically advantageous or disadvantageous, so life must always be in flux. Although internal cellular processes and molecular signaling are of prime importance, there are other consequential influences implicit to life on Earth such as the orbital cycles of the Earth, or the gravitational cycles of the Moon. These are part of the epigenetic construct that influences the cell and have churned biology since its inception. These full effects are speculative, yet some basic forces are known, such as the lapping of water on a shore generating

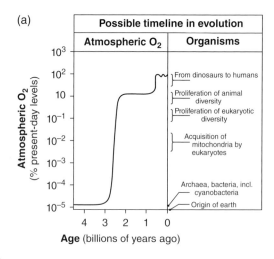

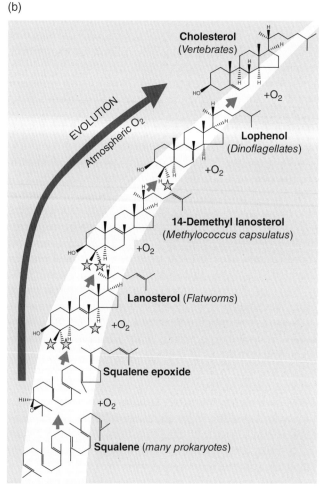

Figure 18.3 Cholesterol arose in response to a rising oxygen level in the atmosphere. Large amounts of oxygen are necessary for the biosynthesis of cholesterol. **a)** Atmospheric oxygen over the last 5 billion years regressed against corresponding life forms. **b)** The relationship between atmospheric oxygen, and the step-wise evolution of cholesterol biosynthesis. (*See insert for color representation of the figure.*)

primitive lipid bubbles, or micelles that are an essential component of communicative life. Further, historic increases in the mineral content of the oceans have enhanced the biologic productivity of the seas. Moreover, an excess of certain elements, even crucial ones, subjects organisms to physiologic stress. As a result, the cellular systems that determine homeostasis have evolved counterbalancing, epistatic mechanisms to survive the ongoing and ever-changing challenges posed by the environment. Calcium is of particular interest in this regard because the cell has to maintain a critical level of this mineral at all times, and any changes in calcium content, either increases or decreases beyond limits, severely impair the cell and eventually will kill it if left unchecked (apoptosis). Hence it is not surprising that calcium is an essential component for the functioning of the nervous system in multicellular organisms, development of the vertebrate embryo, and cell migration.

From the beginnings of life, there has been tension between calcium and lipid homeostasis (Figure 18.4), alleviated by the formation of calcium channels by exploiting those self-same lipids, yielding a common evolutionary strategy. The subsequent rise in atmospheric carbon dioxide, generating carbonic acid when dissolved in water, enhanced calcium leaching from rock. Calcium is essential for all metabolism, and it is through calcium-based mechanisms that the inception of life is marked by a calcium spark first kindled by sperm fertilization of the ovum.

Further, although cholesterol and other related sterols dramatically modified the physical properties of cell membranes in a way that was essential for the further evolution of eukaryotic cells to their multicellular forms, that would have been impossible without concomitant calcium regulation. The lipid composition of the plasma membranes of eukaryotic cells includes substantial amounts of cholesterol, whereas prokaryotes are universally devoid of cholesterol. Konrad Bloch discovered the cholesterol biosynthetic pathway during the 1950s, hypothesizing that if one could determine the

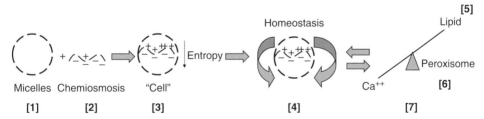

Figure 18.4 The polycyclic hydrocarbons that were contained within asteroid "snowballs" that pelted the early Earth, devoid of oxygen to burn them up upon entry into the atmosphere, were suspended in the early nascent oceans. Lipid emulsions in an aqueous medium spontaneously generate micelles [1], semi-permeable membrane-bound spheres. Within these spheres, chemiosmosis [2] – the partitioning of positively and negatively charged ions – would have generated energy to reduce the endogenous entropy of the protocell [3]. Homeostatic control of this system [4] would have sustained the negative entropy of the cell far from equilibrium. Increasing amounts of carbon dioxide in the atmosphere dissolved in water to form carbonic acid, which leached calcium from the ocean floor. The subsequent rise in oceanic calcium threatened the existence of early life on Earth since calcium ions gum up lipids, proteins, and nucleotides, the essence of life. In response, lipids [5] were employed to form calcium channels to regulate the flow of calcium into and out of the cell. Subsequently, peroxisomes [6] evolved to buffer the deleterious effects of increased cytoplasmic calcium caused by physiologic stress. Over biologic evolution the epistatic balancing of calcium by lipids [7] has become the fundament of vertebrate physiology and pathophysiology.

contemporary sterol pathway, a directed evolutionary process operating on a small molecule could be discerned. His search was for a step-by-step sequence producing a molecule functionally superior to its precursor as a form of molecular evolution. This has proven elusive, but it could be demonstrated that cholesterol evolved in response to the appearance of oxygen in the atmosphere, facilitated by the cytochrome P450 family of enzymes necessary for cholesterol synthesis. Bloch speculated that the biologic advantage associated with cholesterol might have been due to the "reduced fluidity" or "increased microviscosity" that the addition of cholesterol imparts to the membranes of the liquid crystalline state of phospholipid bilayers. Myer Bloom and Ole Mouritsen thought that the biosynthesis of cholesterol in an aerobic atmosphere removed a fundamental constraint on the evolution of eukaryotic cells. This proposed role for the physical properties of biomembranes in the evolution of eukaryotes is compatible with Thomas Cavalier-Smith's characterization of the evolution of eukaryotic cells, in which he identifies "twenty-two characters universally present in eukaryotes that are totally absent from prokaryotes." He presents detailed arguments for the advent of exocytosis and endocytosis as the most likely properties to have provided the driving force for the evolution of eukaryotic cells into their present form, which is dependent on the boundary layer of a cell membrane as a critical aspect.

The further importance of membranes as boundary elements for cells was strengthened when Mouritsen and Bloom hypothesized that, in addition to influencing the cohesive strength of biomembranes, the main role of cholesterol in this evolutionary step was to relax an important constraint on membrane thickness imposed by the biologic necessity for membrane fluidity. The introduction of cholesterol into the phospholipid bilayer membrane (Figure 18.5) increases the orientational order, but does not increase microviscosity. Such fluid-like properties would allow for large membrane curvatures without abnormal increases in permeability. As a result, with the appearance of large amounts of molecular oxygen in the Earth's atmosphere sometime between 2.3 and 1.5 billion years ago, a bottleneck in the evolution of eukaryotic cells was removed by the resultant incorporation of sterols into the plasma membrane. This functional interrelationship between cholesterol, membrane thickness, and endo- and exocytosis for metabolism driven by oxygenation is recapitulated through the evolution of the surfactant system from the swim bladder of fish to the vertebrate lung, increasing gas exchange for increased feeding efficiency in the former, and for adaptation to life on land in the latter. Such a reprise of a trait that served one purpose in evolution, only to serve a homologous purpose later in evolution, is referred to as an exaptation; if the reprised trait is heterologous, it is referred to as pleiotropy.

Cholesterol represents a molecular phenotypic trait that has been positively selected for, beginning with unicellular organisms (Figure 18.5) all the way up through to the complex physiologic properties of lung surfactant, cell–cell signaling via G-protein-coupled receptors, and endocrine regulation of physiology. All of these traits are catalyzed by cytochrome P450 enzymes. Commencing with the advent of cholesterol synthesis, its appearance in the cell membrane of unicellular eukaryotes fostered the evolution of vertebrates. Cholesterol facilitated metabolism, respiration, and locomotion, the three basic elements of vertebrate evolution. The subsequent specialization of cholesterol in lipid rafts within the evolving cell membrane formed the basis for cell–cell signaling since the receptors for such interactions reside within the rafts, where they trigger intracellular

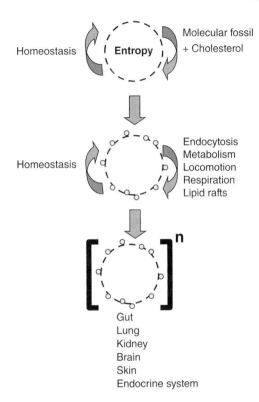

Figure 18.5 Cholesterol in the unicellular cell membrane and evolution of metazoans. The advent of cholesterol in the cell membrane fostered eukaryotic unicellular evolution, followed by multicellular evolution by giving rise to lipid rafts and the endocrine system. (*See insert for color representation of the figure.*)

second messengers for cellular growth, differentiation, and metabolism. Ultimately, these localized paracrine cell–cell signaling mechanisms evolved into the endocrine system of complex vertebrates.

The evolution of respiration was particularly important as the driver of metabolism, both functionally and conceptually. Following the cellular-molecular progression of biologic traits that facilitated efficient gas exchange across the cell membrane expedited our understanding of overall vertebrate physiologic evolution. Cholesterol is the most primitive of lung surfactants, being expressed in the swim bladder epithelial lining cells of fish (Figure 18.6) where it acts as a lubricant to prevent the walls of the bladder from sticking together. As the swim bladder of fish evolved into the terrestrial lung, adapting to the drying up of water bodies some 400 million years ago, changes in the surfactant lipid composition allowed for the progressive decrease in the size of the gas exchange unit to increase the surface area-to-blood volume ratio. At the cellular level, this was accomplished through epithelial–mesenchymal interactions mediated by soluble growth factors and their cognate receptors, remodeling the gas-exchange surface and alveolar wall thickness.

The nature of these evolutionary modifications of lung alveolar structure and function in adaptation to environmental demands is revealed by both the ontogeny and phylogeny of the vertebrate lung, from fish to human. By regressing the molecular

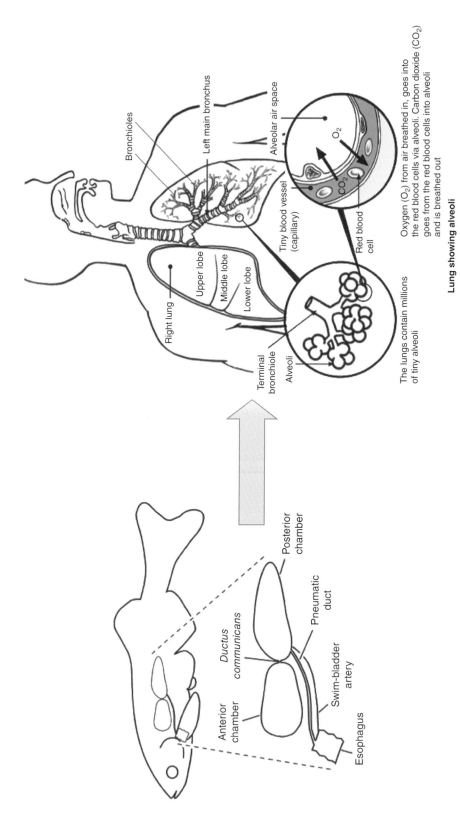

Figure 18.6 Evolution of the lung from the swim bladder. Cholesterol, the most primitive surfactant, is produced by the gas gland epithelial lining cells. Lung surfactant contains cholesterol and other constituents that facilitated lung evolution in amphibians, reptiles, mammals, and birds.

Lung showing alveoli

Oxygen (O_2) from air breathed in, goes into the red blood cells via alveoli. Carbon dioxide (CO_2) goes from the red blood cells into alveoli and is breathed out

Alveolar air space

O_2

CO_2

Red blood cell

The lungs contain millions of tiny alveoli

Alveoli

Terminal bronchiole

Tiny blood vessel (capillary)

Left main bronchus

Bronchioles

Right lung

Upper lobe

Middle lobe

Lower lobe

Anterior chamber

Ductus communicans

Posterior chamber

Pneumatic duct

Swim-bladder artery

Esophagus

mechanisms common to both development and phylogeny of the alveolus against specific eons in vertebrate evolution – the increased salinity of the oceans, the water-to-land transition, and the rising levels of atmospheric oxygen – we have found that these cellular-molecular modifications alternate between external and internal stress-mediated mechanisms. This observation raises questions as to how and why environmental changes on other planets could similarly have led to life forms specific to those conditions in a parallel manner. For example, there are experiments that demonstrate the effects of electromagnetic waves on phospholipids. This raises the possibility that life generated under such conditions would be homologous with the immersion of lipids in water and also permit micelle formation as a circumscribed boundary mechanism for negentropic flux and the controlled regulation of chemiosmosis and homeostasis.

A focus on the necessity and utility of lipids in initiating and facilitating the evolution of eukaryotes yields a cohesive, scale-free evolutionary strategy for life on Earth as a reciprocating agency. Importantly, it is not only that lipid, or DNA, or calcium play crucial roles. Life is effectively harnessed when all these requisites are bounded and constrained within the reciprocating environment of the cell membrane. In fostering metabolism, gas exchange, locomotion, and endocytosis/exocytosis, cholesterol in the cell membrane of unicellular eukaryotes formed a crucial element for what was to come. The basic difference between prokaryotes and eukaryotes is the soft, compliant cell membrane of the latter, interacting with the external environment and adapting to it. The process is then further internalized by the cellular endomembrane system as an extension of the cell membrane. This iterative process was set in motion in competition with prokaryotes, which can emulate multicellular behaviors with phenomena such as biofilm formation and quorum sensing. All crucial cellular features, such as peroxisomes, mechanotransduction, mechanisms of water-to-land transition, lipofibroblasts, or basic processes that support endothermy/homeothermy are functional fractals of the originating principle of lipids in service to the evolution of eukaryotes. Furthermore, all of these attributes are based on intracellular and cell–cell signaling mechanisms, thereby underlying all aspects of the developmental path of physiology on this planet.

Following the course of vertebrate physiology from its unicellular origins instead of its overt phenotypic appearances and functional associations provides a robust, predictive portrayal of the means by which complex physiology evolved from unicellular organisms. Further, it centers evolutionary development squarely within the cell, and its connections to other cells, which continues to this day in an unbroken continuum from prokaryote to eukaryote. This same process gives rise to both differentiated tissues and hologenomes. And the same pathway rationalizes the universal experience of the multicellular realm in its return to its unicellular roots by necessarily re-centering within the unicell zygote. A coherent rationale is provided for embryogenesis and the subsequent stages of life, offering a context in which epigenetic markers are introduced to the genome. At each level, the emphasis is shifted from the limiting perch of self-organization as a theoretical construct to the actual enactment of organization as a principle of self-reinforcement based on the reciprocative agency of communication between cells that have both awareness and an instinct to interact. This basic requirement for communication provides a reasonable concept for what constitutes life as awareness – as a reciprocating and communicative entity. Life began as a dynamic equilibrium of balancing selection between lipids and calcium to sustain negentropy within the cell in defiance of the second law of thermodynamics. Since that is the ultimate mechanism for evolution, is it any wonder that we go back to those first principles during the life cycle, or perhaps we never left them in the first place? That is, the multicellular state is derivative of these base principles that are then used to monitor the ever-changing environment, glean information, subsequently filtering it in the process of reiterating the embryo. This serves as a robust mechanism for either stability or change, depending upon the circumstances.

A cohesive, mechanistically integrated view of physiology has long been sought, going back to Aristotle's concept of *entelechy*. L.L. Whyte described it as the "unitary principle in physics and biology," but the concept lacked a scientifically causal basis, so it remained purely philosophical. But with the advent of growth factor signaling as the mechanistic basis for molecular embryology, Whyte's vision of a highly focused understanding can finally be realized. Viewing physiology as a continuum from unicellular to multicellular organisms provides fundamental insight to ontogeny and phylogeny as an integral whole, directly linking the external physical environment to the cellular. The processes themselves can be reconstructed going forward, once it is realized that there are "first principles of physiology." These processes superimpose on a background of homeostatic drive that is dependent upon intracellular and cell–cell communication.

Life probably began much like the sea foam that can be found on any shoreline subject to wave motion, since similar lipids naturally form primitive "cells" when vigorously agitated in water. Algae, for example, can be as much as 77% lipid by dry weight. Such primitive cells provided a protected space for catalytic reactions that decreased and stabilized the internal energy state within the cell from which life could emerge. Crucially, that cellular space permits the circumvention of the second law of thermodynamics. This principle has typically been considered a manifestation of an essential self-organizing property of life, self-perpetuating, yet always in flux, staying apace with, and yet continually separable from a stressful, ever-changing external environment. When the cell is further considered, however, the concept of self-organizing can be further refined as self-reinforcing and reciprocating interactions among the essential elements that enable and sustain any cell.

Even from the inception of life, rising calcium levels in the ocean have driven a perpetual balancing selection for calcium homeostasis, mediated by lipid metabolism. Metaphorically, this invokes the Greek ouroboros, an ancient symbol depicting a serpent eating its own tail. Ouroboros embodies self-reflexivity or cyclicity, especially in the sense of something constantly recreating itself. Just like the mythological phoenix, it operates in cycles that begin anew as soon as each successive cycle ends. Critically, the basic cell permits the internalization of factors in the environment that would otherwise have destroyed it – oxygen, minerals, heavy metals, gravitational effects, and even bacteria – all facilitated by an internal membrane system that compartmentalized those factors within the cell to render them useful. These membrane interfaces are the fundamental "biologic imperative" that separates life from non-life.

Unicellular organisms have dominated the Earth for most of its existence, and remain an imperative aspect. Far from static, these organisms are constantly adapting. From them, the simplest protists evolved first, producing oxygen and carbon dioxide that modified the nitrogen-filled atmosphere. The rising levels of atmospheric carbon dioxide, largely generated by cyanobacteria, acidified the oceans by forming carbonic acid, progressively leaching more and more calcium from rock into the ocean waters, eventually forcing a proliferation of life from sea to land.

Figure 18.7 The explicate and implicate orders as two sides of a Möbius strip. David Bohm has written that we see reality through our subjective senses, whereas the true reality lies elsewhere as a "force field."

The existence of a protected space within primitive "cells" allowed for the formation of the endomembrane system, giving rise to chemiosmosis, or the generation of bioenergy through the partitioning of ions within the cell, much like a battery. This is only a single manifestation of a crucial aspect of life. The philosopher David Bohm considered two "orders" of reality, the implicate and explicate, depicted in Figure 18.7 as if they were on either side of a Möbius strip as a continuous function seen in two different realms; in the cell, this is the extensive set of potential responses to stress as opposed to the full range of exhibited actions and reactions. At all stages, this is part of the homeostatic balancing mechanisms that maintain a cell within living boundaries. And at every point, the range of reactions is governed by awareness and communication.

How Life Progressed

Early in evolutionary progression, the otherwise toxic ambient calcium concentrations within primitive cells had to be lowered by the formation of calcium channels, composed of lipids embedded within the cell membrane. This also led to the complementary formation of the endoplasmic reticulum, an internal membrane system for the compartmentalization of intracellular calcium. Ultimately, the advent of cholesterol synthesis led to the incorporation of cholesterol into the cell membrane of eukaryotes, differentiating them (our ancestors) from prokaryotes (bacteria), which are devoid of cholesterol. This process was contingent on an enriched oxygen atmosphere, since it takes five oxygen molecules to synthesize one cholesterol molecule (see Figure 18.3). The cholesterol-containing cell membrane thinned out, critically increasing oxygen transport and enhancing motility through increased cytoplasmic streaming; this was also conducive to endocytosis, or cell eating.

All of these processes are the primary characteristics of vertebrate evolution. At some point in this progression of cellular complexity, impelled by oxygen-promoting metabolic drive, the evolving physiologic load on the system resulted in endoplasmic reticulum stress, periodically causing the release of toxic calcium into the cytoplasm of the cell. The counterbalancing, or epistatic mechanism, was the "invention" of the peroxisome, an organelle that utilizes lipids to buffer excess calcium. That mechanism became homeostatically fixed, further promoting the movement of ions into and out of the cell. Importantly, the internalization of the external environment by this mechanism reciprocally conveyed functional biologic information about the external surroundings, and promoted intracellular communication.

Walter B. Cannon later formulated the concept that biologic systems are designed to "trigger physiological responses to maintain the constancy of the internal environment in face of disturbances of external surroundings," which he termed homeostasis. He emphasized the need for reassembling the data being amassed for the components of biologic systems into the context of whole organism function. Hence, Ewald Weibel and Richard Taylor tested their theory of "symmorphosis," the hypothesis that physiology has evolved to optimize the economy of biologic function.

Harold Morowitz is a proponent of the concept that the energy that flows through a system also helps organize that system. This is better reformulated to reflect the realm in which cells live; communication flowing through systemic awareness is the essential organizing aspect.

As eukaryotes thrived, they experienced increasing pressure for metabolic efficiency in competition with their prokaryotic cousins. They ingested bacteria via endocytosis, which were assimilated as mitochondria, providing more bioenergy to the cell for maintaining homeostatic flux. Eventually, eukaryotic metabolic cooperativity between cells gave rise to multicellular organisms, which were effectively able to compete with prokaryotes. As Simon Conway Morris has archly noted, "Look! Once there was bacteria, now there is New York." There are reasons to consider this proposition. The unicellular realm can obviously engineer complex environments through behavioral traits such as quorum sensing and biofilm formation. Such actions are purposeful, based on awareness invested within all cells, an ability to act and react, and to communicate reaction and intent to others. In so doing, even at the most primitive state of our evolutionary path on this planet, the unicellular environment demonstrates characteristics of a multicellular organization in an organized confederacy. Communication is the enabling component of that process. And from this more loosely affiliated structural architecture, evolution iterates in a pattern best characterized as essentially fractal. The same basic impulses impel each stage. Evolution can best be envisioned as the consequence of cellular growth factors and their signal-mediating receptors in counterbalance to selection at every level, most particularly at that of the collaborative collection of cellular constituents that affect localized ecologies that become hologenomes. This is the frame within which our vertebrate ancestors facilitated cell–cell signaling, forming the basis for metazoan evolution. It is this same process that is recapitulated each time the organism undergoes embryogenesis.

This cellular focus on the process of evolution serves a number of purposes. First, it regards the mechanism of evolution from its unicellular origins as the epitome of the integrated genotype and phenotype. Multicellular organisms evolved from the starting place of the unicellular cell membrane and its cellular partners as the common origin for all evolved complex traits. Second, it offers a discrete direction for experimentally determining the constituents of evolution based on the ontogeny and phylogeny of cellular precursors. For example, it is commonplace for evolution scientists to emphasize the fact that any given evolved trait had its antecedents in an earlier phylogenetic species as a pre-adapted, or exapted trait. These ancestral traits can then subsequently be cobbled together to form a novel structure and/or function. Inescapably, if this reasoning is followed to its logical conclusion, all metazoan traits must have evolved from their unicellular origins. Indeed, Nicole King *et al.* have shown empirically that the entire metazoan "toolkit" is present in the unicellular form.

The Origin of Complex Physiologic Traits

How might complex traits such as those like our own, and in which we are vitally interested, have arisen? Physiologic stress must have been the primary force behind such a generative process, transduced by changes in the homeostatic control mechanisms, affected via cellular communication. For example, when physiologic stress occurs in any complex organism, it increases blood pressure, causing vascular wall shear stress, most pronouncedly in the microvascular beds of visceral organs. Such shear stress generates reactive oxygen species (ROS), specifically at points of greatest vascular wall friction. ROS are known to damage DNA, causing mutations and duplications, particularly at those sites most affected physiologically by the prevailing stress. This can result in context-specific gene mutations, and even gene duplications, all of which can profoundly affect the process of evolution. However, understanding the biochemical processes undergirding the genetic ones equips a profound and testable mechanism for understanding the entire aggregate of genetic changes as both modifications of prior genetic lineages, and yet "fit enough to survive" in their new form.

Over evolutionary time, such context-specific varying of modifications to structure and function would iteratively have altered various internal organs (Figure 18.8). These divergences would either successfully conform to the conditions at hand, or failing to do so, cause yet another round of damage repair. Either an existential solution was found or the lineage became extinct; either way, such physiologic changes would have translated into both phylogenetic and ontogenetic evolution. Such an evolutionary process need not be unidirectional. In the forward direction, developmental mechanisms recapitulate phylogenetic structures and functions, culminating in homeostatically controlled processes. And in the reverse direction, the best illustration lies with the genetic changes that occur under conditions of chronic disease, usually characterized by simplification of structure and function. For example, all scarring mechanisms are typified by fibroblastic reversion to their primordial signaling pathway. This formation of scar tissue sustains the integrity of the tissue or organ, albeit suboptimally, thereby allowing the organism to reproduce before being overwhelmed by the ongoing injury repair.

Nowhere are such mechanisms of molecular evolution more evident than during the water-to-land transition. Rises in oxygen and carbon dioxide in the Phanerozoic atmosphere over the course of the last 500 million years partially dried up the oceans, lakes, and rivers, opening new niches to be exploited. Importantly, all complex organisms are hologenomes, shifting microbial realms that might be expected to be among the more responsive players in ecological shifts, with the potential to engage in new and novel partnerships. This changing nature of the types of hologenomic partnerships becomes a reciprocating mechanism that results in the cellular engineering of novelty for mutual advantages, and as a result of selection pressures. In all such circumstances, cell–cell communication is the primary faculty that would permit and sustain new cellular combinations.

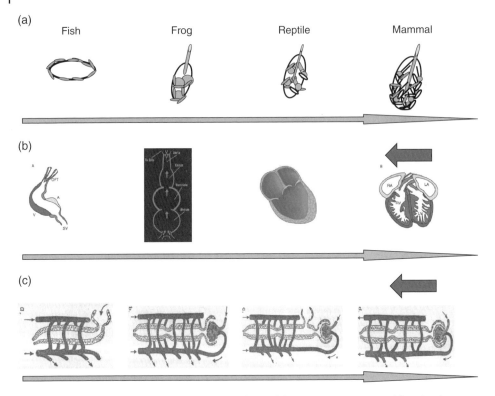

Figure 18.8 Over evolutionary time context-specific modifications to structure and function have altered internal organs. As a result of physiologic stress, tissue-specific genetic mutations have fostered internal adaptation by visceral organs, giving rise to physiologic phylogenetic changes in **a)** the lung, **b)** the heart, and **c)** the kidney.

There were two known gene duplications (Figure 18.9) that occurred during this period of terrestrial adaptation – the parathyroid hormone-related protein receptor (PTHrPR), and the β-adrenergic receptor (βAR). The cause of these gene duplications can be deduced from their effects on vertebrate physiology. PTHrP is necessary for a variety of traits relevant to land adaptation – ossification of bone, skin barrier development, and the formation of alveoli in the lung. Bone had to ossify to maintain the integrity of skeletal elements under the stress of higher gravitational forces on land compared to relative buoyancy in water. PTHrP signaling is necessary for calcium incorporation into bone. It is known from the fossil record that there were at least five attempts to breach land by aquatic ancestors, so a successful biologic combination had to be sought. The effect of shear stress on PTHrP-expressing organs like bone, lung, skin, and kidney may have precipitated the duplication of the PTHrP receptor; this amplification enabled land adaptation, and so is an illustration of such a solution.

As a result of such positive selection pressure for PTHrP signaling (see Figure 1.3), its genetic expression ultimately evolved in both the pituitary and adrenal cortex, further stimulating adrenocorticotropic hormone (ACTH) and corticoids, respectively, in response to the manifold stresses of land adaptation. This pituitary-adrenal hormonal cascade would have amplified the production of adrenaline (epinephrine) in the adrenal

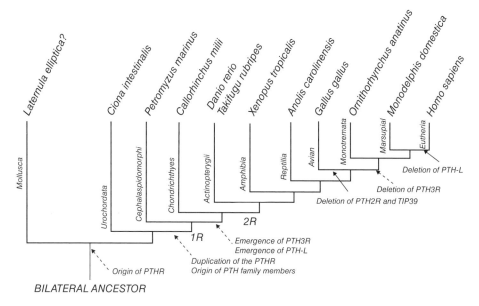

Figure 18.9 Gene duplications and mutations in the water-to-land transition. There were two gene duplications (1R to 2R) involving the parathyroid hormone-related protein receptor (PTHrPR) – with the later phylogenetic loss of the PTHrP homologs fish parathyroid hormone-like peptide (PTH-L) and TIP39 – and the β-adrenergic receptor (βAR), plus mutation of the mineralocorticoid receptor to the glucocorticoid receptor (GR), all of which facilitated vertebrate land adaptation.

medulla, since corticoids produced in the adrenal cortex pass through the microvascular arcades of the medulla on their way to the systemic bloodstream. The passage of corticoids through this medullary labyrinth enzymatically stimulates the rate-limiting step in adrenaline synthesis, phenylethanolamine-N-methyltransferase (PNMT). Positive selection pressure for this functional trait may have resulted from cyclic periods of hypoxic stress. Episodes of intermittent large increases and decreases in atmospheric oxygen over geologic time, known as the Berner hypothesis, may have precipitated lapses in the capacity of the lung to oxygenate efficiently, demanding alternating antagonistic adaptations to hyperoxia and hypoxia as a result. Crucially, we divide the microbial realm into aerobic and anaerobic categories, and we ourselves, just like all other complex organisms, have both types among our obligatory cellular partnerships. Our gut is inhabited by trillions of anaerobic inhabitants crucial for our digestive, metabolic, and immune systems. The creation of these forms of dualities in partnership permits complex organisms to have a literal foot in each door of oxic and hypoxic capacities. The search for solutions among cellular participants and selection pressures were the mutual drivers of multicellular terrestrial success.

Intermittent episodes of relative hypoxia would have been alleviated by the increased adrenaline production, stimulating lung alveolar surfactant secretion, transiently increasing gas exchange by facilitating the distension of the existing alveoli. The increased distension of the alveoli, in turn, would have fostered the generation of more alveoli by stimulating stretch-regulated PTHrP secretion, which is both mitogenic, and angiogenic for the alveolar capillary bed. This would accelerate the evolution of the alveolar bed in the interim through positive selection pressure for those members of the species most capable of increasing their PTHrP secretion.

Increased amounts of PTHrP flowing through the adrenal may also have been responsible for the evolution of the capillary system of the medulla, since PTHrP is angiogenic. Such pleiotropic effects are emblematic of the importance across tissues of cell–cell signaling, which, along with positive selection, would have driven the evolutionary process.

The gene duplication described was not unique to PTHrP. A similar genetic duplication occurred for the βARs. The increase in their density within the alveolar capillary bed relieved a major constraint during the evolution of the lung in the adaptation to life on land. The βARs are required as a parsimonious mechanism for independent blood pressure control in both the lung alveoli and in the systemic circuit. The pulmonary system has a limited capacity to withstand the swings in blood pressure to which other visceral organs are subjected given the thinning of the alveolar wall for optimal gas exchange. PTHrP is a potent vasodilator, so it had the capacity to compensate for the blood pressure constraint in the interim. But eventually the βARs evolved to coordinately accommodate both the systemic and local blood pressure control within the alveolar space.

As a further example of the iterative nature of these processes in response to the stress of land adaptation, the glucocorticoid (GC) receptor evolved from the mineralocorticoid (MC) receptor during this same period through an amino acid modification of the mineralocorticoid receptor (MR). Since blood pressure would have tended to increase during the vertebrate adaptation to land in response to gravitational demands, there would have been positive selection pressure to reduce the vascular stress caused by blood pressure elevation by the MC aldosterone during this phase of land vertebrate evolution. The evolution of the GC receptor would have placed positive selection on GC regulation by reducing the hypertensive effect of the MCs, diverting steroidogenesis toward cortisol production. In turn, the positive selection for the GC cortisol would have stimulated βAR expression, hypothetically explaining how and why the βARs superseded the blood pressure-reducing function of PTHrP. It is these ad hoc existential interactions that promoted land adaptation through independent local blood pressure regulation within the alveolus. This integration of blood pressure control in the lung and periphery by catecholamines represents allostatic evolution, and crucially depends on exquisite reciprocal communication mechanisms.

Moreover, increased episodes of adrenaline production in response to stress may have fostered the evolution of the central nervous system. Peripheral adrenaline limits blood flow through the blood–brain barrier, which would have caused increased adrenaline and noradrenaline (norepinephrine) production within the evolving brain; noradrenaline promotes neuronal development. One might even speculate that this cascade led to human creativity and problem solving as an evolved expression of that same axis as an alternative to fight or flight, since it is well known that learning requires a modicum of stress.

The duplication of the βAR gene may also have been instigated by the same intermittent cyclical hypoxia resulting from the process of lung adaptation, subsequently facilitating independent blood pressure regulation within the alveolar microvasculature – both of these mechanisms would have been synergized by the evolution of glucocorticoids during this transition, given the interrelationship between stress and glucococorticoid stimulation of βARs.

The processes of evolution can be illustrated in this manner as dependent upon molecular signaling pathways such as those that evolved in service to the water-to-land transition – the PTHrP receptor, the βAR, and the GC receptor – which all aided and abetted the evolution of the vertebrate lung, the rate-limiting step in land adaptation. The synergistic interactions of the lung and pituitary-adrenal axis producing adrenaline relieved the constraint on the lung through increased PTHrP production, fostering the ultimate adaptive formation of more alveoli. Perhaps this is the reason why the lung has excess physiologic capacity as the ultimate determinant of land vertebrate adaptation – it was either that, or become extinct. In each circumstance, the pattern of finding solutions via gene duplications, genetic interchange, and cellular cooperation is reiterated in a fractal-like manner. The linkage among all these steps is rooted within base cellular awareness and cellular communication, with development then proceeding over a broad arc of tissues that are in communication with each other.

To illustrate, it was the advent of cholesterol that catalyzed the evolution of eukaryotes from prokaryotes. But both prokaryotes and eukaryotes utilize polycyclic hydrocarbons under the control of oxygen for homeostasis. In the case of prokaryotes these hydrocarbons are hopanoids, whereas in the case of eukaryotes they are sterols, but in either case such lipids benefit the physiology of the organism. And the polycyclic hydrocarbons themselves may have come along with the source of water on Earth as co-travelers on comets and asteroids, bombarding the nascent Earth to form the oceans, lakes, and rivers where life evolved (Figure 18.10). The lipids have acted to foster physiology as a continuous process from bacteria to eukaryotes, from the cell membrane to cell–cell signaling via receptors embedded in lipid rafts, to the endocrine mechanisms that facilitated the water-to-land transition (see above). The ancient nature of this process is attested to by the concerted effects of microgravity on cell polarity and replication in even the most primitive single-celled eukaryotes.

Figure 18.10 The asteroid origins of water and polycyclic hydrocarbons on Earth. Prokaryotes and eukaryotes share commonalities, with the exception of the cell wall/membrane.

Cellular Communication

It is worthwhile now considering the means by which cells communicate. Not so long ago the only established mechanism consisted of cell wall receptors and molecular signaling. However, it is now known that microorganisms communicate by varied means, and that the communication process is intense. As Karen Visick and Clay Fuqua note, "It is clear that chatter among microorganisms is extensive and pervasive." Its quantitative importance is evidenced by the estimate that 6–10% of all genes in the bacterium *Pseudomonas aeruginosa* are devoted to cell–cell signaling systems.

The varied mechanisms by which this communication proceeds are currently being extensively researched, yielding surprising results. Kenneth Nealson had documented a previously unknown electrical communication at a distance between bacteria in differing layers of sediment in the Aarhus Bay in Denmark. Gyanendra Dubey and Sigal Ben-Yehuda described sophisticated intercellular nanotubes as pathways of communication between microbes, which permit the interchange of content. Other ready means are now known, including mechanical signals in the form of mechanotransduction, mechanical load, or mechanoelectrical phenomena as exhibited by hair cells. However, cells can communicate with each other without physical contact or the diffusion of molecules, in ways that are only beginning to be explored.

For example, a potential higher form of intercellular communication relating to underlying calcium signaling has been demonstrated by cells that are physically separated from one another. The exact means is not yet understood. Receptor-mediated interactions based on electromagnetic waves have been documented. Guenter Albrecht-Buehler has proposed that centrosomes are infrared detectors. Ashkan Farhadi cites research that supports the conclusion that cells can generate electromagnetic waves and communicate via electromagnetic signals at a distance, as does Maxim Trushin. Felix Scholkmann *et al.*, in a comprehensive review, discuss a broad range of experiments documenting non-chemical and non-contact cell-to-cell communication. Daniel Fels discusses photons as information carriers, in the form of biophotons that are emitted at wavelengths of 200–650 nm, spanning both the ultraviolet and visible spectrums. The necessary implications of a means of non-local correlations in animal physiology are well known, though the mechanisms involved are not. The instantaneous reaction of groups of tissues to stimuli over widely dispersed tissues and enormously varied scales is incompatible with only a molecular, or even a cell–cell electrical evoked response trigger. Rita Pizzi *et al.* have presented data that strongly suggest the validity of non-local properties in biologic systems by an as yet undetermined path. Therefore, intercellular communication is vast and proceeds by a range of mechanisms that are only now being researched.

This variety of means of communication between cells lends further weight as to why all complex organisms return to the unicellular zygotic state. Fidelity of those patterns of communication in a life cycle that imposes a large number of epigenetic experiences might clearly be considered of paramount importance. In all likelihood, ontogeny takes a path toward the recapitulation of phylogeny in order to vouchsafe the integrity of all of the homeostatic mechanisms that support communication among the cellular parties – whether to facilitate cellular integrity, on the one hand, or to drive development, on the other. Without such a fail-safe mechanism for the foundational principles of life, there would inevitably have been drift from the fundaments of cellular faculties, putting the core process of evolution at risk in response to environmental changes.

One implication of this perspective on evolution – starting from the unicellular state phylogenetically, being recapitulated ontogenetically – is the likelihood that it is the unicellular state that is actually the object of selection. The multicellular state – which Stephen Jay Gould and Richard Lewontin called "spandrels" – is merely a biologic probe for monitoring the environment between unicellular stages in order to register and facilitate adaptive changes. This consideration can be based on both *a priori* and empiric data. Regarding the former, emerging evidence for epigenetic inheritance demonstrates that the environment can cause heritable changes in the genome, but they only take effect phenotypically in successive generations. This would suggest that selection actually operates at the level of the germ cells of the offspring. There is some observational evidence to support this; the starvation model of metabolic syndrome may illustrate this experimentally, for example. Maternal dietary restriction can cause obesity, hypertension, and diabetes in the offspring. But the offspring also mature sexually at an earlier stage due to the excess amount of body fat. Though seemingly incongruous, this may represent the primary strategy to accelerate the genetic transfer of information to the next generation (positive selection), effectively overarching the expected paucity of food in favor of the next environment. The concomitant obesity, hypertension, and diabetes are unfortunate side effects of this otherwise adaptive process in the adults, resulting from enhanced bioenergetics allocation for reproduction. Under these circumstances, one can surmise that it is the germ cells that are the explicit object of selection; in other words, the adults are disposable, as Thomas Kirkwood has opined, the difference being that now there is a testable mechanism.

Hologenomic evolution theory provides yet another mechanism for selection emerging from the unicellular state. According to that theory, all complex organisms actually are vast collaborations of linked, co-dependent, cooperative, and competitive localized environments and ecologies functioning as a unitary organism toward the external environment. These co-linked ecologies are comprised of both the innate cells of that organism, and all of the microbial life that is cohabitant with it. The singular function of these ecologies is to maintain the homeostatic preferences of their constituent cells. In this theory, evolutionary development is the further expression of cooperation, competition, and connections among the cellular constituents in each of those linked ecologies in successive iterations as they successfully sustain themselves against a hostile external genetic environment. Ontogeny would then recapitulate phylogeny since the integrity of the linked environments that constitute a fully developed organism can only be maintained by reiterating those environmental ecologies in succession toward their full expression in the organism as a whole.

There is a further justification for thinking that the unicellular state is the actual object of selection. This primacy is focused within calcium signaling as an initiating event for all of biology. There is experimental evidence that the increases in carbon dioxide during the Phanerozoic eon caused acidification of the oceans, with consequent leaching of calcium from the ocean floor. The rise in calcium levels can be causally linked to the evolution of the biota, and is intimately involved with nearly all biologic processes. For example, fertilization of the ovum by sperm induces a wave of calcium that triggers embryogenesis. The same sorts of processes continue throughout the life cycle, until the organism dies. There seems to be a disproportionately high investment in the zygote from a purely biologic perspective. However, given the prevalence of calcium signaling at every stage, on the one hand, and the participation of the gonadocytes in epigenetic inheritance, on the other, the reality of the vectorial trajectory of the life cycle becomes apparent – it cannot be static, it must move either toward or away from change. As Wallace Arthur has taught us, the embryo is "biased."

Where Might we Place Our Emphasis?

This unicellular-centric vantage point provides an impulse to shift our gaze toward our true place in the biologic realm. This is occurring in medicine as it is increasingly apparent that our growing knowledge of the obligate interactions with our immense microbiome requires such an adjustment. We are learning that the unicellular world has surprising mechanisms and capacities. The proper understanding is a byway through the "Big Bang" of the cell forward, with all its faculties and strictures. By concentrating on cellular dynamics, an entirely coherent path is revealed. Tennyson's line about "Nature, red in tooth and claw" is the merest simplification of evolution. We and all other creatures evolved from unicellular organisms through cooperation, codependence, collaboration, and competition. These are all archetypical cellular capacities. It is not surprising then that we ourselves are only examples of cellular fractal reiterations extending forward from these base capacities, and thereby embody selfsame and similar behaviors.

How might these insights direct us toward any search for extraterrestrial life? By analogy, our inquiry must be centered on awareness, reciprocation, and boundary conditions. It is by these means that communication clearly flourishes on our planet. Reciprocal mechanisms that are conclusively enacted as awareness are then epitomized in the cell as the fundamental unit of life. It is likely that the property of awareness is not definitively invested exclusively within the cellular unit insofar as viruses, prions and other proteins also possess a property of awareness and discriminative preference. Therefore, as any extraterrestrial life could reasonably be expected to be quite different from our own macro-organic sensibilities and faculties, then any exploration of our own inner life, and an inquisitive search for dialogue with our own cellular companions would be a fruitful means of uncovering a mode of communication with alien life outside of our own planet. Cellular mechanisms have not been our focus, since our prior assumption had rested on the belief that cellular communication was based on direct molecular interactions. However, research is proving that alternatives exist. What antenna or radio array has been erected to communicate with our hologenomic cellular partners? Communication among these obligate constituents of ourselves is active and abundant, yet ignored by us. Are they not an alien life form, at least with respect to our own faculties? Do they not have their own means of making their intentions clear to others? How might our search for intelligent life elsewhere be altered if we determined the advantage and had the disposition to communicate with the "alien" life within us and on us? This is life that is so intimately entwined with our own that it enables our survival and reproduction on this planet. However, even as we enthusiastically engage in a broad search for alien life beyond our planetary limits, there is no similar drive to communicate with the alien bacterial life with which we are intimately associated.

Life on Earth is likely due to the uniquely combined presence of water and lipids on its surface, and the effects of the Sun, Moon, gravity, carbon dioxide, oxygen, and the availability of polycyclic hydrocarbons. All of these have been put to service by the cell.

So the proper focus in the search for a hospitable celestial body might shift toward an exploration of the combinations of physical components that might lead to the formation of micelles that serve as a protected space within which negentropy, chemiosmosis, and homeostasis could have occurred. Others have shown, for example, that electromagnetic force can alter the configuration of phospholipid membranes, and that van der Waals forces will affect the structure of such membranes.

Although the fractal pattern in geometry is a powerful metaphor for biologic mechanisms, their emergence still requires a further impulse. What then is that iterative means? That answer is that cells communicate within their boundaries and among each other. In the unicellular environment, the reciprocating relationships between lipids, DNA, and calcium, are amplified and reinforced within those boundaries. Herein lies a powerful paradox. Boundaries reinforce creativity in biology. Our creativity might be unlocked by looking within our own boundaries based on a thorough understanding of all those mechanisms that exist to enable communications within and among cells.

Adam Frank and Woodruff Sullivan have recently explored concepts of a sustainable human civilization in astrobiologic terms using the Drake equation as a vehicle. The Drake equation relates to a series of estimates of the number of civilizations in our galaxy that might support radio communication as a proxy for intelligent life. The probability determination is based upon a series of assumptions about the numbers of planets that might potentially support life, the number that eventually do so, the fraction of those that might develop technology that releases signs of their existence, and the length of life of those civilizations. The case is made that the search for extraterrestrial life could be rewarded by evaluating habitable planets based on an ensemble of markers for species with energy-intensive technology (SWEIT) on our planet. That trajectory, as appraised on Earth through its impact on climate change, would then bear on both issues of Earthly sustainability and the search for extraterrestrial civilizations. By modeling SWEIT through energy consumption rates, and population and planetary systems, the authors propose that there might be some discrimination between "natural" and SWEIT-derived consequences on climate on this planet that might be generalized to inform our search for alien life. It is pertinent that emphasis is directed toward the most important variable in the Drake equation, namely "L," or the length of time that a civilization survives to emit detectable signals. Although the Drake equation relates to estimates of the probability of intelligent life that employs technologies of our human kind, its underlying principles could be enlarged to include other attributes of the biologic realm. The cellular world utilizes its technologies according to its own capacities. Its substrate is biologic material. Its tools include a wide range of physical phenomena and energy usage that are all forms of communication, including genetic interchange. Its language is perhaps abstruse, but could be made accessible if deeply explored. Importantly, the prokaryotic realm has been continuous on this planet, the only species to be so sustained. There are many scientists who feel that the species divisions that are made among prokaryotes are artificial and that they are in fact only a single species with separable breeding fractions. Therefore, the Drake equation can be further empowered by two means. First, by freeing our imagination about what constitutes technology, and secondly, by concentrating on the alternate methods by which it is employed for communication by prokaryotes, the only continuous biologic set on this planet, one that has done so to sustain itself for well over three billion years. This latter particularly satisfies the consequential "L" in the equation. This combination then represents a proper supplemental focus for astrobiologic inquiry.

Conclusion

An evolutionary focus on the cell provides a novel bridge toward devising a means by which extraterrestrial life might be sought, or how communication with it might be achieved. Until recently, the cognitive capacity of unicellular life had not been appreciated. Yet its primacy is apparent. It is the most enduring evolutionary participant on our planet, having been the only life form for the first several billion years, and recapitulates through multicellular organisms for the exploitation of environmental niches to cope with epiphenomena. The mechanisms by which it sustains this success are iterative biologic forces reciprocating with physical phenomena intrinsic to our planet. Understanding the capacities and limitations of the unicell permits the identification of basic principles of evolutionary development that devolved as a series of coordinate interactions between nucleobases, lipids, and calcium within the bounding constraints of semi-permeable membranes. This fortuitous reciprocity yields negentropic self-reinforcement as a "first principle" of evolution, from which other fundamental paths radiate. Evolutionary creativity on our planet can then be viewed as a paradoxical product of boundary conditions that permit homeostatic moments of varying length and amplitude that can productively absorb a variety of epigenetic impacts to meet environmental challenges. This is what our planet does too, in its own form and by its own means.

On Earth then, the cell as an elemental fractal unit utilizes all of its own intrinsic processes to reiterate and sustain itself in the variety of forms that can be appraised across the planet. With this in mind, it is not surprising that all multicellular complex organisms return to the unicellular state as a requisite means of recalibrating themselves for consistent evolutionary development. It therefore represents the ultimate unit of selection. However, underlying all such development is intracellular and intercellular communication, which involves numerous mechanisms. Without this replete dialogue, development would cease. Therefore, the exchange of information is the fundament of life, and research into the means and mechanisms of that exchange at the cellular level here on Earth should be foundational both for cell biology and astrobiology.

Consider the slime mold *Dictyostelium discoideum*. It exists in either an amoeboid or a colonial form, depending upon food abundance. Under high food abundance conditions it is in the free-swimming form, leading to the supposition that it evolved in the amoeboid state. Importantly, *Dictyostelium* clearly exhibits epigenetic inheritance. This suggests that the colonial form, as in metazoans, is a derivative means of acquiring information from the environment under stressful conditions. More recently, it has also been shown that *Dictyostelium* expresses the target of rapamycin (*TOR*) gene, which senses nutrients and regulates AKT signaling for cytoskeletal polarization. The particular expressed molecule, mTOR (mechanistic target of rapamycin), is a highly complex and flexible microbial protein affecting a vast number of unicellular pathways, and is also crucial for eukaryotic development and metabolism. For example, an important group of mTOR factors are primarily involved in regulating the use of lipids for energy in the cell. It is not surprising then that, in our human selves, mTOR dysfunction has been linked to a diverse set of disorders such as diabetes, cancer, tumors, epilepsy, degenerative brain disorders, depression, and autism. This clearly illustrates the unicellular developmental influences that reverberate across the vastly differentiated tissues that compromise all complex organisms.

These forms of intimate reciprocation are also mirrored in the interactions between physical phenomena on our planet and biologic systems. An example is the effect of microgravity on yeast. Both polarity and budding are lost in $0 \times g$, indicating that eukaryotes adapt to gravity, likely due to its effect on the cytoskeleton via signaling through TOR and AKT. This then further interrelates to reproductive processes mitigated through a calcium spike. The cytoskeleton functions as a critical element of the cellular "state," whether during homeostasis, mitosis, or meiosis. The fundamental relationships between ion flux, cell division, and adaptation to gravity extend throughout all organisms, including reproduction, indicating that there is a greater scope to that process than just the generation of progeny. Reproduction is a critical means of monitoring the environment that permits adaptation by sorting epigenetic impacts through reciprocation with basic cellular mechanisms via the unicell as the primary focus of selection.

These same effects then correspond within the multicellular realm. Gravitational effects translated through the cell induce mechanotransduction in multicellular organisms, as illustrated for bone plasticity according to Wolff's law, or through fetal lung development in response to the stretching effects of fluid distension *in utero*, and even uterine physiology itself during gestation. The interrelationships between these physiologic traits can be adduced through the vertical integration of the parathyroid hormone-related protein (PTHrP) signaling that initiates within physiologic stress on the cell membrane for metabolism, locomotion, and respiration. Through this same path, land adaptation was fostered by the duplication of the PTHrP receptor, for skeletal, pulmonary, renal, dermal, and cerebral development. All of these physiologic properties contribute to the evolutionary success of the organism, but are accomplished at the level of the cell in service to it.

What might that mean for astrobiology? When we humans, as aggregated cellular entities, grasp the means by which we might communicate with our own cellular partners and competitors, with those constituencies that are innately part of us through cellularly enacted collaborations, an elucidating path toward understanding alien life will emerge. By comprehending the dialogue and information exchange among our own intimate cellular companions, a productive means of communicating with forms of alien life inherently divergent from our own cognitive sphere might be grasped. Further, whatever life can be found will likely be at such a distance as to preclude us, in our human form, from ever being in proximate contact with it. Perhaps too, through the cell, we might determine the means by which those distances might be overcome by utilizing an embodiment of life more enduring than our own. How might this elusive search be pursued? As a first iteration, through a directed attempt to dialogue with our own fundamental selves.

This Chapter has demonstrated how the principles of physiology and evolution on Earth can provide a logical way of thinking about extraterrestrial life. Chapter 19, entitled "Pleiotropy reveals the mechanism for evolutionary novelty," provides a way of thinking about how the return to the unicellular state during the life cycle offers the opportunity for the reallocation of genes to generate novel physiologic traits.

19

Pleiotropy Reveals the Mechanism of Evolutionary Novelty

The Ghost in the Machine

Based on the current state of an organism's physiology, pleiotropy is superficially seen as the same gene randomly utilized for functionally unrelated traits. George Williams exploited this perspective for pleiotropy to explain why senescence occurs as the cost paid for Darwinian reproductive advantage, characterizing it as antagonistic pleiotropy – when one gene controls more than one trait, with one trait being beneficial to the organism's reproductive fitness earlier in life, the other being detrimental to it later in life. In the aggregate, this was Williams' explanation for aging.

Alternatively, perhaps pleiotropy occurs deterministically rather than by chance, based on specific physiologic principles, possibly revealing the underlying nature of physiology and the evolutionary principle. Maybe pleiotropy was fostered by evolution through interactions between the "first principles of physiology" and the ever-changing environment. Pleiotropic novelties may emerge through cellular-molecular recombinations and permutations based on both past and present conditions, in service to the future needs of the organism.

Rubik's Cube and Pleiotropic Evolution

Erno Rubik invented his eponymous 'cube' (see Figure 19.1) to teach his architectural students about spatial relationships. By manipulating the cube, you can generate 4×10^{19} permutations and combinations of green, yellow, white, orange, red, and blue squares in space and time. By analogy, as an embryo "twists and turns" developmentally in biologic space and time it generates hundreds of different cell-types in forming itself; and those cell-types comprise tissue-specific, homeostatic interactions that constitute integrated functional structures and functions. Yet the genes of each cell are all the same, but are all phenotypically different, both within and between tissues, posing a biologic "puzzle."

Pleiotropy is defined as the expression of a single gene that generates two or more distinct phenotypic traits – much like twisting a Rubik's cube and generating various seemingly random permutations and combinations of colors. But they are actually interrelated based on field theory. In the case of the biologic process, such permutations and recombinations originate from the zygote to generate the various cellular pheno-types that compose the body, with equally varied homeostatic interactions. If this process is followed phylogenetically and ontogenetically, it provides insights to the

Evolution, the Logic of Biology, First Edition. John S. Torday and Virender K. Rehan.
© 2017 John Wiley & Sons, Inc. Published 2017 by John Wiley & Sons, Inc.

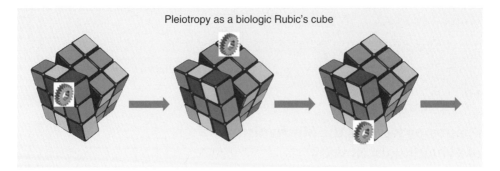

Pleiotropy as a biologic Rubic's cube

Figure 19.1 Pleiotropy as a Rubik Cube. (*See insert for color representation of the figure.*)

mechanisms of evolution, just as unicellular organisms gave rise to multicellular organisms under the iterative, interactive influences of both internal and external environmental selection pressures. We use the Rubik's cube as a device for understanding how and why the mechanism of pleiotropy helps to understand how one gene can affect multiple phenotypes. To enforce this simile, there are images of cells on the faces of the cube (see Figure 12.5) associated with different colors. When the cube is reconfigured to generate new color combinations, those cellular images are repermutated and recombined. The inference is that the cellular phenotypic traits are modified in much the same way as they were during the process of evolutionary adaptation. The reallocation of genes and phenotypic traits is not due to "random selection." Instead, it is determined by homeostatic constraints within each iterative cellular niche. Those constraints are derivative of the unicellular bauplan, and each subsequent generation must remain faithful to those homeostatic constraints at every scale, phylogenetically, developmentally, and physiologically; if they do not, they can be compensated for by other genetic motifs, "silenced," or they can be embryonically lethal. It is this self-organizing, self-referential process, determined by homeostatic principles constraining cell–cell interactions predicated on the genes being faithfully expressed by specific germline cells (endoderm, mesoderm, ectoderm) that explains why physiologic traits are pleiotropically distributed. More importantly, the contingent and emergent properties of this process provide internally consistent mechanisms for evolutionary novelty, since pleiotropy offers the opportunity to "repurpose" pre-existing genetic traits for different phenotypic functions, as needed.

In our prequel to this book, entitled *Evolutionary Biology, Cell-Cell Communication, and Complex Disease*, we used this pleiotropic property of biology to explain the evolutionary mechanisms of both physiology and pathophysiology. In the case of the former, we demonstrated how the alveolus of the lung and the glomerulus of the kidney are virtually the same functionally at the cell-molecular level, even though they nominally seem to be unrelated based on the descriptive, top-down or bottom-up perspective, one mediating gas exchange between the environment and the circulation, the other mediating fluid and electrolyte balance in the systemic circulation. However, they both functionally sense and transduce pressure signals, and thereby regulate homeostasis through cellular cross-talk between the epithelial cell source for parathyroid hormone-related protein (PTHrP) and neighboring mesodermal fibroblast receptor target. In the case of the lung, distension of the alveoli causes PTHrP produced by the epithelial type II cell to signal to its receptor on the lipofibroblast to regulate lung surfactant production, reducing surface tension to maintain alveolar homeostasis. In the kidney, distension of the glomerulus causes PTHrP produced by endodermally derived podocytes that surround the fluid-filled space within the glomerulus to regulate the mesodermally derived mesangium, the thin mesodermal membrane supporting the glomerular capillary loops, homeostatically monitoring and regulating fluid and electrolyte balance for the systemic circulation.

The Lung as the Archetypal Pleiotropic Mechanism

As previously mentioned, the evolution of the lung was existential for the survival of land-dwelling vertebrates, since the rise in atmospheric temperature due to the green-house effect caused by accumulating carbon dioxide forced vertebrates to adapt to land. The physico- chemically determined, integrated developmental and phylogenetic cell–cell interactions regulating lung surfactant offer a way of understanding the ontogenetic and phylogenetic structural-functional interrelationships that facilitated the decrease in alveolar diameter to increase lung surface area for gas exchange, thus avoiding the otherwise fatal increase in surface tension resulting in alveolar collapse, or atelectasis. And suffice it to say that increasing surface area is the only biologic strategy for increasing oxygenation. Beginning with the fish swim bladder as a biologic mechanism for adapting to water buoyancy – inflating to float, deflating to sink – fish have successfully utilized gases to optimize their adaptations to water. All of the key molecular features of the mammalian lung as a reciprocating gas exchanger were already pre-adapted in the fish swim bladder – surfactant phospholipid and protein to prevent the walls of the bladder from sticking together, PTHrP functioning during swim bladder development, and the β-adrenergic receptor regulating the filling and flushing of the swim bladder with gas absorbed from or secreted to the circulation in physoclistous fish. These pieces of the evolutionary "puzzle" repermutated and recombined within the physiologic constraints of the existing structure and function to form the lung. The only additional "trick" to be learned was neutral lipid trafficking (NLT), the active mobilization mediated by adipocyte differentiation related protein (ADRP), a member of the PAT (perilipin, ADRP, TIP47) family of lipid transport and storage proteins ubiquitously expressed wherever lipids are stored. That molecular pathway likely evolved from the lipofibroblast phenotype, initially utilizing neutral lipids to protect the lung against oxidant injury, followed by its role in NLT as a means of more efficiently producing surfactant in response to the ever-increasing excursions of the alveolar wall in response to metabolic demand phylogenetically.

The Lung as a Homolog of the Plasmalemma, Skin, and Brain

Developmentally, the lung emanates from the foregut as an expansion of the surface area of the alimentary tract. As a homolog of the gut, the lung also acts as an interface between the internal and external "environments" of the body. But the homology goes much deeper since the stratum corneum of the skin forms a lipid-antimicrobial barrier much like the alveolar surfactant forming tubular myelin as a "membrane barrier" – in both the skin and lung, the epithelium secretes lamellar bodies composed of lipid-protein complexed with antimicrobial peptides. And the skin and brain are structurally-functionally homologous, both phylogenetically and pathophysiologically – the nervous system of the skin in worms gave rise to the central nervous system of vertebrates, referred to as the "skin-brain." Pathophysiologically, both the skin and brain exhibit lipodystrophies in common in such neurodegenerative diseases as Niemann–Pick, Tay–Sachs, and Gaucher's diseases. It has been speculated by some that this is a reflection of "too much of a 'good thing' going bad." Viewed as a coevolutionary process, the protective effects of the skin barrier may have come at the price of pleiotropic abnormal brain function.

For example, the functional homology between the lung alveolus and kidney glomerulus is that both are mechanotransducers for the physiologic stretching of their mutual walls. In the case of the lung, PTHrP signals to increase surfactant production, which prevents alveolar collapse due to increased surface tension. In the case of the kidney, the podocytes that line the glomerulus also secrete PTHrP, which then signals to the mesangium to regulate water and electrolyte economy as a function of fluid distension. In either case, the calcium-regulatory activity of PTHrP, which is ubiquitously expressed in all epithelial cells, has been embellished due to its myriad functionally evolved properties. And due to its angiogenic properties, PTHrP promotes capillary formation for gas exchange in the alveolar bed, and for fluid and electrolyte exchange in the glomeruli. Phylogenetically, in the fish kidney the growth of the primitive filtering capillaries of the glomus were stimulated locally by PTHrP, culminating in expansion of the capillary network to form glomeruli, thereby increasing the efficiency of water and electrolyte homeostasis in service to land adaptation.

NKX2-1 and Thyroid, Pituitary, and Lung Pleiotropy

The foregut is a plastic structure from which the thyroid, lung, and pituitary arose through the NKX2-1/TTF-1 gene regulatory pathway. Evolutionarily, this is consistent with the concept of *terminal addition*, since the deuterostome gut develops from the anus to the mouth. Moreover, when NKX2-1/TTF-1 is deleted in embryonic mice, the thyroid, lung, and pituitary fail to form during embryogenesis, providing experimental evidence that it is the genetic common denominator for all three organs. Their phylogenetic relationship has been traced back to *Amphioxus*, and to cyclostomes, since the larval endostyle (a longitudinal ciliated groove on the ventral wall of the pharynx for gathering food particles) is the structural homolog of the thyroid gland.

Phylogeny of the Thyroid

Phylogenetically, the lamprey has a follicular thyroid gland, which is an evolved endostyle. The expression of an endostyle in developing lampreys is not direct evidence for descent of lampreys from protochordates, but rather that the evolutionary history of the lamprey is deep and ancient in origin, and that it shares the common feature of having a filter-feeding mechanism during its larval stage of development. Notably, the other extant agnathan, the hagfish, possesses thyroid follicles before hatching. Since hagfish evolution is considered to be conservative, going back 550 million years, this suggests that thyroid follicles could also be considered to have an ancient history.

Evolutionarily Vertical Integration of the Thyroid

The increased bacterial load within the endostyle may have stimulated the cyclic AMP-dependent protein kinase A (PKA) pathway, since bacteria produce endotoxin, a potent PKA agonist. This structure-function relationship may have evolved into thyroid-stimulating hormone (TSH) regulation of the thyroid since TSH regulates the thyroid via the cAMP-dependent PKA signaling pathway. This mechanism hypothetically generated such novel structures as the thyroid, lung, and pituitary, all of which are induced by the PKA-sensitive NKX2-1/TTF-1 pathway. The brain–lung–thyroid syndrome, in which infants with NKX2-1/TTF-1 mutations develop hypotonia, hypothyroidism, and respiratory distress syndrome, or surfactant deficiency disease, is further evidence for the coevolution of the lung, thyroid, and pituitary.

Embryologically, the thyroid evaginates from the foregut in the embryonic mouse beginning on day 8.5, about one day before the lung and pituitary emerge, suggesting that the thyroid may have been a molecular prototype for the lung during evolution, providing a testable and refutable hypothesis. The thyroid rendered molecular iodine in the environment bioavailable by binding it to threonine to synthesize thyroid hormone, whereas the lung made molecular oxygen bioavailable, first by inducing fat cell-like lipofibroblasts as cytoprotectants, which then stimulated surfactant production by producing leptin, relieving the constraint on the blood–gas barrier by making the alveoli more distensible. This, in turn, would have facilitated the use by the metabolic system of rising oxygen concentration in the atmosphere, placing further selection pressure on the alveoli, and giving rise to the stretch-regulated surfactant system mediated by PTHrP and leptin. Subsequent selection pressure on the cardiopulmonary system may have facilitated liver evolution, since the progressively increasing size of the heart may have induced precocious liver development, fostering increased glucose regulation. The brain serves as a glucose sink, and there is experimental evidence that increasing glucose during pregnancy increases the size of the developing brain. Further evolution of the brain, specifically the pituitary, would have served to further the evolution of complex physiologic systems.

The thyroid and lung have played important roles in physiologically accommodating toxic substances in the environment during vertebrate evolution. The thyroid has assimilated environmental iodine, and the lung has assimilated the rising, fluctuating oxygen levels. Moreover, the thyroid and the lung have interacted synergistically to facilitate vertebrate evolution. Thyroid hormone stimulates embryonic lung morphogenesis during development, while also accommodating the increased lipid metabolism needed for surfactant production by driving fatty acids into muscle to increase motility, as opposed to oxidation of circulating lipids to form toxic lipoperoxides. The selection pressure for metabolism was clearly facilitated by the synergy between these foregut derivatives.

A Retrospective Understanding of Evolution

Like, pentimenti, the telltale signs that an artist has painted over an old canvas, the cellular-molecular structure and function of the mammalian alveolus (see Figure 12.1) reveal the signature for phylogenetic traits that facilitated the evolution of land vertebrates in a step-wise fashion. Figure 19.2 depicts (at the far left) the transition from prokaryotes to eukaryotes, which likely resulted from rising oxygen tension in the atmosphere stimulating hopanoid production under the control of hypoxia-inducible factor-1 (HIF-1). This scenario would resolve the age-old controversy as to whether evolution was gradual or saltatory – it was both. Since this resolution is a prime example for understanding mechanistic evolution, we will expound on it as follows. Historically, Darwin thought that evolution was a slow, gradual process. He did not think that this process was "smooth," but rather, "stepwise," with species evolving and accumulating small variations over long periods of time. Darwin further speculated that if evolution were gradual there would be fossil evidence for small incremental change within species. Yet he and his supporters have been unable to find most of these hypothesized "missing links." Darwin thought

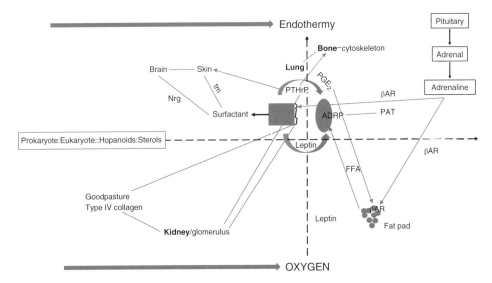

Figure 19.2 Origins of vertebrate physiologic homologies. Sterols (*far left*) under the control of hypoxia inhibitory factor-1 (HIF-1) link prokaryotes and eukaryotes together functionally; thus the major vertebrate physiologic homologies are linked through the induction of endothermy (top solid arrow) by atmospheric oxygen (bottom solid arrow). In the center of the schematic is the molecular regulation of alveolar surfactant production. It is homologous with brain (neuregulin, Nrg) and skin (tubular myelin, tm), both of which are under parathyroid hormone-related protein (PTHrP) regulation. Lung and bone share functional homology through PTHrP stretch-regulated metabolism. Prostaglandin E_2 (PGE$_2$), leptin, adipocyte differentiation related protein (ADRP) and β-adrenergic receptor (βAR) share homologies with the fat pad free fatty acid (FFA) regulation. The lung and kidney share functional homologies through PTHrP and type IV collagen ("Goodpasture"). Periods of hypoxia due to shortfalls in the saltatory process of lung evolution caused physiologic stress, stimulating the pituitary-adrenal axis production of adrenaline (epinephrine), which both relieved the constraint on the alveoli stimulating surfactant secretion, and stimulated peripheral fat cell secretion of FFAs, causing increased metabolic activity and body heat = endothermy.

that the lack of fossil evidence for gradualism was due to the low likelihood that the intermediate steps in such transitions would have been preserved.

Then, in 1972 evolutionary biologists Stephen Jay Gould and Niles Eldredge suggested that the "missing links" in the fossil record were real, representing periods of stasis in morphology, calling this mode of evolution "punctuated equilibrium." This infers that species are generally morphologically stable, changing little for millions of years. This slow pace is "punctuated" by rapid bursts of change resulting in new species. According to this theory, changes leading to new species do not result from slow, incremental changes in the mainstream population. Instead, changes occur in populations living on the periphery, or in isolated populations where their gene pools vary more widely due to slightly different environmental conditions. When the environment changes, such "peripheral" or "isolated" species possess variations in morphology that might allow them an adaptive advantage.

Perhaps the kinds of mechanisms that have been invoked for pleiotropy would reconcile for both gradualism and punctuated equilibrium. Consistent with Darwin's thinking, evolution could have occurred on a continuous basis molecularly in response to physiologic stress as previously described, but only leaving fossil evidence when form reached a macro-scale, making it superficially appear as though evolution had occurred in bursts. These two scenarios are all the more reasonable when one considers the episodic increases and decreases in atmospheric oxygen that have been documented over the last 500 million years, referred to as the Berner hypothesis. The hyperoxic increases in oxygen caused concomitant increases in the size of land animals; however, the hypoxic decreases have not been considered up until now, yet ironically would have had profound effects on vertebrate evolution given that hypoxia is the most potent effector of complex physiologic systems. Elsewhere, hypoxia has been invoked as the mechanism underlying the evolution of endothermy/homeothermy. This hypothetical perspective is validated by the pleiotropic effects resulting from the gene duplications for the PTHrP receptor, the β-adrenergic receptor, as well as the mutation of the glucocorticoid receptor, all of which occurred during the water-to-land transition. These events corroborate the repurposing of pre-existing genes for novel phenotypic adaptations.

Even earlier in vertebrate evolution, hopanoids, under positive control of HIF-1, may have "liquefied" the bacterial cell wall due to rising levels of oxygen in the atmosphere. That event would have marked the phenotypic transition from prokaryotes to eukaryotes, the former having hard exterior walls, the latter having compliant cell membranes. That transition may have been further catalyzed by the novel synthesis of cholesterol, also under positive control by HIF-1, catapulting the evolution of eukaryotes (see Figure 19.2). The two emboldened arrows for "Endothermy" and "Oxygen" in Figure 19.2 represent the major drivers of vertebrate evolution. All three of these processes – prokaryote/eukaryote evolution, oxygen, and endothermy – have acted synergistically to promote vertebrate evolution, indicated by the dotted arrows that interconnect them.

Denouement

Based on descriptive biology, pleiotropy lacks any biologic significance. However, when pleiotropy is seen as the result of positive selection for atavistic cell–cell signaling events over the ontogenetic and phylogenetic history of the organism, the repurposing of genes constrained by structure and function at the cellular-molecular level is highly relevant to our understanding of evolution. Thus, such deep, unobvious pleiotropic homologies transcend the superficialities of comparative anatomy, only being revealed by knowledge of functional molecular-physiologic motifs. The deepest of these are related to the physiologic effects of stretching, or mechanotransduction, on surfactant metabolism, which refers all the way back to adaptation to gravitational force, the most ancient, omnipresent, and constant of all environmental effectors of evolution.

For example, the alveolar type II (ATII) cells produce prostaglandin E_2 (PGE_2), particularly when they are distended, causing secretion of lipid substrate from lipofibroblasts for lung surfactant phospholipid production by the ATII cells; without PGE_2, the lipids would remain bound up within the lipofibroblasts. This physiologic property is common with the effect of PGE_2 on the secretion of free fatty acids (FFAs) from peripheral fat cells, a trait that hypothetically evolved as a functional homology to the evolution of endothermy (see Chapter 17). To relieve the constraint on the evolving alveolar bed, adrenaline (epinephrine) stimulated surfactant secretion to increase gas exchange transiently until the indigenous PTHrP mechanism would generate more alveoli. The pleiotropic coevolution of the PGE_2 mechanism facilitating FFA utilization in both the lung and fat pad was not a chance event; it was synergistic when seen within the context of the evolving lung, both within and between organs. In further support of this hypothesis, the role of the lung in the evolution of endothermy is further evidence for the causal evolutionary interrelationship between the pulmonary and neuroendocrine systems. The release of excess FFAs from the fat pad would otherwise have been toxic, but instead adaptively increased body temperature, complementing the evolution of dipalmitoylphosphatidylcholine, which is three times more surface-active at $37\,°C$ than at $25\,°C$.

The etiology of Goodpasture syndrome similarly reveals the interrelationship between genetic mutation and evolution. The disease is caused by an autoimmune reaction to an evolved isoform of type IV collagen. Phylogenetically, alpha 3(IV) NC1 type IV collagen is absent from worms and flies, but appears in fish, in which, however, it does not generate the Goodpasture syndrome antibody. It is ubiquitous in amphibians, reptiles, birds, and mammals. Alpha 3(IV) NC1 has the evolutionarily relevant physico-chemical characteristic of being more hydrophobic than other type IV collagens, giving it a functional role in preventing water loss across the lung and kidney epithelial surfaces in adaptation to land. The evolution of this specific type IV collagen isoform during the process of land adaptation is unlikely to have occurred merely by chance, given its ability to prevent water loss on land. More likely, it evolved empirically by trial and error until the right configuration evolved.

Thus, not unlike chemistry and physics, biology is also founded on first principles that can be understood ontologically and epistemologically rather than through dogmatic teleologic mechanisms and tautologic concepts. George Williams' antagonistic pleiotropy hypothesis for aging was alluded to above. In large part, this perspective is reflective of the systematic error authored by Ernst Mayr that there are dichotomous proximate and ultimate mechanisms of evolution that must be dissociated from one another based on Darwinian principles of mutation and selection. That dictum was formulated more than 60 years ago, and theorists who offered differing perspectives, such as Ernst Haeckel, Hans Spemann, and Jean Baptiste Lamarck, have all been discounted. However, in the interim a great deal more about biology has been learned that re-energizes some previously discarded principles toward understanding evolutionary development. This is particularly true for cell biology, where pathways can be identified that inform us of a continuum between the proximate and ultimate mechanisms of evolution – Mayr exemplified his proximate/ultimate principle for evolution theory using bird migration, which was too complex to be understood as one continuous process in 1952. However, we now know how seasonal changes in ambient light affect the neuroendocrine system to foster reproductive migratory behavior.

Using the insights gained from seeing pleiotropy mechanistically as the repurposing of the same genetic signaling cascade to form new phenotypes, heterochrony can also be understood mechanistically, particularly since the mechanism of heterchrony has never been provided before. Haeckel used the concept of heterochrony as a way of expressing how development could facilitate evolutionary change. To this day, no one has described heterochrony as a mechanism for reallocating cell–cell signaling to accommodate adaptive change, yet it is the premise we have used throughout this book.

The key to understanding this mechanism lies in the transition from the blastula to the gastrula, since this is the stage of embryogenesis when the unilamellar cell membrane becomes the endoderm, mesoderm, and ectoderm, the three germ layers that generate the rest of the organism – it is during this phase that the epigenetic information must be transmitted functionally to the germ layers. Lewis Wolpert, a leader in the field of developmental biology, has rightfully stated that "It is not birth, marriage, or death, but gastrulation, which is truly the most important time in your life."

The author of *The Structure of Scientific Revolutions*, Thomas Kuhn, famously said that an indicator of a paradigm shift was a change in the language; going from a descriptive to a mechanistic way of expressing pleiotropy and heterochrony would be emblematic of such a paradigm shift.

The above resolution of the significance of pleiotropy is tantamount to Niels Bohr's eloquent explanation for the duality of light as both wave and particle based on principles of quantum mechanics. In his Complementarity lecture at Lake Como, Switzerland, in 1927 he resolved this seeming paradox by explaining that it was an artifact of the way in which the light was measured. The cell is similarly both genes and phenotypes depending upon the metric, yet in reality it is only as an integral whole that it exists and its fate is determined by the ever-transcendent evolutionary mechanisms that perpetuate it. In his groundbreaking book *Wholeness and the Implicate Order*, the physicist David Bohm explains how our subjective senses cloud our perception of the reality that lies beneath, leading to such dichotomous thinking. It is this realization that allows us to understand the mechanistic basis for pleiotropy, and that of evolution itself in the process.

Cells solve problems; they use the tools that they have or can generate. Many generations of scientists have attempted to discern the puzzle of evolutionary development, yet they have lacked the tools that can be productively employed today. What we have now learned is in many ways unexpected. Contrary to our expectation, what was old can again become new. In that sense, this book is dedicated to those who have labored before us. Their efforts can now be married to compelling research. Through this combination, a new paradigm for evolutionary development unfurls that is congruent with the dominant truth that can be asserted about our physiologic path from the "first principles of physiology" based on the cell membrane of unicellular organisms. It is clearly evident that all complex organisms must unavoidably return to their unicellular roots.

The physiologic pathways and the cellular communication mechanisms that underscore it explain the imperative for this immutable recapitulation.

This chapter has demonstrated how a mechanistic understanding of pleiotropy as descriptive, dogmatic biology provides insight to the process of evolution. Chapter 20, entitled "Meta-Darwinism," provides examples of the power of the cellular-molecular approach to evolution.

Selected Readings

Kuhn TS. (1962) *The Structure of Scientific Revolutions.* University of Chicago Press, Chicago.

MacDonald BA, Sund M, Grant MA, Pfaff KL, Holthaus K, Zon LI, Kalluri R. (2006) Zebrafish to humans: evolution of the alpha3-chain of type IV collagen and emergence of the autoimmune epitopes associated with Goodpasture syndrome. *Blood* **107**:1908–1915.

Torday JS, Rehan VK. (2012) *Evolutionary Biology, Cell-Cell Communication and Complex Disease.* John Wiley & Sons, Inc., Hoboken, NY.

20

Meta-Darwinism

The unique power of science is in its ability to predict. It is the only way we have of knowing what we do not know. Both chemistry and physics originated as descriptive, non-mechanistic, non-predictive disciplines, respectively alchemy and astrology. But once their "first principles" were realized, these disciplines emerged as chemistry and physics. In contrast to that, biology remains descriptive, though its practitioners continue to think that it is mechanistic for want of "first principles." The mere fact that the sequencing of the human genome has not led to any breakthroughs in curing disease over the course of the past 16 years, since the draft sequence was made public, is practical evidence for the lack of predictive power. The utility of the Human Genome Project (HGP) should allow us to understand health deterministically, instead of defining it relativistically as the absence of disease. Clearly, we are using the wrong logic since genomics has not succeeded in conquering disease. The fact that many of the major discoveries in biology – penicillin, various nitric oxide applications, steroids for the prevention of respiratory distress syndrome, pathologic hypothermia – were serendipitous observations speaks to the lack of predictive value in the current descriptive paradigm. In contrast to that, the approach that has been expounded in this book is predictive because it is predicated on the origins and causal nature of biology. We will now cite aspects of this approach that exemplify its predictive value.

Compartmentation of the Life Cycle in Service to Epigenetics

Perhaps the biggest breakthrough in twenty-first century biology and medicine is the realization that environmental factors can directly affect genetic inheritance, referred to as epigenetics. There have now been several experimental demonstrations of this principle. Its importance is reflected by the fact that only approximately 3% of human genetic diseases are Mendelian, leaving an unknown etiology for the vast majority of human diseases. Given the myriad ways in which environmental factors are known epidemiologically to affect biologic systems, it is reasonable to speculate that their nature is epigenetic, given that it is the only other known mechanism of inheritance, boding well for their effective treatment and possible eradication in the foreseeable future. For example, our laboratory studies the effects of cigarette smoke on childhood asthma. It had long been known that there was an association between parental

Evolution, the Logic of Biology, First Edition. John S. Torday and Virender K. Rehan.
© 2017 John Wiley & Sons, Inc. Published 2017 by John Wiley & Sons, Inc.

smoking and the incidence of childhood asthma, but it was difficult to show causality until we examined the effect of nicotine, a proxy for smoking, on lung development in rats. Treatment of the mother rat with nicotine caused asthma in the offspring for at least three generations due to the formation of epigenetic "marks," or chemical modifications of DNA that cause alterations in the deciphering of its code biologically. This effect of nicotine is insidious because it remains active in the environment, and its photochemical adducts are 1000 times more bioactive than the parent compound. Beyond this, it is now thought that beehive collapses worldwide are due to the effect of neonicotinoids, nicotine derivatives used as insecticides. Neonicotinoids are orders of magnitude more toxic in insects than in humans, so it is thought that they are relatively harmless in humans. However, these compounds are lipid soluble, and cross the placenta during pregnancy, potentially explaining such diseases as asthma and autism. For example, nicotine stimulates specific nicotinic acetylcholine receptors in the upper airway of the lung, and in the brain; these same receptors are associated with autism. And the liberal use of neonicotinoid insecticides worldwide might help to explain why asthma has reached epidemic proportions and autism is on the rise at an alarming rate.

Children are more susceptible to the nicotine in cigarette smoke because it settles on surfaces, particularly floors, where it can be inhaled, ingested, or acquired transdermally via adherence to skin on hands and knees as children crawl around. This life stage-specific nature of exposure to an epigenetic modifier transcends the acquisition of asthma. In an earlier chapter we had mentioned the compartmentation of epigenetic inheritance during meiosis and embryologic development as a way of biologically selecting for the acquisition of epigenetic marks. Compartmentation may also apply postnatally since the nominal stages of the life cycle – infancy, childhood, adolescence, adulthood, old age – also represent ways in which age-specific physiology and behaviors determine environmental exposures. We have already mentioned the relationship between crawling and the increased exposure to nicotine; during infancy the child is exposed to mother's milk while breast feeding, and to the microbiome of the mother's skin; adolescence is associated with "risk-taking" and the onset of puberty; adulthood has its own characteristic exposures to cigarettes, alcohol, sexual interactions, and novel environments due to increased mobility; in contrast, during old age we tend to become more socially and biologically isolated.

These associations between life cycle stages and environmental exposures are biologically determined by the endocrine system. Such interrelationships may seem teleologic at first glance, but there is scientific evidence that the endocrine system itself is under epigenetic control by the environment, suggesting causation rather than mere association and correlation. One experimental model that demonstrates this complex interrelationship biologically and evolutionarily is the maternal food restriction model for metabolic syndrome. If mother rats are deprived of 50% of their normal food intake at mid-gestation the offspring are born smaller and develop hypertension, obesity, and type II diabetes, or what is referred to clinically as metabolic syndrome. The biomedical research community has been exploiting this model of chronic disease because of its potential clinical use to predict and prevent metabolic syndrome, yet they may be missing an important "big picture" aspect of this phenomenon – the offspring become sexually active earlier in their life cycle due to precocious adrenarche, the increased production of androgens by the adrenal cortex, leading to precocious puberty. This may actually be the primary selection pressure in response to low food abundance since it would speed up reproduction, hastening the exposure of the next generation to a potentially higher food-abundance environment. This scenario is likely correct since the consequences of maternal food restriction only prevail if the offspring are exposed to a normal environment; if they are reared in a low food environment, as anticipated by the intrauterine exposure to low food abundance, they do not develop metabolic syndrome. By offering the offspring the opportunity to enter a normal food environment precociously, during the subsequent pregnancy the food environment will prepare the offspring appropriate to the food environment to come. This exemplifies the importance of focusing on the evolutionary mechanism rather than on the consequent disease because the latter is an epiphenomenon. By focusing on the primary selection mechanism, the consequences can be understood from their fundament both for prediction and prevention, rather than devising pharmacologic "band-aids" to alleviate the symptoms of the overt disease.

Rational Drug Design

Historically, in medical science new drugs have been discovered through trial-and-error, or simply through luck. However, as demand for new and more effective drugs has increased, a new method for drug development called rational drug design has begun to replace the old methods. In rational drug design, biologically active compounds are specifically designed or chosen to affect a specific target. This method often involves the use of molecular design software, which researchers use to create three-dimensional models of drugs and their biologic targets. As a result this process is also referred to as computer-aided drug design.

A biologic drug target is usually one of two kinds. The first is a molecule in the human body that causes disease when it is defective. The second is a molecule produced by a disease-causing microorganism. Rational drug development involves designing new chemical compounds that interact with these targets in a beneficial way, such as by interacting with cholesterol to rid the body of it or by interacting with a microorganism to disable or kill it.

Older methods of drug development have shortcomings that make drug discovery and development costly. The easiest and fastest method of developing a new drug is simply to discover, through sheer luck, that a certain compound is biologically active against a drug target of interest. The classic example is the serendipitous discovery of penicillin by Alexander Fleming in 1928. The microbiologist discovered the first antibiotic when some bacterial cultures he was working with became contaminated with a bactericidal fungus. Of course, this type of chance discovery does not happen very often, and luck is not something that drug companies can rely on for the development of new medications.

The usual method for developing new medications is an arduous, large-scale process called combinatorial library screening. In this process, large numbers of chemical compounds are created and then screened for biologic activity. If any given compound shows signs of interacting with a biologic target of interest, it is further investigated and might be developed into a new drug. This process can take years and large sums of money, and even at the end of the development period, the drug might not be effective or safe enough for human use.

Rational drug design is a streamlined process that requires careful analysis of both the target of the drug, and the drug itself. This method of drug design uses special equipment to examine the three-dimensional structure of a drug target in order to then design a compound that will hypothetically interact with the target. This process requires in-depth knowledge of chemistry and biology, because chemical interactions between drugs and their targets are what determine whether a drug is biologically active.

Compounds can be located for testing in two ways. The first involves combinatorial library screening, but in this case the process is streamlined because researchers using rational drug design methods will screen the library for compounds of a shape that is specific enough to interact with the drug target of interest. The second method involves the actual design of a compound that can interact with the target. This requires consideration of the chemical makeup of the compound and knowledge of what chemical groups the compound might require in order to be capable of interacting with the drug target.

The combined approach of rational drug design and the targeting of homeostatic mechanisms, particularly those that have evolved over eons from the unicellular state, would offer an opportunity to avoid the loss of such metabolic control due to disease, decreasing or eliminating the morbidities and comorbidities caused by the conventional medical practice of waiting for the patient to become symptomatic. Moreover, if the subject has a constitutive or genetic predisposition to some disease process he/she could be treated prophylactically. In either case, the combined effects of early intervention would truly constitute preventive medicine, saving both pain and suffering as well as healthcare costs for society

Formulation of a New Society

Modern social systems are being bombarded daily with huge organizational challenges such as climate change, drought, overpopulation, poverty, famine, and general infrastructural weaknesses. Our traditional approach to such profound problems has been ad hoc coping strategies reflective of our self-perception as beings, going all the way back to the dawn of time. However, there is a means of re-envisioning modern communities that would be in harmony with their inhabitants based on our deepest understanding of our own physiology. This can be accomplished through an emerging understanding of human physiology that integrates its unicellular origins with processes that originated with the Big Bang of the Universe (Figure 20.1). This novel approach assumes that social structures should ideally foster and optimize human existence. Extending this understanding of our deepest physiology from its atomic origins can suggest how this knowledge could be exploited to optimize and epitomize the universal human social condition.

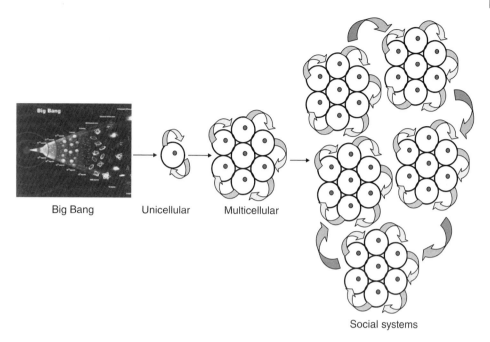

Big Bang   Unicellular   Multicellular

Social systems

Figure 20.1 From the Big Bang to social systems. On the far left is the Big Bang of the Universe, which scattered the elements based on their atomic mass, creating an informatic hierarchy. Biology exploited the physical environment to generate autonomous cells that used homeostasis to maintain negentropy. Competition between prokaryotes and eukaryotes gave rise to multicellular organisms, which ultimately formed social systems, all based on the "first principles of physiology."

Anthropomorphisms Subvert the Natural Biologic Imperative to Cooperate

Nowadays, biologists are militant about dissuading us from thinking hierarchically about the evolution of species. Each has its own set of traits that allow it to adapt to its environmental niche, including humans. A classic example of how our highly evolved central nervous system misguides us is the "anthropic principle," that the Earth environment is "just right" – oxygen in the atmosphere, ambient temperature, water, minerals, and so forth. That perception is very deceptive because it suggests some sort of higher power placing us and all other biota on Earth, when in fact we have evolved from the physical environment. For example, by regressing the genetic pathway for the evolution of the mammalian lung against major epochs in vertebrate evolution – salinization of the oceans, the water-to-land transition, and the Phanerozoic oxygen fluctuations – a pattern of alternating internal and external selection pressures mediated by genetic mechanisms consistent with specific physiologic developmental and phylogenetic adaptations emerges (see Figure 14.1). This observation is more in keeping with Spencerian environmental adaptation than with Darwinian survival of the fittest. Moreover, this perspective is mechanistically consistent with Gaia theory, proposing that organisms interact with their inorganic surroundings on Earth to form a self-regulating, complex system that contributes to maintaining the conditions for life on the planet.

Euphysiology

Up until now, social communities have been founded on basic human needs for food, shelter, education, religious institutions, trade, and government. Towns and cities were constructed on bodies of water both for agricultural and sanitation requirements. For example, Roman fortresses were built on the principle of bilateral symmetry, with entrances at all four compass points, not unlike Da Vinci's idealized portrayal of Vitruvian Man. The ultimate size of these entities was pragmatically determined by need, constrained by capacity, and formulated using the base 10 to emulate the number of fingers and toes – this is rapidly changing with our efforts to translate everything into the base 2, which would allow the application of the binary system to understand physiology from a novel perspective.

We have always considered our own physiology from its ends instead of its means. The conventional view has examined the physiology of complex organisms as an association of parts and generally linked steps, when in fact physiology is a highly integrated process that has evolved intact from our unicellular origins starting some 500 million years ago. Physiology only seems to be complex because we have been reasoning after the fact within a conceptually limiting teleological frame of reference. In reality, physiology is quite simple if examined from its first principles moving forward, having been established by relatively simple unicellular eukaryotic organisms. From this point of origin, physiology is properly assessed as the culmination of adaptations to the environment in support of epigenetic inheritance.

Unicellular eukaryotes (eukaryotes are defined as those organisms with a nuclear envelope) evolved from bacteria, or prokaryotes, some two billion years ago (see Figure 1.3). With the advent of cholesterol synthesis, and its incorporation into the eukaryotic cell membrane, eukaryotes were able to efficiently perform the three key functional tasks that characterize vertebrate physiology – locomotion, endocytosis/exocytosis, and respiration. Each of these traits evolved as a direct result of the physicochemical thinning of the cell membrane caused by the insertion of cholesterol.

Prokaryotes and eukaryotes continually compete with one another. Prokaryotes evolved the capacities for forming biofilms and for quorum sensing, which are pseudomulticellular properties. But eukaryotes actually evolved the capacity to form truly multicellular organisms as a result of competition with prokaryotes over the course of the last billion years.

Physiologic stress was a major driver for vertebrate evolution, epitomized by the water-to-land transition (WLT). As a result of the epic ecologic selection pressure entailed by the biota increasing the amount of carbon dioxide in the atmosphere – causing lakes and rivers to evaporate (Romer hypothesis) – there were three known gene duplications that facilitated specific land-adaptive traits as a result of their amplification. These duplications supported many disparate processes or their functioning organs: skeletal structural adaptations to increased gravitational force on land; lung physiology for air breathing; the kidneys for water and electrolyte regulation; the skin for barrier function; and the brain to integrate all of this newly acquired complex physiology. The three gene duplications – parathyroid hormone-related protein receptor (PTHrPR), the β-adrenergic receptor (βAR), and the glucocorticoid receptor (GR) – were all instrumental in facilitating the evolution of all these traits.

There were at least five known attempts by vertebrates to breach land during the WLT based on the fossil record; these involved crucial skeletal and concomitant visceral organ adaptations based on the gene duplications. The PTHrPR, which is essential for bone remodeling, is also necessary for the development of the lung and skin, and is indirectly involved in the development of the kidney. Physiologic stress would have caused shearing of the microvasculature, particularly in key tissues and organs necessary for adaptation to land (skeleton, lung, skin, kidney), consequently generating radical oxygen species known to cause gene duplications. The over-expression of the βAR gene due to duplication overcame the constraint caused by the shared regulation of blood pressure in both the lung alveoli and the peripheral circulation; and in turn, glucocorticoid signaling in response to physiologic stress would have facilitated βAR over-expression. These physiologic adaptations may all ultimately have been facilitated by the evolution of the mammalian lung, during which intermittent phases of hypoxia would have stimulated the pituitary-adrenal axis. This would have increased the production of adrenaline (epinephrine) by the adrenal medulla, culminating in relief of the hypoxic constraint by increasing surfactant secretion into the alveoli, making the alveoli more distensible due to reduced surface tension on the alveolar wall. As a consequence, there would have been increased production of PTHrP, fostering alveolarization and vascularization of newly formed alveoli. Such positive selection for PTHrP signaling may have fostered the aforementioned expression of PTHrP in both the pituitary and adrenal cortex, further amplifying the stimulation of adrenaline production. This stress-mediated mechanism thus both enhanced alveolarization and caused release of free fatty acids from peripheral fat cells. This would have resulted in increased overall metabolism, body temperature, and surfactant bioactivity, since the surfactant is 300% more active at 37 °C than at 25 °C. Thus, endothermy evolved as a result of these positive adaptations, mediated by the genes known to have been duplicated during the WLT in complicated physiologically directed linkages.

As added evidence of this evolutionary mechanism for the adaptation of vertebrate visceral organs during the WLT, the adrenal medulla of mammals formed vascular arcades during this period. These further amplified the production of adrenaline due to the increased microvascular surface area. These vascular arcades may have been generated by PTHrP secreted by the adrenal cortex since PTHrP is angiogenic. In fish, the adrenal cortex and medulla are separated, lacking this amplification mechanism. However, under such functionally mediated positive selection pressure, the adrenal cortex and medulla evolved into one integrated structure sharing a common vasculature. That positive selection pressure for adrenaline amplification may have been a balancing selection for hypoxic stress due to pulmonary insufficiency. This might explain why land-dwelling vertebrates have glomerular capillary complexes that make fluid and electrolyte regulation more efficient for land habitation. PTHrP expression within the renal artery may have fostered the evolution of the glomerulus from the glomus, a much simpler vascular kidney invagination, since PTHrP is expressed in the podocytes lining the glomeruli, signaling to the mesangium for regulation of fluid and electrolytes. Therefore, the internal and external selection pressures for skeletal remodeling, air breathing, neuroendocrine stimulation, and kidney evolution were all positively benefited by the evolution of PTHrP signaling from fish to humans due to the PTHrPR gene duplication.

Critically, these particular gene duplications for vertebrate land adaptation are the very same genetic adaptations involved in the evolution of unicellular eukaryotes. Facilitated by cholesterol, linked physiologic drivers from the unicellular state yield metabolic complexity, locomotion, and respiration. So positive selection for these attributes should not come as a surprise, given the deep phylogenetic "history" of these biologic traits, referring all the way back to the Big Bang of the Cosmos. All of these were part of the continuing attempt every organism must exert to maintain its homeostatic equipoise.

How might such complex, interrelated, fundamental physiologic mechanisms and evolutionary strategies bear on social systems? Even more importantly, what might be gleaned from these deeply rooted physiologic pathways that could productively relate to our all too "human" interactions? Obviously, civilizations have developed to support our physiologic needs as organisms for water, food, shelter, and mobility. These physiologic adaptations evolved in support of homeostasis as the prevailing mechanism of evolution by such deeply linked mechanisms as illustrated above. In contrast to the ad hoc nature of human cohabitation in large groups – whether village, city, state, country, nation state, or world community – social systems might in future be designed to effectively support human homeostasis in ways that would optimize physiology simultaneously on multiple scales, thereby maximizing our human potential. Through contemporary technological tools such as social networking, people would be enabled to provide biofeedback that would be used to fine-tune and "servo-regulate" such social systems in real time. Data strings could be used to both monitor and modify the social system in order to maintain societal equipoise as a thriving construct, avoiding "clogged arteries" and social decay; perhaps even more insidious, it is known that physiologic stress can cause psychological depression transgenerationally, causing social dysfunction for multiple generations.

Such an organic construct would synergize human activity, empowering individuals to grow and flourish within their environments in concert with their own genetic makeup, referring all the way back to their unicellular origins. The critical point is that our physiologic mechanisms are profoundly interlocking and constantly monitored and assessed within us as biologic organisms, yet our human responses to our physiologic stresses are never systemically assayed in real time. Might that not benefit society if we had the ready means? Why would we leave such useful reciprocal feedback mechanisms to stimuli to the whim of such top-down entities as advertisers and governments? Or to myth and custom, perpetuating racial, gender, and ageist biases.

Witness Mark Twain's *Huckleberry Finn*, which is about social pathology. In Azar Nafisi's *The Republic of Imagination*, she states that "*Huck Finn* shows us that everything that is accepted as the norm, as respectable, is in essence not normal or respectable. It is a book in which 'educated' people are the most ignorant, stealing is 'borrowing,' people with 'upbringings' are scoundrels, goodness is heartless, respectability stands for cruelty, and danger lurks, most especially at home." Twain wrote the book as a way of making us aware of the pathology. Elsewhere, Nafisi states that "Ignorance of the heart, in this book, is the greatest sin." Entraining such metaphoric physiology in all of society is what we are alluding to in this chapter.

Any model such as this could be designed to effectively determine how the by-products of our living interactions might effectively be incorporated into our social structure, or discouraged in order to maximally benefit its inhabitants. It is clear that there are agencies within the environment responsible for disease and pathology. For example, smoking directly afflicts the smoker, but also causes deposition of nicotine in the environment, affecting newborns and toddlers by causing asthma. At all stages of the life cycle, deleterious agents may epigenetically affect any individual. Conversely, there are organic substances in the environment that are known to be beneficial, or might be shown to be so. These might be productively identified and husbanded for our benefit. With an appropriate feedback system, deliberate systematic inclusion and exclusion of a variety of substances could initially be based on experimental evidence, but could also be monitored by an ongoing biofeedback-based mechanism, since there may be subtle effects not predicted by the model.

Importantly, the "physiologic first principles" model allows for monitoring of biologic systems based on homeostatic principles, instead of "input/output" metrics. It can be imagined just as one might consider a patient in the intensive care unit recovering from a heart attack. The physician measures fluid and electrolytes in urine to try to bring the patient back into homeostatic balance. Because the heart is in failure, the lungs are filling with fluid, and the kidneys have shut down due to shock. The hope is that the patient will reset his homeostatic mechanisms by normalizing outputs downstream of the regulatory mechanisms. Care is concentrated on assessing fluid inputs and outputs that are only indirect biomarkers of renal, cardiac, and ultimately lung function. Yet both the alveolus and glomerulus are "pressure transducers," which utilize endodermal PTHrP to regulate physiology by signaling to specialized fibroblasts in both structures. When these signaling mechanisms fail, the fibroblasts in both conditions default to the molecular Wingless/int (Wnt) pathway. In response, peroxisome proliferator activated receptor gamma (PPARγ) agonists inhibit Wnt, attempting to normalize the homeostatic pathways of both the lung and kidney. This is the means by which physiology actually facilitates recovery of homeostasis. Certainly, a deeper level of understanding offers unique opportunities to intervene at an effective and direct level as compared to any indirect assessment means. Such a true mechanistic understanding of physiology allows for higher-order regulation and correction based on fundamental operating principles.

The philosopher Willard Quine, and his predecessor Pierre Duhem, had expressed concern that science was "underdetermined," sensing a lack of competency, leading to subjective conclusions instead of deterministic results. Some of this lack of clarity may have stemmed from misunderstanding our own physiologic makeup, leading instead to default decisions based on custom or subjective opinion rather than data-driven principles. Just like the patient in the ICU whose care could be rewardingly directed by a deep understanding of physiology, an alternative social system based on physiologic realities using contemporary feedback tools could empower a society that is both empathetic and genuinely enlightened. Instead of the artificial mind-body duality of Descartes, we would have a totally integrated model of physiology on which to build social systems directly reflective of humans, not of their environment. What then are such deeply rooted, cell-based physiologic principles? They are the means by which cells govern themselves and have evolved: close collaboration, partnership, reciprocity, as well as competition, and, most importantly, adaptation. A better human society, one that systematically avoids stigma and suffering, has to be based on enacting and amplifying these cell-based physiologic principles. The means to do so by creatively utilizing modern feedback systems is a method that is finally within our grasp, though not yet in hand.

Merging of Evolution Theory and Ecology

Niche construction theory (NCT) postulates that animals create their own environments, increasing their ability to adapt to their surroundings. This is one of the first formal ways in which evolution and ecology have been merged, to great fanfare as these disciplines have been estranged. And now, with the re-emergence of epigenetics, NCT is even more relevant since the two concepts naturally reinforce one another mechanistically.

Gaia Theory

Gaia theory envisions the Earth as a self-regulating complex system involving the biosphere, the atmosphere, the hydrosphere, and the pedosphere, tightly coupled together in an evolving system. As a holistic entity, Gaia seeks a physical and chemical environment optimal for contemporary life.

Gaia evolves through a cybernetic feedback system mediated by the biota, leading to stable conditions for habitability in full homeostasis. The Earth's surface is essential for the optimal conditions of life, depending on interactions between life forms, especially microorganisms, with inorganic elements. These processes determine a global control system that regulates the Earth's surface temperature, atmospheric composition, and ocean salinity, driven by the force of the global thermodynamic disequilibrium of the Earth system.

Planetary homeostasis fostered by living forms had previously been recognized in the field of biogeochemistry, and it is also being investigated in other fields such as Earth system science. Gaia theory is novel because it relies on the precept that homeostatic balance is an active process with the goal of maintaining the optimal conditions for life.

A Universal Operating Database for all Natural Sciences

In his book *Consilience: The Unity of Knowledge*, E.O. Wilson speculates on the possibility of a universal database, given that all knowledge is being reduced to "ones and zeroes." That is a big idea that goes all the way back in history to the ancient Greek philosopher Heraclitus. One of us (J.T.) published a position paper in 2004 speculating about the feasibility of a "periodic table of biology" – an algorithm based on the "first principles of physiology." If and when that is achieved, it could be merged with the mathematical expression of the chemical periodic table, which would provide a working database for all of the natural sciences. Merging that with the existing databases of knowledge would constitute the vision of consilience expressed in E.O. Wilson's book by the same name, and the dawn of a new age of integrated, predictive knowledge.

Out-of-the-Box

Any credible philosophy, particularly one claiming predictive value, should be comprehensive. In that spirit we would like to address some topics that may be resolved by the evolutionary approach we have suggested. The physicist David Bohn authored a book entitled *Wholeness and the Implicate Order*. In it, he hypothesized that the way we observe reality is determined by our evolved senses, which he referred to as the "explicate order." He contrasts that with the true nature of reality, which he calls the "implicate order." In that spirt, we hope that the approach we have proposed will allow us to understand ourselves as derivative of the physical environment, allowing us to "factor out" the human subjectivity from our perception of reality, offering the opportunity to understand the true physical realm.

Philosophers like Duhem and Quine have questioned the ability of science to test hypotheses, since an empiric test of the hypothesis requires one or more background assumptions. And the failure of the great physicist Ilya Prigogine and the mathematician-philosopher Michael Polanyi to solve the riddle of "life" reinforces the unknowable complexity of evolution as all of biology, as Theodosius Dobzhansky phrased it. Yet simply introducing the cellular principle into the mix has allowed us to gain novel insights about biology never before tenable.

Microbiome as Immortality

We now know that we live in an intimate relationship with bacteria, which we acquire from the womb during development. When we die, our microbiome re-enters the soil and aquifer, where it can be assimilated by flora and fauna, rendering us immortal.

Ethics Based on Biologic Principles

When protocells formed from water and lipids, self-organizing and self-referential, entraining entropy to defy the second law of thermodynamics, aided by homeostasis, we gained the essence of free will for the first time. Some 3–4 billion years later multicellular organisms evolved from prokaryotes. In a profile of the pre-eminent ethicist Derek Parfit, entitled "How to be Good," Parfit pondered the paradox of Darwinian survival of the fittest and empathy, epitomizing the fundamental difference between the descriptive and mechanistic approaches to evolution, the former ending in confusion, the latter generating a robust platform for further inquiry and resolution.

Mind

If, as put forward in this book, all of physiology is fractal, emanating from the unicellular state as the ultimate unity, then the ways in which we and paramecia perceive the external environment are one and the same. In paramecia, consciousness is expressed as calcium pulses; in us the same holds true, though the calcium pulses are transmitted from neuron to neuron. That calcium pulse begins with the sperm fertilizing the egg, and does not cease until we die. It is the essence of mind.

Selected Readings

Bohm D. (1980) *Wholeness and the Implicate Order*. Routledge & Kegan Paul, London.

Polanyi M. (1968) Life's irreducible structure. Live mechanisms and information in DNA are boundary conditions with a sequence of boundaries above them. *Science* **160**:1308–1312.

Prigogine I, Stengers I. (1984) *Order Out of Chaos*. Bantam Books, New York.

Wilson E.O. (1998) *Consilience: The Unity of Knowledge*. Alfred A. Knopf, New York.

The content of this book represents a novel, mechanistic, testable, refutable approach to the question of "how and why" evolution? This book is dedicated to that sea change.

Index